Introduction to Materials Management

THIRD EDITION

J. R. Tony Arnold, CFPIM, CIRM

PRENTICE HALL
Upper Saddle River, New Jersey   Columbus, Ohio

Library of Congress Cataloging-in-Publication Data

Arnold, J. R. Tony.
    Introduction to materials management / J.R. Tony Arnold.—3rd ed.
        p.   cm.
    Includes bibliographical references and index.
    ISBN 0-13-862087-3 (hardcover)
    1. Materials management.   I. Title.
TS161.A76   1998
658.7—dc21

                                                    97-30065
                                                    CIP

Editor: Stephen Helba
Production Editor: Rex Davidson
Design Coordinator: Julia Zonneveld Van Hook
Text Designer: Kip Shaw
Cover Designer: Karrie M. Converse
Production Manager: Laura Messerly
Illustrations: Tom Kennedy
Marketing Manager: Debbie Yarnell

This book was set in Dutch 801 by Carlisle Communications, Ltd. and was printed and bound by R. R. Donnelley & Sons Company. The cover was printed by Phoenix Color Corp.

 © 1998, 1996, 1991 by Prentice-Hall, Inc.
Simon & Schuster/A Viacom Company
Upper Saddle River, New Jersey 07458

Printed in the United States of America

10 9 8 7 6 5 4

ISBN: 0-13-862087-3

Prentice-Hall International (UK) Limited, *London*
Prentice-Hall of Australia Pty. Limited, *Sydney*
Prentice-Hall of Canada, Inc., *Toronto*
Prentice-Hall Hispanoamericana, S. A., *Mexico*
Prentice-Hall of India Private Limited, *New Delhi*
Prentice-Hall of Japan, Inc., *Tokyo*
Simon & Schuster Asia Pte. Ltd., *Singapore*
Editora Prentice-Hall do Brasil, Ltda., *Rio de Janeiro*

Preface

Introduction to Materials Management is an introductory text designed for students in community colleges and university programs. It is used in technical programs such as industrial engineering or manufacturing engineering, and in business programs. The text has also proved suitable for those already in industry, whether or not they are working in materials management.

This text has been widely adopted by colleges and universities not only in North America but in other parts of the world. It is listed in the American Production and Inventory Control Society (APICS) *CPIM Exam Content Manual* as the text reference for the *Basics of Supply Chain Management* (BSCM) CPIM certification examination. It is used by production and inventory control societies in other countries, such as South Africa, Australia, New Zealand, and France. As well, it is used by consultants in presenting in-house courses to their customers.

While the second edition covered most of the content of the BSCM examination, there were some gaps. These gaps have been addressed in the third edition. Two new chapters have been added, Products and Processes (Chapter 14), and Total Quality Management (Chapter 16). Materials management must react to whatever processes exist, and processes are largely determined by the design of products. The chapter on products and process design examines some of the basic factors and concepts in designing products for ease of manufacture, and in designing and selecting processes. Total Quality Management has become a major concept in operations management. This chapter explores the concepts and philosophy of total quality management.

Materials management means different things to different people. In this text, materials management includes all activities in the flow of materials from the supplier through to the consumer. Such activities include physical supply, operations planning and control and physical distribution. Other terms sometimes used are business logistics and supply chain management. Often the emphasis in business logistics is on transportation and distribution systems with little concern for what goes on in the factory. While there are chapters in this text devoted to transportation and distribution, most emphasis is placed on operations planning and control.

Distribution and operations are managed by planning and controlling the flow of materials through them and by utilizing the system's resources to achieve a desired customer service level. These activities are the responsibility of materials management, and affect every department in a manufacturing business. If the materials management system is not well designed and operated, the distribution and manufacturing system will be less effective and more costly. Anyone working in manufacturing or distribution should have a good basic understanding of the factors influencing materials flow. This text aims to provide that understanding.

The American Production and Inventory Control Society has defined the body of knowledge, the concepts, and the vocabulary used in production and inventory control. This is important, not only in developing an understanding of production and inventory control, but in making clear communication possible. Where applicable, the definitions and concepts in the text subscribe to APICS vocabulary and concepts.

The first six chapters of this text cover the basics of production planning and control. Chapter 7 discusses the important factors in purchasing; Chapter 8 is on forecasting. Chapters 9, 10, and 11 look at the fundamentals of inventory management. Chapter 12 discusses physical inventory and warehouse management, and Chapter 13 examines the elements of distribution systems including transportation, packaging, and material handling. Chapter 14 discusses the factors influencing product and process design. Chapter 15 looks at the philosophy and the environment of Just-in-Time manufacturing. It explains how operations planning and control systems relate to Just-in-Time. Chapter 16 examines the elements of total quality management.

The text covers all the basics of supply chain management and production and inventory control. The material, examples, questions, and problems lead the student logically through the material. The style is simple and user-friendly. Students who have used the material attest to this.

I have received help and encouragement from a number of valued sources, among them my friends, colleagues, and students at Fleming College. I thank the faculty of other colleges and the many members of APICS chapters who continue to offer their support and helpful advice. I also thank my wife, Vicky, for her assistance and patience throughout the time *Introduction to Materials Management* has been in preparation.

This book is dedicated to those who have taught me the most—my students.

J. R. Tony Arnold
Professor Emeritus
Fleming College
Peterborough, Ontario

Contents

1

Introduction to Materials Management

INTRODUCTION

The wealth of a country is measured by its gross national product—the output of goods and services produced by the nation in a given time. Goods are physical objects, something we can touch, feel, or see. Services are the performance of some useful function such as banking, medical care, restaurants, clothing stores, or social services.

But what is the source of wealth? It is measured by the amount of goods and services produced, but where does it come from? Although we may have rich natural resources in our economy such as mineral deposits, farm land, and forests, these are only potential sources of wealth. A production function is needed to transform our resources into useful goods. Production takes place in all forms of transformation—extracting minerals from the earth, farming, lumbering, fishing, and using these resources to manufacture useful products.

There are many stages between the extraction of resource material and the final consumer product. At each stage in the development of the final product, value is added, thus creating more wealth. If ore is extracted from the earth and sold, wealth is gained from our efforts, but those who continue to transform the raw material will gain more and usually far greater wealth. Japan is a prime example of this. It has very few natural resources and buys most of the raw materials it needs. However, the Japanese have developed one of the wealthiest economies in the world by transforming the raw materials they purchase and adding value to them through manufacturing.

Manufacturing companies are in the business of converting raw materials to a form that is of far more value and use to the consumer than the original raw materials. Logs are converted into tables and chairs, iron ore into steel, and steel into cars and refrigerators. This conversion process, called manufacturing or production, makes a society wealthier and creates a better standard of living.

To get the most value out of our resources, we must design production processes that make products most efficiently. Once the processes exist, we need to manage their operation so they produce goods most economically. Managing the operation means planning for and controlling the resources used in the process: labor, capital, and material. All are important, but the major way in which management plans and controls is through the flow of materials. The flow of materials controls the performance of the process. If the right materials in the right quantities are not available at the right time, the process cannot produce what it should. Labor and machinery will be poorly utilized. The profitability, and even the existence, of the company will be threatened.

OPERATING ENVIRONMENT

Operations management works in a complex environment affected by many factors. Among the most important are government regulation, the economy, competition, customer expectations, and quality.

Government.   Regulation of business by the various levels of government is extensive. Regulation applies to such areas as the environment, safety, product liability, and taxation. Government, or the lack of it, affects the way business is conducted.

Economy.   General economic conditions influence the demand for a company's products or services and the availability of inputs. During economic recession the demand for many products decreases while others may increase. Materials and labor shortages or surpluses influence the decisions management makes. Shifts in the age of the population, needs of ethnic groups, low population growth, freer trade between countries, and increased global competition all contribute to changes in the marketplace.

Competition.   Competition is severe today.

- Manufacturing companies face competition from throughout the world. They find foreign competitors selling in their markets even though they themselves may not be selling in foreign markets. Companies also are resorting more to worldwide sourcing.

- Transportation and the movement of materials are relatively less costly than they used to be.

- Worldwide communications are fast, effective, and cheap. Information and data can be moved almost instantly halfway around the globe.

Customers. Both consumers and industrial customers have become much more demanding, and suppliers have responded by improving the range of characteristics they offer. Some of the characteristics and selection customers expect in the products and services they buy are

- a fair price
- higher (right) quality products and services
- delivery lead time
- better pre-sale and after-sale service
- product and volume flexibility

Quality. Since competition is international and aggressive successful companies provide quality that not only meets customers' high expectations, but exceeds them. Chapter 16 discusses quality in detail.

Order qualifiers and order winners. Generally a supplier must meet set minimum requirements to be considered a viable competitor in the marketplace. Customer requirements may be based on price, quality, delivery, and so forth and are called **order qualifiers.** For example, the price for a certain type of product must fall within a range for the supplier to be considered. But being considered does not mean winning the order. To win orders a supplier must have characteristics that encourage customers to choose its products and services over competitors'. Those competitive characteristics, or combination of characteristics, that persuade a company's customers to choose its products or services are called **order winners.** They provide a competitive advantage for the firm. Order winners change over time and may well be different for different markets. For example, fast delivery may be vital in one market but not in another. Characteristics that are order winners today probably will not remain so, because competition will try to copy winning characteristics, and the needs of customers will change.

Manufacturing Strategy

A highly market-oriented company will focus on meeting or exceeding customer expectations and on order winners. In such a company all functions must contribute toward a winning strategy. Thus, operations must have a strategy that allows it to supply the needs of the marketplace and provide fast on-time delivery.

Delivery lead time. From the supplier's perspective, this is the time from receipt of an order to the delivery of the product. From the customer's perspective it may also include time for order preparation and transmittal. Customers want delivery lead time to be as short as possible, and manufacturing must design a strategy to achieve this. There are four basic strategies: engineer-to-order, make-to-order, assemble-to-order, and make-to-stock. Customer involvement in the product design, delivery lead time, and inventory state are influenced by each strategy. Figure 1.1 shows the effect of each strategy.

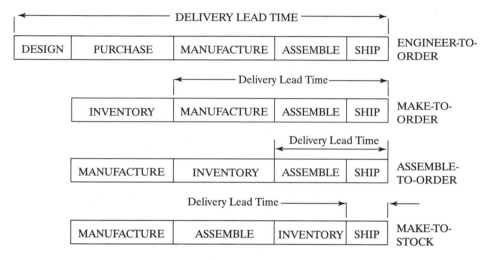

Figure 1.1   Manufacturing strategy and lead time.

Engineer-to-order means that the customer's specifications require unique engineering design or significant customization. Usually the customer is highly involved in the product design. Inventory will not normally be purchased until needed by manufacturing. Delivery lead time is long because it includes not only purchase lead time, but design lead time as well.

Make-to-order means that the manufacturer does not start to make the product until a customer's order is received. The final product is usually made from standard items but may include custom designed components as well. Delivery lead time is reduced because there is little design time required and inventory is held as raw material.

Assemble-to-order means that the product is made from standard components that the manufacturer can inventory and assemble according to a customer order. Delivery lead time is reduced further because there is no design time needed and inventory is held ready for assembly. Customer involvement in the design of the product is limited to selecting the component part options needed.

Make-to-stock means that the supplier manufactures the goods and sells from finished goods inventory. Delivery lead time is shortest. The customer has little direct involvement in the product design.

THE SUPPLY CHAIN CONCEPT

There are three phases to the flow of materials. Raw materials flow into a manufacturing company from a physical supply system, they are processed by manufacturing, and finally finished goods are distributed to end consumers through a physical distribution system. Figure 1.2 shows this system graphically. While this figure shows only one supplier and one customer, usually the supply chain consists of several companies linked in a supply/demand relationship. For example, the customer of one supplier buys product, adds value to it, and supplies yet another customer. Similarly, one customer may have several suppliers and may in turn supply several customers. As long as there is a chain of supplier/customer relationships, they are all members of the same supply chain.

Figure 1.2 Supply-production-distribution system.

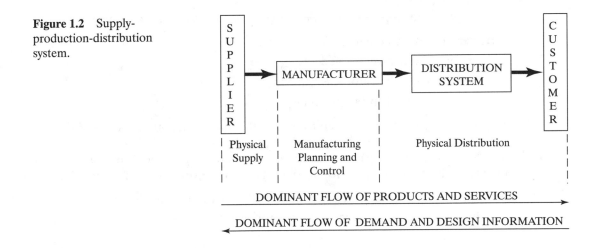

There are a number of important factors in supply chains:

• The supply chain includes all activities and processes to supply a product or service to a final customer.

• Any number of companies can be linked in the supply chain.

• A customer can be a supplier to another customer so the total chain can have a number of supplier/customer relationships.

• While the distribution system can be direct from supplier to customer, depending on the products and markets, it can contain a number of intermediaries (distributors) such as wholesalers, warehouses, and retailers.

• Product or services usually flow from supplier to customer and design and demand information usually flows from customer to supplier. Rarely is this not so.

Although these systems vary from industry to industry and company to company, the basic elements are the same: supply, production, and distribution. The relative importance of each depends on the costs of the three elements.

Conflicts in Traditional Systems

In the past, supply, production, and distribution systems were organized into separate functions that reported to different departments of a company. Often policies and practices of the different departments maximized departmental objectives without considering the effect they would have on other parts of the system. Because the three systems are interrelated, conflicts often occurred. While each system made decisions that were best for itself, overall company objectives suffered. For example, the transportation department would ship in the largest quantities possible so it could minimize shipping costs. However, this increased inventory and resulted in higher inventory-carrying costs.

To get the most profit, a company must have at least four main objectives:

1. Provide best customer service.
2. Provide lowest production costs.
3. Provide lowest inventory investment.
4. Provide lowest distribution costs.

These objectives create conflict among the marketing, production, and finance departments because each has different responsibilities in these areas.

Marketing's objective is to maintain and increase revenue; therefore, it must provide the best customer service possible. There are several ways of doing this:

- Maintain high inventories so goods are always available for the customer.
- Interrupt production runs so that a noninventoried item can be manufactured quickly.
- Create an extensive and costly distribution system so goods can be shipped to the customer rapidly.

Finance must keep investment and costs low. This can be done in the following ways:

- Reduce inventory so inventory investment is at a minimum.
- Decrease the number of plants and warehouses.
- Produce large quantities using long production runs.
- Manufacture only to customer order.

Production must keep its operating costs as low as possible. This can be done in the following ways:

- Make long production runs of relatively few products. Fewer changeovers will be needed and specialized equipment can be used, thus reducing the cost of making the product.

- Maintain high inventories of raw materials and work in process so production is not disrupted by shortages.

These conflicts among marketing, finance, and production center on customer service, disruption of production flow, and inventory levels. Figure 1.3 shows this relationship.

Figure 1.3 Conflicting objectives

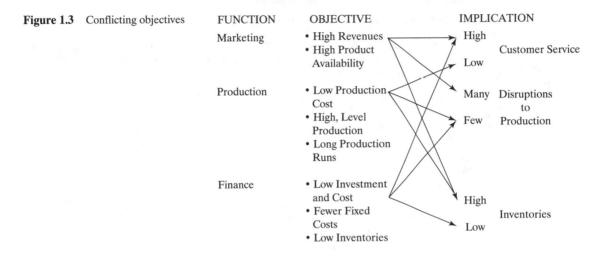

Today the concepts of Just-in-Time (JIT) manufacturing stress the need to supply customers with what they want when they want it and to keep inventories at a minimum. These objectives put further stress on the relationship among production, marketing, and finance. Chapter 15 will discuss the concepts of JIT manufacturing and how it influences materials management.

One important way to resolve these conflicting objectives is to provide close coordination of the supply, production, and distribution functions. The problem is to balance conflicting objectives to minimize the total of all the costs involved and maximize customer service consistent with the goals of the organization. This requires some type of integrated materials management or logistics organization that is responsible for supply, production, and distribution. Rather than having the planning and control of these functions spread among marketing, production, and distribution, they should occur in a single area of responsibility.

WHAT IS MATERIALS MANAGEMENT?

The concept of having one department responsible for the flow of materials, from supplier through production to consumer, is relatively new. Although many companies have adopted this type of organization, there are still a number that have not. If companies wish to minimize total costs in this area and provide a better level of customer service, they will move in this direction.

The name usually given to this function is **materials management.** Other names include **distribution planning and control** and **logistics management,** but the one used in this text is materials management.

Materials management is a coordinating function responsible for planning and controlling materials flow. Its objectives are as follows:

- Maximize the use of the firm's resources.
- Provide the required level of customer service.

Materials management can do much to improve a company's profit. An income (profit and loss) statement for a manufacturing company might look something like the following:

	Dollars	Percent of Sales
Revenue (sales)	$1,000,000	100
Cost of Goods Sold		
Direct Material	$500,000	50
Direct Labor	$200,000	20
Factory Overhead	$200,000	20
Total Cost of Goods Sold	$900,000	90
Gross Profit	$100,000	10

Direct labor and direct material are costs that increase or decrease with the quantity sold. Overhead (all other costs) does not vary directly with sales. For simplicity this section assumes overhead is constant, even though it is initially expressed as a percentage of sales.

If, through a well-organized materials management department, direct materials can be reduced by 10% and direct labor by 5%, the improvement in profit would be:

	Dollars	Percent of Sales
Revenue (sales)	$1,000,000	100
Cost of Goods Sold		
Direct Material	$450,000	45
Direct Labor	$190,000	19
Overhead	$200,000	20
Total Cost of Goods Sold	$840,000	84
Gross Profit	$160,000	16

Profit has been increased by 60%. To get the same increase in profit ($60,000) by increasing revenue, sales would have to increase to $1.2 million.

	Dollars	Percent of Sales
Revenue (sales)	$1,200,000	100
Cost of Goods Sold		
Direct Material	$600,000	50
Direct Labor	$240,000	20
Overhead	$200,000	17
Total Cost of Goods Sold	$1,040,000	87
Gross Profit	$160,000	13

EXAMPLE PROBLEM

a. If the cost of direct material is 60%, direct labor is 10%, and overhead is 25% of sales, what will be the improvement in profit if direct material is reduced 5%?

b. How much will sales have to increase to give the same increase in profit? (Remember, overhead cost is constant.)

Answer

a.	Before Improvement	After Improvement
Revenue (sales)	100%	100%
Cost of Goods Sold		
Direct Material	60%	55%
Direct Labor	10%	10%
Overhead	25%	25%
Total Cost of Goods Sold	95%	90%
Gross Profit	5%	10%

$$
\begin{aligned}
\textbf{b.} \quad \text{Profit} &= \text{sales} - (\text{direct material} + \text{direct labor} + 0.25) \\
&= \text{sales} - (0.6\ \text{sales} + 0.1\ \text{sales} + 0.25) \\
&= \text{sales} - 0.7\ \text{sales} - 0.25 \\
.1 &= 0.3\ \text{sales} - 0.25 \\
0.3\ \text{Sales} &= 0.35 \\
\text{Sales} &= \frac{0.35}{0.3} = 1.17
\end{aligned}
$$

Sales must increase 17% to give the same increase in profit.

Reducing cost contributes directly to profit. Increasing sales increases direct costs of labor and materials so profit does not increase directly. Materials management can reduce costs by being sure that the right materials are in the right place at the right time and the resources of the company are properly used.

There are several ways of classifying this flow of material. A very useful classification, and the one used in this text, is **manufacturing planning and control** and **physical supply/distribution.**

Manufacturing Planning and Control

Manufacturing planning and control is responsible for the planning and control of the flow of materials through the manufacturing process. The primary activities carried out are as follows:

1. *Production planning.*   Production must be able to meet the demand of the marketplace. Finding the most productive way of doing so is the responsibility of production planning. It must establish correct priorities (what is needed and when) and make certain that capacity is available to meet those priorities. It will involve:

   a. Forecasting.

   b. Master planning.

   c. Material requirements planning.

   d. Capacity planning.

2. *Implementation and control.*   These are responsible for putting into action and achieving the plans made by production planning. These responsibilities are accomplished through production activity control (often called **shop floor control**) and purchasing.

3. *Inventory management.*   Inventories are materials and supplies carried on hand either for sale or to provide material or supplies to the production process. They are part of the planning process and provide a buffer against the differences in demand rates and production rates.

Production planning, implementation, control, and inventory management work together. Inventories in manufacturing are used to support production or are the result of production. Only if items are purchased and resold without further processing can inventory management operate separately from production planning and control. Even then, it cannot operate apart from purchasing.

Inputs to the manufacturing planning and control system.   There are five basic inputs to the manufacturing planning and control system:

1. The *product description* shows how the product will appear at some stage of production. *Engineering drawings and specifications* are methods of describing

the product. Another method, and the most important for manufacturing planning and control, is the **bill of material.** As used in materials management, this document does two things:

- Describes the components used to make the product.
- Describes the subassemblies at various stages of manufacture.

2. *Process specifications* describe the steps necessary to make the end product. They are a step-by-step set of instructions describing how the product is made. This information is usually recorded on a route sheet or in a routing file. These are documents or computer files that give information such as the following on the manufacture of a product:

- Operations required to make the product.
- Sequence of operations.
- Equipment and accessories required.
- Standard time required to perform each operation.

3. The *time needed to perform operations* is usually expressed in standard time which is the time taken by an average operator, working at a normal pace, to perform a task. It is needed to schedule work through the plant, load the plant, make delivery promises, and cost the product. Usually, standard times for operations are obtained from the routing file.

4. *Available facilities.* Manufacturing planning and control must know what plant, equipment, and labor will be available to process work. This information is usually found in the work center file.

5. *Quantities required.* This information will come from forecasts, customer orders, orders to replace finished-goods inventory and the material requirements plan.

Physical Supply/Distribution

Physical supply/distribution includes all the activities involved in moving goods, from the supplier to the beginning of the production process, and from the end of the production process to the consumer.

The activities involved are as follows:

- Transportation.
- Distribution inventory.
- Warehousing.
- Packaging.
- Materials handling.
- Order entry.

Materials management is a balancing act. The objective is to be able to deliver what customers want, when and where they want it, and do so at minimum cost. To achieve this objective, materials management must make tradeoffs between the level of customer service and the cost of providing that service. As a rule, costs rise as the service level increases, and materials management must find that combination of inputs to maximize service and minimize cost. For example, customer service can be improved by establishing warehouses in major markets. However, that causes extra cost in operating the warehouse and in the extra inventory carried. To some extent, these costs will be offset by potential savings in transportation costs if lower cost transportation can be used.

By grouping all those activities involved in the movement and storage of goods into one department, the firm has a better opportunity to provide maximum service at minimum cost and increase profit. The overall concern of materials management is the balance between priority and capacity. The marketplace sets demand. Materials management must plan the firm's priorities (what goods to make and when) to meet that demand. Capacity is the ability of the system to produce or deliver goods. Priority and capacity must be planned and controlled to meet customer demand at minimum cost. Materials management is responsible for doing this.

SUMMARY

Manufacturing creates wealth by adding value to goods. To improve productivity and wealth, a company must first design efficient and effective systems for manufacturing. It must then manage these systems to make the best use of labor, capital, and material. One of the most effective ways of doing this is through the planning and control of the flow of materials into, through, and out of manufacturing. There are three elements to a material flow system: supply, manufacturing planning and control, and physical distribution. They are connected and what happens in one system affects the others.

Traditionally, there are conflicts in the objectives of a company and in the objectives of marketing, finance, and production. The role of materials management is to balance these conflicting objectives by coordinating the flow of materials so customer service is maintained and the resources of the company are properly used.

This text will examine some of the theory and practice considered to be part of the "body of knowledge" as presented by the American Production and Inventory Control Society. Chapter 15 will study the concepts of Just-in-Time manufacturing to see how they affect the practice of materials management.

QUESTIONS

1. What is wealth, and how is it created?
2. What is value added, and how is it achieved?
3. Name and describe four major factors affecting operations management.
4. What is an order qualifier and an order winner?
5. Describe the four primary manufacturing strategies. How does each affect delivery lead time?
6. What is a supply chain? Describe five important factors in supply chains.
7. What must manufacturing management do to manage a process or operation? What is the major way in which management plans and controls?
8. Name and describe the three main divisions of supply, production, and distribution systems.
9. What are the four objectives of a firm wishing to maximize profit?
10. What is the objective of marketing? What three ways will help it achieve this objective?
11. What are the objectives of finance? How can these objectives be met?
12. What are the objectives of production? How can these objectives be met?
13. Describe how the objectives of marketing, production, and finance are in conflict over customer service, disruption to production, and inventories.
14. What is the purpose of materials management?
15. Name and describe the three primary activities of manufacturing planning and control.
16. Name and describe the inputs to a manufacturing planning and control system.
17. What are the six activities involved in the physical supply/distribution system?
18. Why can materials management be considered a balancing act?

PROBLEMS

1.1 If the cost of manufacturing (direct material and direct labor) is 50% of sales and profit is 10% of sales, what would be the improvement in profit if, through better planning and control, the cost of manufacturing was reduced from 50% to 45%?

Answer. Profits would improve 50%.

1.2 In problem 1.1, how much would sales have to increase to provide the same increase in profits?

Answer. Sales would have to increase 10%.

1.3 On the average, a firm has 10 weeks of work-in-process, and annual cost of goods sold is $10 million. Assuming that the company works 50 weeks a year:

a. What is the dollar value of the work-in-process?

b. If the work-in-process could be reduced to 5 weeks and the annual cost of carrying inventory was 20% of the inventory value, what would be the annual saving?

Answer. **a.** $2,000,000

b. $200,000

1.4 On the average, a company has 12 weeks of work-in-process and annual cost of goods sold of $36 million. Assuming that the company works 50 weeks a year:

   a. What is the dollar value of the work-in-process?

   b. If the work-in-process could be reduced to 5 weeks and the annual cost of carrying inventory was 20% of the inventory value, what would be the annual saving?

1.5 Amalgamated Fenderdenter's cost of goods sold is $10 million. The company spends $4 million for purchase of direct materials and $2 million for direct labor; overhead is $3.5 million and profit is $500,000. Direct labor and direct material vary directly with the cost of goods sold, but overhead does not. The company wants to increase profit to $1 million.

   a. By how much should the firm increase sales?

   b. By how much should the firm decrease material costs?

   c. By how much should the firm decrease labor cost?

2

Production Planning System

INTRODUCTION

This chapter introduces the manufacturing planning and control system. First, it deals with the total system and then with some details involved in production planning. Subsequent chapters will discuss master production scheduling, material requirements planning, capacity management, production activity control, purchasing, and forecasting.

Manufacturing is complex. Some firms make a few different products, while others make many products. However, each uses a variety of processes, machinery, equipment, labor skills, and material. To be profitable, a firm must organize all these factors to make the right goods at the right time at top quality and do so as economically as possible. It is a complex problem, and it is essential to have a good planning and control system.

A good planning system must answer four questions:

1. What are we going to make?
2. What does it take to make it?
3. What do we have?
4. What do we need?

These are questions of priority and capacity.

Figure 2.1   Priority–capacity relationship.

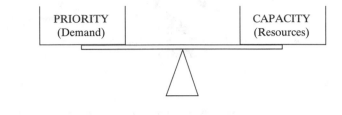

Priority relates to what products are needed, how many are needed, and when they are needed. The marketplace establishes the priorities. Manufacturing is responsible for devising plans to satisfy the market demand if possible.

Capacity is the capability of manufacturing to produce goods and services. Eventually it depends on the resources of the company—the machinery, labor, and financial resources, and the availability of material from suppliers. In the short run, capacity is the quantity of work that labor and equipment can perform in a given period. The relationship that should exist between priority and capacity is shown graphically in Figure 2.1.

In the long and short run, manufacturing must devise plans to balance the demands of the marketplace with its resources and capacity. For long-range decisions, such as the building of new plants or the purchase of new equipment, the plans must be made for several years. For planning production over the next few weeks, the time span will be days or weeks. This hierarchy of planning, from long range to short range, is covered in the next section.

MANUFACTURING PLANNING AND CONTROL SYSTEM

There are five major levels in the manufacturing planning and control (MPC) system:

- Strategic business plan.
- Production plan (sales and operations plan).
- Master production schedule.
- Material requirements plan.
- Purchasing and production activity control.

Each level varies in purpose, time span, and level of detail. As we move from strategic planning to production activity control, the purpose changes from general direction to specific detailed planning, the time span decreases from years to days, and the level of detail increases from general categories to individual components and workstations.

Since each level is for a different time span and for different purposes, each differs in the following:

- Purpose of the plan.
- Planning horizon—the time span from now to some time in the future for which the plan is created.

Figure 2.2 Manufacturing planning and control system.

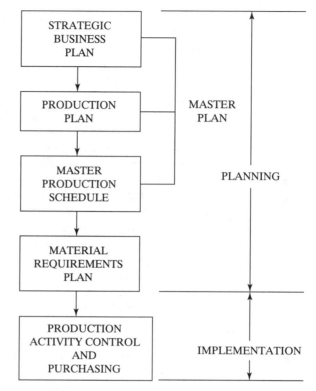

- Level of detail—the detail about products required for the plan.
- Planning cycle—the frequency with which the plan is reviewed.

At each level, three questions must be answered:

1. What are the priorities—how much of what is to be produced and when?
2. What is the available capacity—what resources do we have?
3. How can differences between priorities and capacity be resolved?

Figure 2.2 shows the planning hierarchy. The first four levels are planning levels. The result of the plans is authorization to purchase or manufacture what is required. The final level is when the plans are put into action through production activity control and purchasing.

The following sections will examine each of the planning levels by purpose, horizon, level of detail, and planning cycle.

The Strategic Business Plan

The strategic business plan is a statement of the major goals and objectives the company expects to achieve over the next two to ten years or more. It is a statement of the broad direction of the firm and shows the kind of business—product lines, markets, and so on—the firm wants to do in the future. The plan gives general direction about how the company hopes to achieve these objectives. It is based on long-range forecasts and includes participation from marketing, finance, production, and engineering. In turn, the plan provides direction and coordination among the marketing, production, financial, and engineering plans.

Marketing is responsible for analyzing the marketplace and deciding the firm's response: the markets to be served, the products supplied, desired levels of customer service, pricing, promotion strategies, and so on.

Finance is responsible for deciding the sources and uses of funds available to the firm, cash flows, profits, return on investment, and budgets.

Production must satisfy the demands of the marketplace. It does so by using plants, machinery, equipment, labor, and materials as efficiently as possible.

Engineering is responsible for research, development, and design of new products or modifications to existing ones. Engineering must work with marketing and production to produce designs for products that will sell in the marketplace and can be made most economically.

The development of the strategic business plan is the responsibility of senior management. Using information from marketing, finance, and production, the strategic business plan provides a framework that sets the goals and objectives for further planning by the marketing, finance, engineering, and production departments. Each department produces its own plans to achieve the objectives set by the strategic business plan. These plans will be coordinated with one another and with the strategic business plan. Figure 2.3 shows this relationship.

Figure 2.3   Business plan.

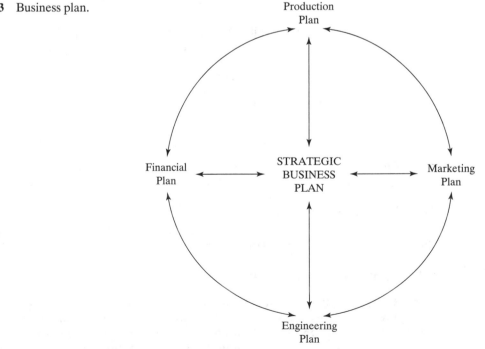

The level of detail in the strategic business plan is not high. It is concerned with general market and production requirements—total market for major product groups, perhaps—and not sales of individual items. It is often stated in dollars rather than units.

Strategic business plans are usually reviewed every six months to a year.

The Production Plan

Given the objectives set by the strategic business plan, production management is concerned with the following:

- The quantities of each product group that must be produced in each period.
- The desired inventory levels.
- The resources of equipment, labor, and material needed in each period.
- The availability of the resources needed.

The level of detail is not high. For example, if a company manufactures children's bicycles, tricycles, and scooters in various models, each with many options, the production plan will show major product groups, or families: bicycles, tricycles, and scooters.

Production planners must devise a plan to satisfy market demand within the resources available to the company. This will involve determining the resources needed to meet market demand, comparing the results to the resources available, and devising a plan to balance requirements and availability.

This process of determining the resources required and comparing them to the available resources takes place at each of the planning levels and is the problem of capacity management. For effective planning, there must be a balance between priority and capacity.

Along with the market and financial plans, the production plan is concerned with implementing the strategic business plan. The planning horizon is usually six to 18 months and is reviewed perhaps each month or quarter.

The Master Production Schedule

The master production schedule (MPS) is a plan for the production of individual end items. It breaks down the production plan to show, for each period, the quantity of each end item to be made. For example, it might show that 200 Model A23 scooters are to be built each week. Inputs to the MPS are the production plan, the forecast for individual end items, sales orders, inventories, and existing capacity.

The level of detail for the MPS is higher than for the production plan. Whereas the production plan was based upon families of products (tricycles), the master production schedule is developed for individual end items (each model of tricycle). The planning horizon usually extends from three to 18 months but primarily depends on the purchasing and manufacturing lead times. This will be discussed in Chapter 3 in the section on master production scheduling. Usually, the plans are reviewed and changed weekly or monthly.

The Material Requirements Plan

The material requirements plan (MRP) is a plan for the production and purchase of the components used in making the items in the master production schedule. It shows the quantities needed and when manufacturing intends to make or use them. Purchasing and production activity control use the MRP to decide the purchase or manufacture of specific items.

The level of detail is high. The material requirements plan establishes when the components and parts are needed to make each end item.

The planning horizon is at least as long as the combined purchase and manufacturing lead times. As with the master production schedule, it usually extends from three to 18 months.

Purchasing and Production Activity Control

Purchasing and production activity control (PAC) represent the implementation and control phase of the production planning and control system. Purchasing is responsible for establishing and controlling the flow of raw materials into the factory. Production activity control is responsible for planning and controlling the flow of work through the factory.

The planning horizon is very short, perhaps from a day to a month. The level of detail is high since it is concerned with individual components, workstations, and orders. Plans are reviewed and revised daily.

Figure 2.4 shows the relationship among the various planning tools, planning horizons, and level of detail.

The levels discussed in the preceding sections will be examined in more detail in later chapters. This chapter discusses production planning. Later chapters deal with master production scheduling, material requirements planning, and production activity control.

Figure 2.4   Level of detail versus planning horizon.

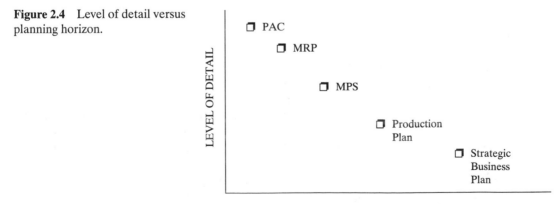

Capacity Management

At each level in the manufacturing planning and control system, the priority plan must be tested against the available resources and capacity of the manufacturing system. Chapter 5 will describe some of the details of capacity management. For now, it is sufficient to understand that the basic process is one of calculating the capacity needed to manufacture the priority plan and of finding methods to make that capacity available. There can be no valid, workable production plan unless this is done. If the capacity cannot be made available when needed, then the plans must be changed.

Determining the capacity required, comparing it to available capacity, and making adjustments (or changing plans) must occur at all levels of the manufacturing planning and control system.

Over several years, machinery, equipment, and plants can be added to or taken away from manufacturing. However, in the time spans involved from production planning to production activity control, these kinds of changes cannot be made. Some changes, such as changing the number of shifts, working overtime, subcontracting the work, and so on, can be accomplished in these time spans.

SALES AND OPERATIONS PLANNING (SOP)

The strategic business plan integrates the plans of all the departments in an organization and is normally updated annually. However, these plans should be updated as time progresses so that the latest forecasts and market and economic conditions are taken into account. Sales and operations planning (SOP) is a process for continually revising the strategic business plan and coordinating plans of the various departments.

While the strategic business plan is updated annually, sales and operations planning is a dynamic process in which the company plans are updated on a regular basis, usually at least monthly. The process starts with the sales and marketing departments, which compare actual demand with the sales plan, assess market potential, and forecast future demand. The updated marketing plan is then communicated to manufacturing, engineering, and finance, which adjust their plans to support the revised marketing plan. If these departments find they cannot accommodate the new marketing plan, then the marketing plan must be adjusted. In this way the strategic business plan is continually revised throughout the year and the activities of the various departments are coordinated. Figure 2.5 shows the relationship between the strategic business plan and the sales and operations plan.

Figure 2.5 Sales and operations planning.

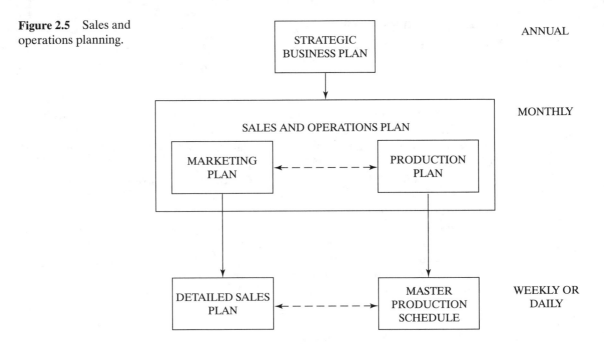

Sales and operations planning is medium range and includes the marketing, production, engineering, and finance plans. Sales and operations planning has several benefits:

- It provides a means of updating the strategic business plan as conditions change.
- It provides a means of managing change. Rather than reacting to changes in market conditions or the economy after they happen, the SOP forces management to look at the economy at least monthly and places it in a better position to plan changes.
- Planning ensures the various department plans are realistic, coordinated, and support the business plan.
- It provides a realistic plan that can achieve the company objectives.
- It permits better management of production, inventory, and backlog.

MANUFACTURING RESOURCE PLANNING (MRP II)

Because of the large amount of data and the number of calculations needed, the manufacturing planning and control system will probably have to be computer based. If a computer is not used, the time and labor required to make calculations manually is extensive and forces a company into compromises. Instead of scheduling requirements through the planning system, the company may have to extend lead times and build inventory to compensate for the inability to schedule quickly what is needed and when.

The system is intended to be a fully integrated planning and control system that works from the top down and has feedback from the bottom up. Strategic business planning integrates the plans and activities of marketing, finance, and production to create plans intended to achieve the overall goals of the company. In turn, master production scheduling, material requirements planning, production activity control, and purchasing are directed toward achieving the goals of the production and strategic business plans and, ultimately, the company. If priority plans have to be adjusted at any of the planning levels because of capacity problems, those changes should be reflected in the levels above. Thus, there must be feedback throughout the system.

The strategic business plan incorporates the plans of marketing, finance, and production. Marketing must agree that its plans are realistic and attainable. Finance must agree that the plans are desirable from a financial point of view, and production must agree that it can meet the required demand. The manufacturing planning and control system, as described here, is a master game plan for all departments in the company. This fully integrated planning and control system is called a **manufacturing resource planning,** or **MRP II,** system. The phrase "MRP II" is used to distinguish the "manufacturing resource plan" (MRP II) from the "materials requirement plan" (MRP).

MRP II provides coordination between marketing and production. Marketing, finance, and production agree on a total workable plan expressed in the production plan. Marketing and production must work together on a weekly and daily basis to adjust the plan as changes occur. Order sizes may need to be changed, orders canceled, and delivery dates adjusted. This kind of change is made through the master production schedule. Marketing managers and production managers may change master production schedules to meet changes in forecast demand. Senior management may adjust the production plan to reflect overall changes in demand or resources. However, they all work through the MRP II system. It provides the mechanism for coordinating the efforts of marketing, finance, production, and other departments in the company. MRP II is a method for the effective planning of all resources of a manufacturing company.

Figure 2.6 shows a diagram of an MRP II system. Note the feedback loops that exist.

MAKING THE PRODUCTION PLAN

We have looked briefly at the purpose, planning horizon, and level of detail found in a production plan. This section will discuss some details involved in making production plans.

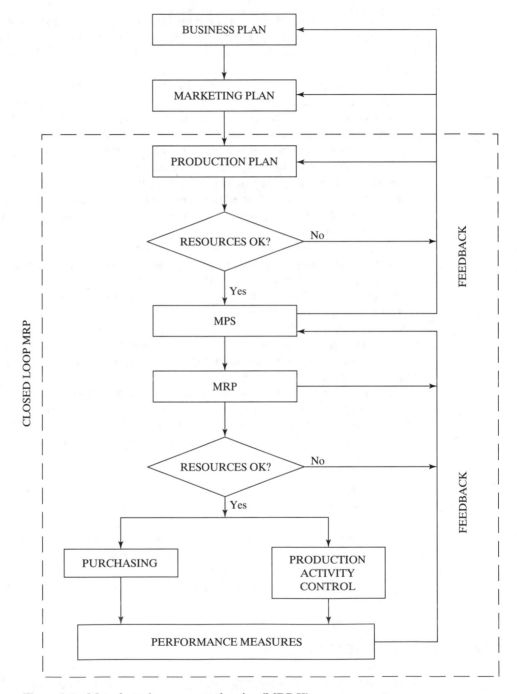

Figure 2.6   Manufacturing resource planning (MRP II).

Based on the market plan and available resources, the production plan sets the limits or levels of manufacturing activity for some time in the future. It integrates the capabilities and capacity of the factory with the market and financial plans to achieve the overall business goals of the company.

The production plan sets the general levels of production and inventories over the planning horizon. Its prime purpose is to establish production rates that will accomplish the objectives of the strategic business plan. These include inventory levels, backlogs (unfilled customer orders), market demand, customer service, low-cost plant operation, labor relations, and so on. The plan must extend far enough in the future to plan for the labor, equipment, facilities, and material needed to accomplish it. Typically, this is a period of 6 to 18 months and is done in monthly and sometimes weekly periods.

The planning process at this level ignores such details as individual products, colors, styles, or options. With the time spans involved and the uncertainty of demand over long periods, the detail would not be accurate or useful, and the plan would be expensive to create. For planning purposes, a common unit or small number of product groups is what is needed.

Establishing Product Groups

Firms that make a single product, or products that are similar, can measure their output directly by the number of units they produce. A brewery, for instance, might use barrels of beer as a common denominator. Many companies, however, make several different products, and a common denominator for measuring total output may be difficult or impossible to find. Product groups need to be established. While marketing naturally looks at products from the customers' point of view of functionality and application, manufacturing looks at products in terms of processes. Thus, firms need to establish product groups based on the similarity of manufacturing processes.

Manufacturing must provide the capacity to produce the goods needed. It is concerned more with the demand for the specific kinds of capacity needed to make the products than with the demand for the product.

Capacity is the ability to produce goods and services. It means having the resources available to satisfy demand. For the time span of a production plan, it can be expressed as the time available or, sometimes, as the number of units or dollars that can be produced in a given period. The demand for goods must be translated into the demand for capacity. At the production planning level, where little detail is needed, this requires identifying product groups, or families, of individual products *based on the similarity of manufacturing process.* For example, several calculator models might share the same processes and need the same kind of capacity, regardless of the variations in the models. They would be considered as a family group of products.

Over the time span of the production plan, large changes in capacity are usually not possible. Additions or subtractions in plant and equipment are impossible or very difficult to accomplish in this period. However, some things can be altered, and it is the responsibility of manufacturing management to identify and assess them. Usually the following can be varied:

- People can be hired and laid off, overtime and short time can be worked, and shifts can be added or removed.
- Inventory can be built up in slack periods and sold or used in periods of high demand.
- Work can be subcontracted or extra equipment leased.

Each alternative has its associated benefits and costs. Manufacturing management is responsible for finding the least-cost alternative consistent with the goals and objectives of the business.

Basic Strategies

In summary, the production planning problem typically has the following characteristics:

- A time horizon of 12 months is used, with periodic updating perhaps every month or quarter.
- Production demand consists of one or a few product families or common units.
- Demand is fluctuating or seasonal.
- Plant and equipment are fixed within the time horizon.
- A variety of management objectives such as low inventories, efficient plant operation, good customer service, and good labor relations.

Suppose a product group has the demand forecast shown in Figure 2.7. Note that the demand is seasonal.

Figure 2.7   Hypothetical demand curve.

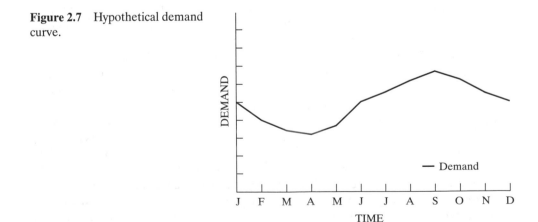

There are three basic strategies that can be used in developing a production plan:

1. Chase strategy.
2. Production leveling.
3. Subcontracting.

Chase (demand matching) strategy. **Chase strategy** means producing the amounts demanded at any given time. Inventory levels remain stable while production varies to meet demand. Figure 2.8 shows this strategy. The firm manufactures just enough at any one time to meet demand exactly. In some industries, this is the only strategy that can be followed. Farmers, for instance, must produce in the growing season. The post office must process mail over the Christmas rush and in slack seasons. Restaurants have to serve meals when the customers want them. These industries cannot stockpile or inventory their products or services and must be capable of meeting demand as it occurs.

Figure 2.8 Demand matching strategy.

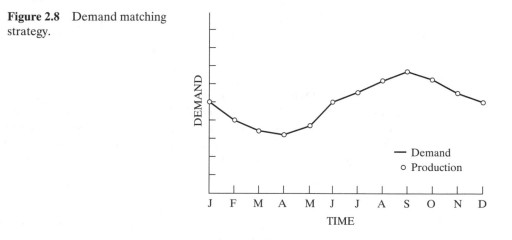

In these cases, the company must have enough capacity to be able to meet the peak demand. Farmers must have sufficient machinery and equipment to harvest in the growing season although the equipment will lie idle in the winter. Companies have to hire and train people for the peak periods and lay them off when the peak is past. Sometimes they have to put on extra shifts and overtime. All these changes add cost.

The advantage to the chase strategy is that inventories can be kept to a minimum. Goods are made when demand occurs and are not stockpiled. Thus, the costs associated with carrying inventories are avoided. Such costs can be quite high, as is shown in Chapter 9 on inventory fundamentals.

Production leveling.   **Production leveling** is continually producing an amount equal to the average demand. This relationship is shown in Figure 2.9. Companies calculate their total demand over the time span of the plan and, on the average, produce enough to meet it. Sometimes demand is less than the amount produced and an inventory builds up. At other times demand is greater and inventory is used up.

Figure 2.9   Production leveling strategy.

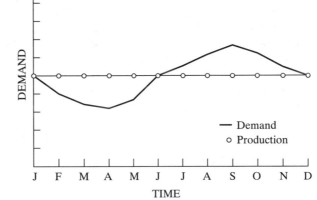

The advantage of a production leveling strategy is that it results in a smooth level of operation that avoids the costs of changing production levels. Firms do not need to have excess capacity to meet peak demand. They do not need to hire and train workers and lay them off in slack periods. They can build a stable work force. The disadvantage is that inventory will build up in low demand periods. This inventory will cost money to carry.

Production leveling means the company will use its resources at a level rate and produce the same amount each day it is operating. The amount produced each month (or sometimes each week) will not be constant because the number of working days varies from month to month.

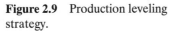

EXAMPLE PROBLEM

A company wants to produce 10,000 units of an item over the next three months at a level rate. The first month has 20 working days; the second, 21 working days; and the third, 12 working days because of an annual shutdown. On the average, how much should the company produce each day to level production?

Answer

$$\text{Total production} = 10{,}000 \text{ units}$$
$$\text{Total working days} = 20 + 21 + 12 = 53 \text{ days}$$
$$\text{Average daily production} = \frac{10{,}000}{53} = 188.7 \text{ units}$$

For some products for which demand is very seasonal, such as Christmas tree lights, some form of production leveling strategy is necessary. The costs of idle capacity, of hiring, training, and laying off, are severe if a company employs a chase strategy.

Subcontracting. As a pure strategy, **subcontracting** means always producing at the level of minimum demand and meeting any additional demand through subcontracting. Subcontracting can mean buying the extra amounts demanded or turning away extra demand. The latter can be done by increasing prices when demand is high or by extending lead times. This strategy is shown in Figure 2.10.

The major advantage of this strategy is cost. Costs associated with excess capacity are avoided, and because production is leveled, there are no costs associated with changing production levels. The main disadvantage is that the cost of purchasing (item cost, purchasing, transportation, and inspection costs) may be greater than if the item were made in the plant.

Few companies make everything or buy everything they need. The decision about which items to buy and which to manufacture depends mainly on cost, but there are several other factors that may be considered.

Firms may manufacture to keep confidential processes within the company, to maintain quality levels, and to maintain a workforce.

They may buy from a supplier who has special expertise in design and manufacture of a component, to allow the firm to concentrate on its own area of expertise, or to provide known and competitive prices.

Figure 2.10 Subcontracting.

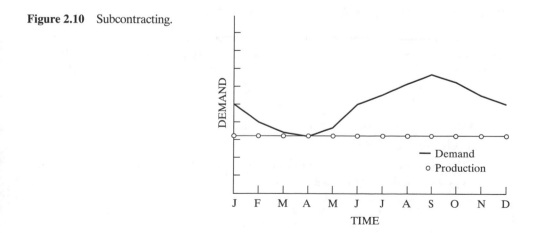

For many items, such as nuts and bolts or components that the firm does not normally make, the decision is clear. For other items that are within the specialty of the firm, a decision must be made whether to subcontract or not.

Hybrid strategy.   The three strategies discussed so far are pure strategies. Each has its own set of costs: equipment, hiring/layoff, overtime, inventory, and subcontracting. In reality, there are many possible hybrid or combined strategies a company may use. Each will have its own set of cost characteristics. Production management is responsible for finding the combination of strategies that minimizes the sum of all costs involved, providing the level of service required, and meeting the objectives of the financial and marketing plans.

Figure 2.11 shows a possible hybrid plan. Demand is matched to some extent, production is partially smoothed, and in the peak period, some subcontracting takes place. The plan is only one of many that could be developed.

Figure 2.11   Hybrid strategy.

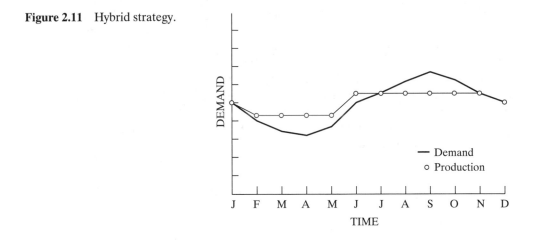

Developing a Make-to-Stock Production Plan

In a make-to-stock environment, products are made and put into inventory before an order is received from a customer. Sale and delivery of the goods are made from inventory. Off-the-rack clothing, frozen foods, and bicycles are examples of this kind of manufacturing.

Generally firms make to stock when:

- Demand is fairly constant and predictable.
- There are few product options.
- Delivery times demanded by the marketplace are much shorter than the time needed to make the product.
- Product has a long shelf life.

The information needed to make a production plan is as follows:

- Forecast by period for the planning horizon.
- Opening inventory.
- Desired ending inventory.
- Any past-due customer orders. These are orders that are late for delivery and are sometimes called *back orders.*

The objective in developing a production plan is to minimize the costs of carrying inventory, changing production levels, and stocking out (not supplying the customer what is wanted when it is wanted).

This section develops a plan for leveling production and one for chase strategy.

Level production plan. Following is the general procedure for developing a plan for level production.

1. Total the forecast demand for the planning horizon.
2. Determine the opening inventory and the desired ending inventory.
3. Calculate the total production required as follows:

$$\text{Total Production} = \text{total forecast} + \text{back orders} + \text{ending inventory} - \text{opening inventory}$$

4. Calculate the production required each period by dividing the total production by the number of periods.
5. Calculate the ending inventory for each period.

▼ ──

EXAMPLE PROBLEM

Amalgamated Fish Sinkers makes a product group of fresh fish sinkers and wants to develop a production plan for them. The expected opening inventory is 100 cases, and they want to reduce that to 80 cases by the end of the planning period. The number of working days is the same for each period. There are no back orders. The expected demand for the fish sinkers is as follows:

Period	1	2	3	4	5	Total
Forecast (cases)	110	120	130	120	120	600

a. How much should be produced each period?

b. What is the ending inventory for each period?

c. If the cost of carrying inventory is $5 per case per period based on ending inventory, what is the total cost of carrying inventory?

d. What will be the total cost of the plan?

Answer

a. Total production required $= 600 + 80 - 100 = 580$ cases

$$\text{Production each period} = \frac{580}{5} = 116 \text{ cases}$$

b. Ending inventory $=$ opening inventory $+$ production $-$ demand

Ending inventory after the first period $= 100 + 116 - 110 = 106$ cases

Similarly, the ending inventories for each period are calculated as shown in Figure 2.12. The ending inventory for period 1 becomes the opening inventory for period 2:

Ending inventory (period 2) $= 106 + 116 - 120 = 102$ cases

c. The total cost of carrying inventory would be:

$$(106 + 102 + 88 + 84 + 80)(\$5) = \$2300$$

d. Since there were no stockouts and no changes in the level of production, this would be the total cost of the plan.

Period		1	2	3	4	5	Total
Forecast (cases)		110	120	130	120	120	600
Production		116	116	116	116	116	580
Ending Inventory	100	106	102	88	84	80	

Figure 2.12   Level production plan: make-to-stock.

Chase strategy.   Amalgamated Fish Sinkers makes another line of product called "fish stinkers." Unfortunately, they are perishable, and the company cannot build inventory for sale later. They must use a chase strategy and make only enough to satisfy demand in each period. Inventory costs will be a minimum, and there should be no stockout costs. However, there will be costs associated with changing production levels.

Let us suppose in the preceding example that changing the production level by one case costs $20. For example, a change from 50 to 60 would cost

$$(60 - 50) \times \$20 = \$200$$

The opening inventory is 100 cases, and the company wishes to bring this down to 80 cases in the first period. The required production in the first period would then be:

$$110 - (100 - 80) = 90 \text{ cases}$$

Assuming that production in the period before period 1 was 100 cases, Figure 2.13 shows the changes in production levels and in ending inventory.

The cost of the plan would be:

$$\text{Cost of changing production level} = (60)(\$20) = \$1200$$
$$\text{Cost of carrying inventory} = (80 \text{ cases})(5 \text{ periods})(\$5) = \$2000$$
$$\text{Total cost of the plan} = \$1200 + \$2000 = \$3200$$

Developing a Make-to-Order Production Plan

In a make-to-order environment, manufacturers wait until an order is received from a customer before starting to make the goods. Examples of this kind of manufacture are custom-tailored clothing, machinery, or any product made to customer specification. Very expensive items are usually made to order. Generally, firms make to order when:

- Goods are produced to customer specification.
- The customer is willing to wait while the order is being made.
- The product is expensive to make and to store.
- Several product options are offered.

Period	0	1	2	3	4	5	Total
Demand (cases)		110	120	130	120	120	600
Production	100	90	120	130	120	120	580
Change in Production		10	30	10	10	0	60
Ending Inventory	100	80	80	80	80	80	

Figure 2.13 Demand matching plan: make-to-stock.

Assemble to order.   Where several product options exist, such as in automobiles, and where the customer is not willing to wait until the product is made, manufacturers produce and stock standard component parts. When manufacturers receive an order from a customer, they assemble the component parts from inventory according to the order. Since the components are stocked, the firm needs only time to assemble before delivering to the customer. Examples of assemble-to-order products include automobiles and computers. Assemble to order is a subset of make to order.

The following information is needed to make a production plan for make-to-order products:

- Forecast by period for the planning horizon.
- Opening backlog of customer orders.
- Desired ending backlog.

Backlog.   In a make-to-order environment, a company does not build an inventory of finished goods. Instead, it has a backlog of unfilled customer orders. The backlog normally will be for delivery in the future and does not represent orders that are late or past due. A custom woodwork shop might have orders from customers that will keep it busy for several weeks. This will be its backlog. If individuals want some work done, the order will join the queue or backlog. Manufacturers like to control the backlog so that they can provide a good level of customer service.

Level production plan.   Following is a general procedure for developing a level production plan:

1. Total the forecast demand for the planning horizon.
2. Determine the opening backlog and the desired ending backlog.
3. Calculate the total production required as follows:

   Total production = total forecast + opening backlog − ending backlog

4. Calculate the production required each period by dividing the total production by the number of periods.
5. Spread the existing backlog over the planning horizon according to due date per period.

EXAMPLE PROBLEM

A local printing company provides a custom printing service. Since each job is different, demand is forecast in hours per week. Over the next five weeks, the company expects that demand will be 100 hours per week. There is an existing backlog of 100 hours, and at the end of five weeks, the company wants to reduce that to 80 hours. How many hours of work will be needed each week to reduce the backlog? What will be the backlog at the end of each week?

Answer

$$\text{Total production} = 500 + 100 - 80 = 520 \text{ hours}$$
$$\text{Weekly production} = \frac{520}{5} = 104 \text{ hours}$$

The backlog for each week can be calculated as:

$$\text{Projected backlog} = \text{old backlog} + \text{forecast} - \text{production}$$

For week 1: Projected backlog $= 100 + 100 - 104 = 96$ hours
For week 2: Projected backlog $= 96 + 100 - 104 = 92$ hours

Figure 2.14 shows the resulting production plan.

Period		1	2	3	4	5	Total
Sales Forecast		100	100	100	100	100	500
Planned Production		104	104	104	104	104	520
Projected Backlog	100	96	92	88	84	80	

Figure 2.14 Level production plan: make-to-order.

Resource Planning

Once the preliminary production plan is established, it must be compared to the existing resources of the company. This step is called resource requirements planning or resource planning. Two questions must be answered:

1. Are the resources available to meet the production plan?
2. If not, how will the difference be reconciled?

If enough capacity to meet the production plan cannot be made available, the plan must be changed.

A tool often used is the **resource bill.** This shows the quantity of *critical* resources (materials, labor, and "bottleneck" operations) needed to make one average unit of the product group. Figure 2.15 shows an example of a resource bill for a company that makes tables, chairs, and stools as a three-product family.

Figure 2.15   Resource bill.

Product	Wood (board feet)	Labor (standard hours)
Tables	20	1.31
Chairs	10	0.85
Stools	5	0.55

If the firm planned to make 500 tables, 300 chairs, and 1500 stools in a particular period, they could calculate the quantity of wood and labor that will be needed. For example, the amount of wood needed is:

Tables:	$500 \times 20 =$	10,000 board feet
Chairs:	$300 \times 10 =$	3000 board feet
Stools:	$1500 \times 5 =$	7500 board feet
Total wood required	$=$	20,500 board feet

The amount of labor needed is:

Tables:	$500 \times 1.31 =$	655 standard hours
Chairs:	$300 \times 0.85 =$	255 standard hours
Stools:	$1500 \times 0.55 =$	825 standard hours
Total labor required	$=$	1735 standard hours

The company must now compare the requirements for wood and labor with the availability of these resources. For instance, suppose the labor normally available in this period is 1600 hours. The priority plan requires 1735 hours, a difference of 135 hours, or about 8.4%. Extra capacity must be found, or the priority plan must be adjusted. In this example, it might be possible to work overtime to provide the extra capacity required. If overtime is not possible, the plan must be adjusted to reduce the labor needed. This might involve shifting some production to an earlier period or delaying shipments.

SUMMARY

Production planning is the first step in a manufacturing planning and control system. The planning horizon usually extends for a year. The minimum horizon depends on the lead times to purchase materials and make the product. The level of detail is not high. Usually, the plan is made for families of products based on the similarity of manufacturing process or on some common unit.

Three basic strategies can be used to develop a production plan: chase, leveling production, or subcontracting. Each has its operational and cost advantages and disadvantages. It is the responsibility of manufacturing management to select the best combination of these basic plans so total costs are minimized and customer service levels are maintained.

A make-to-stock production plan determines how much to produce in each period to meet the following objectives:

- Achieve the forecast.
- Maintain the required inventory levels.

Although demand must be satisfied, the plan must balance the costs of maintaining inventory with the cost of changing production levels.

A make-to-order production plan determines how much to produce in each period to meet the following objectives:

- Achieve the forecast.
- Maintain the planned backlog.

The cost of a backlog that is too large equals the cost of turning away business. If customers have to wait too long for delivery, they might take their business elsewhere. As with a make-to-stock production plan, demand must be satisfied, and the plan must balance the costs of changing production levels with the cost of a backlog that is larger than desired.

QUESTIONS

1. What are the four questions a good planning system must answer?

2. Define capacity and priority. Why are they important in production planning?

3. Describe each of the following plans in terms of their purpose, planning horizon, level of detail, and planning cycle:

   a. Strategic business plan.

   b. Production plan.

   c. Master production schedule.

   d. Material requirements plan.

   e. Production activity control.

4. Describe the responsibilities and inputs of the marketing, production, finance, and engineering departments to the strategic business plan.

5. Describe the relationship among the production plan, the master production schedule, and the material requirements plan.

6. What is the difference between strategic business planning and sales and operations planning (SOP)? What are the major benefits of SOP?

7. What is closed-loop MRP?

8. What is MRP II?

9. In the short run, how can capacity be changed?

10. When making a production plan, why is it necessary to select a common unit or to establish product families?

11. On what basis should product groups (families) be established?

12. What are five typical characteristics of the production planning problem?

13. Describe each of the three basic strategies used in developing a production plan. What are the advantages and disadvantages of each?

14. What is a hybrid strategy? Why is it used?

15. Describe four conditions under which a firm would make to stock or make to order.

16. What information is needed to develop a make-to-stock production plan?

17. What are the steps in developing a make-to-stock production plan?

18. What is the difference between make to order and assemble to order? Give an example of each.

19. What information is needed to develop a make-to-order production plan? How does this differ from that needed for a make-to-stock plan?

20. What is the general procedure for developing a level production plan in a make-to-order environment?

21. What is a resource bill? At what level in the planning hierarchy is it used?

PROBLEMS

2.1 If the opening inventory is 500 units, demand is 800 units, and production is 600 units, what will be the ending inventory?

   Answer. 300 units.

2.2 A company wants to produce 500 units over the next four months at a level rate. The months have 19, 22, 20, and 21 working days, respectively. On the average, how much should the company produce each day to level production?

   Answer. Average daily production = 6.1 units

2.3 A company plans to produce 20,000 units in a three-month period. The months have 22, 24, and 19 working days respectively. What should the average daily production be?

2.4 In problem 2.2, how much will be produced in each of the four months?

   Answer.

   Month 1: 115.9

   Month 2: 134.2

   Month 3: 122

   Month 4: 128.1

2.5 In problem 2.3, how much will be produced in each of the three months?

2.6 A production line is to run at 1000 units per month. Sales are forecast as shown in the following. Calculate the expected period-end inventory. The opening inventory is 500 units. All periods have the same number of working days.

Period	1	2	3	4	5	6
Forecast	800	900	1200	1500	1000	800
Planned Production	1000	1000	1000	1000	1000	1000
Planned Inventory	500						

   Answer. For period 1, the ending inventory is 700 units.

2.7 A company wants to develop a level production plan for a family of products. The opening inventory is 100 units, and an increase to 130 units is expected by the end of the plan. The demand for each period is given in what follows. How much should the company produce each period? What will be the ending inventories in each period? All periods have the same number of working days.

Period		1	2	3	4	5	6	Total
Forecast Demand		100	120	130	140	120	110	
Planned Production								
Planned Inventory	100							

Answer. Total production = 750 units
Period production = 125 units
The ending inventory for period 1 is 125; for period 5, 115.

2.8 A company wants to develop a level production plan for a family of products. The opening inventory is 500 units, and a decrease to 300 units is expected by the end of the plan. The demand for each of the periods is given in what follows. All periods have the same number of working days. How much should the company produce each period? What will be the ending inventories in each period? Do you see any problems with the plan?

Period		1	2	3	4	5	6	Total
Forecast Demand		1200	1200	800	600	800	1000	
Planned Production								
Planned Inventory	500							

2.9 A company wants to develop a level production plan. The beginning inventory is zero. Demand for the next four periods is given in what follows.

   a. What production rate per period will give a zero inventory at the end of period 4?

   b. When and in what quantities will back orders occur?

   c. What level production rate per period will avoid back orders? What will be the ending inventory in period 4?

Period		1	2	3	4	Total
Forecast Demand		10	5	12	9	
Planned Production						
Planned Inventory	0					

Answer. **a.** 9 units
        b. period 1, minus 1
        c. 10 units; 4 units

2.10 If the cost of carrying inventory is $50 per unit per period and stockouts cost $500 per unit, what will be the cost of the plan developed in problem 2.9a? What will be the cost of the plan developed in 2.9c?

Answer. Total cost for plan in question 2.9a = $650
Total cost for plan in question 2.9c = $600

2.11 A company wants to develop a level production plan for a family of products. The opening inventory is 100 units, and an increase to 130 units is expected by the end of the plan. The demand for each month is given in what follows. Calculate the total production, daily production, and production and ending inventory for each month.

Month		May	Jun	Jul	Aug	Total
Working Days		21	19	20	10	
Forecast Demand		105	115	130	140	
Planned Production						
Planned Inventory	100					

Answer. The monthly production for May = 156 units
The ending inventory for May = 151 units

2.12 A company wants to develop a level production plan for a family of products. The opening inventory is 500 units, and a decrease to 300 units is expected by the end of the plan. The demand for each of the months is given in what follows. How much should the company produce each month? What will be the ending inventory in each month? Do you see any problems with the plan?

Month		Jan	Feb	Mar	Apr	May	Jun	Total
Working Days		20	22	20	20	18	19	
Forecast Demand		1200	1300	800	700	700	900	
Planned Production								
Planned Inventory	500							

2.13 Because of its labor contract, a company must hire enough labor for 100 units of production per week on one shift or 200 units per week on two shifts. They cannot hire, lay off, or work overtime. During the fourth week, workers will be available from another department to work part or all of an extra shift (up to 100 units). There is a planned shutdown for maintenance in the second week, which will cut production to half. Develop a production plan. The opening inventory is 200 units, and the desired ending inventory is 300 units.

Week	1	2	3	4	5	6	Total
Forecast Demand	120	160	240	240	160	160	
Planned Production							
Planned Inventory							

2.14 If the opening backlog is 400 units, forecast demand is 600 units, and production is 800 units, what will be the ending backlog?

    Answer. 200 units

2.15 The opening backlog is 800 units. Forecast demand is shown in the following. Calculate the weekly production for level production if the backlog is to be reduced to 400 units.

Period		1	2	3	4	5	6	Total
Forecast Demand		600	700	700	700	600	500	
Planned Production								
Projected Backlog	800							

    Answer. Total production = 4200 units
                  Weekly production = 700 units
                  Backlog at end of week 1 = 700 units

2.16 The opening backlog is 1000 units. Forecast demand is shown in the following. Calculate the weekly production for level production if the backlog is to be increased to 1200 units.

Period		1	2	3	4	5	6	Total
Forecast Demand		1200	1100	1200	1200	1100	1000	
Planned Production								
Projected Backlog	1000							

2.17 For the following data, calculate the number of workers required for level production and the resulting month-end inventories. Each worker can produce 15 units per day, and the desired ending inventory is 9000 units.

Month		1	2	3	4	Total
Working Days		20	24	12	19	
Forecast Demand		28,000	27,500	28,500	28,500	
Planned Production						
Planned Inventory	11,500					

Answer. Workers needed = 98 workers
First month's ending inventory = 12,900 units

2.18 For the following data, calculate the number of workers required for level production and the resulting month-end inventories. Each worker can produce 9 units per day, and the desired ending inventory is 800 units. Why is it not possible to reach the ending inventory target?

Month		1	2	3	4	5	6	Total
Working Days		20	24	12	22	20	19	
Forecast Demand		2800	3000	2700	3300	2900	3200	
Planned Production								
Planned Inventory	1000							

3

Master Production Scheduling

INTRODUCTION

After production planning, the next step in the manufacturing planning and control process is to prepare a master production schedule (MPS). This chapter examines some basic considerations in making and managing an MPS. It is an extremely important planning tool and forms the basis for communication between sales and manufacturing. The MPS is a vital link in the production planning system:

- It forms the link between production planning and what manufacturing will actually build.
- It forms the basis for calculating the capacity and resources needed.
- The MPS drives the material requirements plan. As a schedule of items to be built, the MPS and bills of material determine what components are needed from manufacturing and purchasing.
- It keeps priorities valid. The MPS is a priority plan for manufacturing.

While the production plan deals in families of products, the MPS works with end items. It breaks down the production plan into the requirements for individual end items, in each family, by date and quantity. The production plan limits the MPS. Therefore the total of the items in the MPS should not be different from the total shown on the production plan. For example, if the production plan shows a planned production of 1000 tricycles in a particular week, the total of the individual models planned for by the MPS should be 1000. Within this limit, its objective is to balance the demand (priorities) set by the marketplace with the availability of materials, labor, and equipment (capacity) of manufacturing.

The end items made by the company are assembled from component and subcomponent parts. These must be available in the right quantities at the right time to support the master production schedule. The material requirements planning system plans the schedule for these components based on the needs of the MPS. Thus the MPS drives the material requirements plan.

The master production schedule is a plan for manufacturing. It reflects the needs of the marketplace and the capacity of manufacturing and forms a priority plan for manufacturing to follow.

The MPS forms a vital link between sales and production as follows:

- It makes possible valid order promises. The MPS is a plan of what is to be produced and when. As such, it tells sales and manufacturing when goods will be available for delivery.

- It is a contract between marketing and manufacturing. It is an agreed-upon plan.

The MPS forms a basis for sales and production to determine what is to be manufactured. It is not meant to be rigid. It is a device for communication and a basis to make changes that are consistent with the demands of the marketplace and the capacity of manufacturing.

The information needed to develop an MPS is provided by:

- The production plan.
- Forecasts for individual end items.
- Actual orders received from customers and for stock replenishment.
- Inventory levels for individual end items.
- Capacity restraints.

RELATIONSHIP TO PRODUCTION PLAN

Suppose the following production plan is developed for a family of 3 items:

Week	1	2	3	4	5	6
Aggregate Forecast (units)	160	160	160	160	215	250
Production Plan	205	205	205	205	205	205
Aggregate Inventory (units)	545	590	635	680	670	625

Opening inventories (units) are:

Product A	350
Product B	100
Product C	_50
Total	500

The next step is to forecast demand for each item in the product family.

Week	1	2	3	4	5	6
Product A	70	70	70	70	70	80
Product B	40	40	40	40	95	120
Product C	50	50	50	50	50	50
Total	160	160	160	160	215	250

With these data, the master scheduler must now devise a plan to fit the constraints. The following illustrates a possible solution.

Master Production Schedule

Week	1	2	3	4	5	6
Product A						205
Product B	205	205	205			
Product C				205	205	
Total Planned	205	205	205	205	205	205

Inventory

Week	1	2	3	4	5	6
Product A	280	210	140	70	0	125
Product B	265	430	595	555	460	340
Product C	0	−50	−100	55	210	160
Total Planned	545	640	735	680	670	625

This schedule is satisfactory for the following reasons:

- It tells the plant when to start and stop production of individual items.
- Capacity is consistent with the production plan.

It is unsatisfactory for the following reasons:

- It has a poor inventory balance compared to total inventory.
- It results in a stockout for product C in periods 2 and 3.

EXAMPLE PROBLEM

The Hotshot Lightning Rod Company makes a family of two lightning rods, Models H and I. It bases its production planning on months. For the present month, production is leveled at 1000 units. Opening inventory is 500 units, and the plan is to reduce that to 300 units by the end of the month. The MPS is made using weekly periods. There are four weeks in this month, and production is to be leveled at 250 units per week. The forecast and projected available for the two lightning rods follows. Calculate an MPS for each item.

Answer

Production Plan

Week		1	2	3	4	Total
Forecast		300	350	300	250	1200
Projected Available	500	450	350	300	300	
Production Plan		250	250	250	250	1000

MPS: Model H

Week		1	2	3	4	Total
Forecast		200	300	100	100	700
Projected Available	200	250	200	100	100	
MPS		250	250		100	

MPS: Model I

Week		1	2	3	4	Total
Forecast		100	50	200	150	500
Projected Available	300	200	150	200	200	
MPS				250	150	

DEVELOPING A MASTER PRODUCTION SCHEDULE

The objectives in developing an MPS are as follows:

- To maintain the desired level of customer service by maintaining finished-goods inventory levels or by scheduling to meet customer delivery requirements.
- To make the best use of material, labor, and equipment.
- To maintain inventory investment at the required levels.

To reach these objectives, the plan must satisfy customer demand, be within the capacity of manufacturing, and be within the guidelines of the production plan.

There are three steps in preparing an MPS:

1. Develop a preliminary MPS.
2. Check the preliminary MPS against available capacity.
3. Resolve differences between the preliminary MPS and capacity availability.

Preliminary Master Production Schedule

To show the process of developing an MPS, an example is used that assumes the product is made to stock, an inventory is kept, and the product is made in lots.

A particular item is made in lots of 100, and the expected opening inventory is 80 units. Figure 3.1 shows the forecast of demand, the projected available on hand, and the preliminary MPS.

Period 1 begins with an inventory of 80 units. After the forecast demand for 60 units is satisfied, the projected available is 20 units. A further forecast demand of 60 in period 2 is not satisfied, and it is necessary to schedule an MPS receipt of 100 for week 2. This produces a projected available of 60 units ($20 + 100 - 60 = 60$) at the end of period 2. In period 3, the forecast demand for 60 is satisfied by the projected 60 on

On hand = 80 units
Lot size = 100 units

Period		1	2	3	4	5	6
Forecast			60	60	60	60	60	60
Projected Available	80	20	60	0	40	80	20
MPS			100		100	100	

Figure 3.1 MPS example.

hand, leaving a projected available of zero. In period 4, a further 100 must be received, and when the forecast demand of 60 units is satisfied, 40 units remain in inventory.

This process of building an MPS occurs for each item in the family. If the total planned production of all the items in the family and the total ending inventory do not agree with the production plan, some adjustment to the individual plans must be made so the total production is the same.

Once the preliminary master production schedules are made, they must be checked against the available capacity. This process is called **rough-cut capacity planning.**

▼ ━━━━━━━━━━━━━━━━━━━━━━━━━━━━━━━━━━

EXAMPLE PROBLEM

Amalgamated Nut Crackers, Inc., makes a family of nut crackers. The most popular model is the walnut, and the sales department has prepared a six-week forecast. The opening inventory is 50 dozen (dozen is the unit used for planning). As master planner, you must prepare an MPS. The nutcrackers are made in lots of 100 dozen.

Answer

Week		1	2	3	4	5	6
Forecast Sales		75	50	30	40	70	20
Projected Available	50	75	25	95	55	85	65
MPS		100		100		100	

Rough-Cut Capacity Planning

Rough-cut capacity planning checks whether critical resources are available to support the preliminary master production schedules. Critical resources include bottleneck operations, labor, and critical materials (perhaps material that is scarce or has a long lead time).

The process is similar to resource requirements planning used in the production planning process. The difference is that now we are working with a product and not a family of products. The resource bill, used in resource requirements planning, assumes a typical product in the family. Here the resource bill is for a single product. As before, the only interest is in bottleneck work centers and critical resources.

Suppose a firm manufactures four models of desktop computers assembled in a work center that is a bottleneck operation. The company wants to schedule to the capacity of this work center and not beyond. Figure 3.2 is a resource bill for that work center showing the time required to assemble one computer.

Suppose that in a particular week the master production schedules show the following computers are to be built:

Model D24	200 units
Model D25	250 units
Model D26	400 units
Model D27	100 units

The capacity required on this critical resource is:

$$\text{Model D24 } 200 \times 0.203 = 40.6 \text{ standard hours}$$
$$\text{Model D25 } 250 \times 0.300 = 75.0 \text{ standard hours}$$
$$\text{Model D26 } 400 \times 0.350 = 140.0 \text{ standard hours}$$
$$\text{Model D27 } 100 \times 0.425 = 42.5 \text{ standard hours}$$
$$\text{Total time required } = 298.1 \text{ standard hours}$$

Figure 3.2   Resource bill.

Resource Bill	
Desktop Computer Assembly	
Computer	Assembly Time (standard hours)
Model D24	0.203
Model D25	0.300
Model D26	0.350
Model D27	0.425

EXAMPLE PROBLEM

The Acme Tweezers Company makes tweezers in two models, medium and fine. The bottleneck operation is in work center 20. Following is the resource bill (in hours per dozen).

	Hours per Dozen	
Work Center	Medium	Fine
20	0.5	1.2

The master production schedule for the next four weeks is:

Week	1	2	3	4	Total
Medium	40	25	40	15	120
Fine	20	10	30	20	80

Using the resource bill and the master production schedule, calculate the number of hours required in work center 20 for each of the four weeks. Use the following table to record the required capacity on the work center.

Answer

Week	1	2	3	4	Total
Medium	20	12.5	20	7.5	60
Fine	24	12	36	24	96
Total Hours	44	24.5	56	31.5	156

Resolution of Differences

The next step is to compare the total time required to the available capacity of the work center. If available capacity is greater than the required capacity, the MPS is workable. If not, methods of increasing capacity have to be investigated. Is it possible to adjust the available capacity with overtime, extra workers, routing through other work centers, or subcontracting? If not, it will be necessary to revise the master production schedule.

Finally, the master production schedule must be judged by three criteria:

1. *Resource use.*   Is the MPS within capacity restraints in each period of the plan? Does it make the best use of resources?
2. *Customer service.*   Will due dates be met and will delivery performance be acceptable?
3. *Cost.*   Is the plan economical, or will excess costs be incurred for overtime, subcontracting, expediting, or transportation?

Master Schedule Decisions

The MPS should represent as efficiently as possible what manufacturing will make. If too many items are included, it will lead to difficulties in forecasting and managing the MPS. In each of the manufacturing environments—make to stock, make to order, and assemble to order—master scheduling should take place where the smallest number of product options exists. Figure 3.3 shows the level at which items should be master scheduled.

Make-to-stock products.   In this environment, a limited number of standard items are assembled from many components. Televisions and other consumer products are examples. The MPS is usually a schedule of finished-goods items.

Make-to-order products.   In this environment, many different end items are made from a small number of components. Custom-tailored clothes are an example. The MPS is usually a schedule of the actual customer orders.

Assemble-to-order products.   In this environment, many end items can be made from combinations of basic components and subassemblies. For example, suppose a company manufactures paint from a base color and adds tints to arrive at the final

Figure 3.3   Different MPS environments.

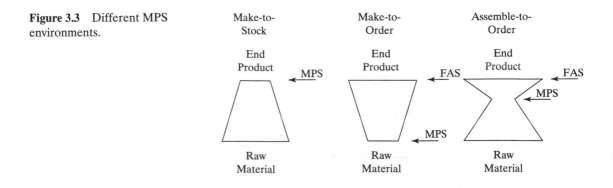

color. Suppose there are ten tints and a final color is made by mixing any three of them with the base. There are 720 possible colors ($10 \times 9 \times 8 = 720$). Forecasting and planning production for 720 items is a difficult task. It is much easier if production is planned at the level of the base color and the ten tints. There are then only ten items with which to deal: the base color and each of the ten tints. Once a customer's order is received, the base color and the required tints can be combined (assembled) according to the order.

Final assembly schedule (FAS). This last step, assembly to customer order, is generally planned using a final assembly schedule. This is a schedule of what will be assembled. It is used when there are many options and it is difficult to forecast which combination the customers will want. Master production scheduling is done at the component level, for example, the base color and tint level. The final assembly takes place only when a customer order is received.

The FAS schedules customer orders as they are received and is based on the components planned in the MPS. It is responsible for scheduling from the MPS through final assembly and shipment to the customer.

Figure 3.4 shows the relationship of the MPS, the FAS, and other planning activities.

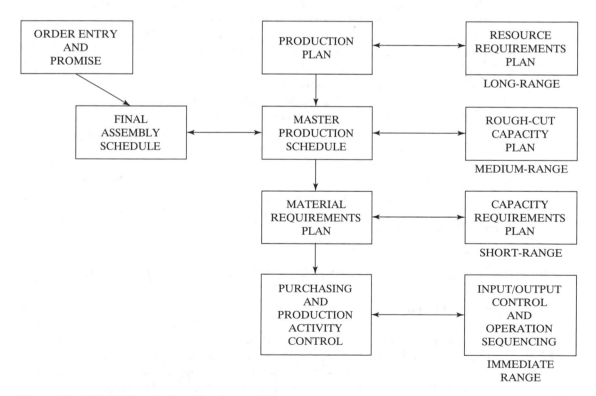

Figure 3.4 MPS, FAS, and other planning activities.

Figure 3.5   Product structure: critical lead time.

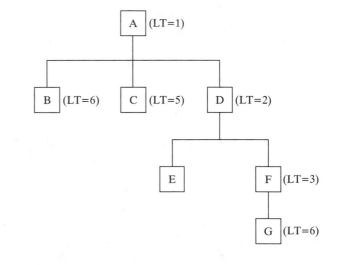

Planning Horizon

The planning horizon is the time span for which plans are made. It must cover a period at least equal to the time required to accomplish the plan. For master production scheduling, the minimum planning horizon is the longest cumulative or end-to-end lead time (LT). For example, in Figure 3.5, the longest cumulative lead time path is A to D to F to G. The cumulative lead time is $1 + 2 + 3 + 6 = 12$ weeks. The minimum planning horizon must be 12 weeks; otherwise, raw material G is not ordered in time to meet delivery.

The planning horizon is usually longer for several reasons. The longer the horizon, the greater the "visibility" and the better management's ability to avoid future problems or to take advantage of special circumstances. For example, firms might take advantage of economical purchase plans, avoid future capacity problems, or manufacture in more economical lot sizes.

As a minimum, the planning horizon for a final assembly schedule must include time to assemble a customer's order. It does not need to include the time necessary to manufacture the components. That time will be included in the planning horizon of the MPS.

PRODUCTION PLANNING, MASTER PRODUCTION SCHEDULING, AND SALES

The production plan reconciles total forecast demand with available resources. It takes information from the strategic business plan and market forecasts to produce an overall plan of what production intends to make to meet forecast. It is dependent on the forecast and, within capacity limits, must plan to satisfy the forecast demand. It is not concerned with the detail of what will actually be made. It is intended to provide a framework in which detailed plans can be made in the MPS.

The MPS is built from forecasts and actual demands for individual end items. It reconciles demand with the production plan and with available resources to produce a plan that manufacturing can fulfill. The MPS is concerned with what items will actually be built, in what quantities, and when, to meet expected demand.

The production plan and the MPS uncouple the sales forecast from manufacturing by establishing a manufacturing plan. Together, they attempt to balance available resources of plant, equipment, labor, and material with forecast demand. However, they are not a sales forecast, nor are they necessarily what is desired. The MPS is a plan for what production *can and will do.*

Figure 3.6 shows the relationship between the sales forecast, production plan, and MPS.

The MPS must be realistic about what manufacturing can and will do. If it is not, it will result in overloaded capacity plans, past-due schedules, unreliable delivery promises, surges in shipments, and lack of accountability.

The master production schedule is a plan for specific end items or "buildable" components that manufacturing expects to make over some time in the future. It is the point at which manufacturing and marketing must agree what end items are going to be produced. Manufacturing is committed to making the goods; marketing, to selling the goods. However, the MPS is not meant to be rigid. Demand changes, problems occur in production, and, sometimes, components are scarce. These events may make it necessary to alter the MPS. Changes must be made with the full understanding and agreement of sales and production. The MPS provides the basis for making changes and a plan on which all can agree.

The MPS and Delivery Promises

In a make-to-stock environment, customer orders are satisfied from inventory. However, in make-to-order or assemble-to-order environments, demand is satisfied from productive capacity. In either case, sales and distribution need to know what is available to satisfy customer demand. Since demand can be satisfied either from inventory or from scheduled receipts, the master production schedule provides a

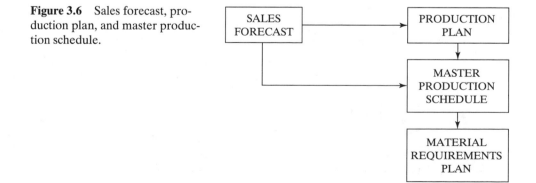

Figure 3.6 Sales forecast, production plan, and master production schedule.

Figure 3.7 The MPS and delivery promises.

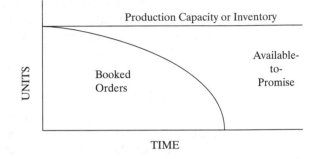

plan for doing either. Figure 3.7 illustrates the concept. As orders are received, they "consume" the available inventory or capacity. Any part of the plan that is not consumed by actual customer orders is available to promise to customers. In this way, the MPS provides a realistic basis for making delivery promises.

Using the MPS, sales and distribution can determine the **available to promise** (ATP). Available to promise is that portion of a firm's inventory and planned production that is not already committed and is available to the customer. This allows delivery promises to be made and customer orders and deliveries to be scheduled accurately.

The ATP is calculated by adding scheduled receipts to the beginning inventory and then subtracting actual orders scheduled before the next scheduled receipt. A **scheduled receipt** is an order that has been issued either to manufacturing or to a supplier. Figure 3.8 illustrates a calculation of an ATP:

$$
\begin{aligned}
\text{ATP for period 1} \;&=\; \text{on hand} - \text{customer orders due before next MPS}\\
&=\; 100 - 80\\
&=\; 20 \text{ units}\\
\text{ATP for period 2} \;&=\; \text{MPS scheduled receipt} - \text{customer orders due before next MPS}\\
&=\; 100 - (10 + 10)\\
&=\; 80 \text{ units}\\
\text{ATP for period 4} \;&=\; 100 - 30 = 70 \text{ units}
\end{aligned}
$$

Inventory on hand: 100 units

Period	1	2	3	4	5
Customer Orders	80	10	10		30
MPS Receipts		100		100	
ATP	20	80		70	

Figure 3.8 Available-to-promise calculation.

This method assumes that the ATP will be sold before the next scheduled receipt arrives. It is there to be sold, and the assumption is that it will be sold. If it is not sold, whatever is left forms an on-hand balance available for the next period.

Continuing with the example problem on page 47, Amalgamated Nut Crackers, Inc., has now received customer orders. Following is the schedule of orders received and the resulting available-to-promise calculation:

Week	1	2	3	4	5	6
Customer Orders	80	45	40	50	50	5
MPS	100		100		100	
ATP	25		10		45	

Sometimes, customer orders are greater than the scheduled receipts. In this case, the *previous* ATP is reduced by the amount needed. Consider the following example:

Week	1	2	3	4	5
Customer Orders	50	20	40	50	
MPS	100		100		
ATP	30		10		

Can the master planner accept an order for another 20 for delivery in week 3? Ten of the units are available from week 3, and ten can be taken from the ATP in week 1, so the order can be accepted.

▼ ━━

EXAMPLE PROBLEM

Calculate the available to promise for the following example. Can an order for 30 more be accepted for delivery in week 5? What will be the ATP if the order is accepted?

Week	1	2	3	4	5
Customer Orders	50	20	30	30	15
MPS	100		100		
ATP	30		25		

Answer

Week	1	2	3	4	5
Customer Orders	50	20	30	30	45
MPS	100		100		
ATP	25		0		

Projected Available

Our calculations so far have based the projected available on the forecast demand. Now there are also customer orders to consider. Customer orders will sometimes be greater than forecast and sometimes less. Projected available is now calculated based on whichever is the greater. For example, if the beginning projected available is 100 units, the forecast is 40 units, and customer orders are 50 units, the ending projected available is 50 units, not 60. Projected available is calculated using *the greater of forecast or customer orders.*

EXAMPLE PROBLEM

Given the following data, calculate the projected available. The order quantity is 120.

Week	1	2	3	4
Forecast	40	70	30	40
Customer Orders	60	60	40	20

Answer

Week		1	2	3	4
Projected Available	100	40	90	50	10
MPS				120	

So far we have considered how to calculate scheduled receipts and available to promise. Using the Amalgamated Nut Cracker, Inc., example, we now combine the two calculations into one record.

Week		1	2	3	4	5	6
Forecast Demand		75	50	30	40	70	20
Customer Orders		80	45	40	50	50	5
Projected Available	50	70	20	80	30	60	40
MPS		100		100		100	
ATP		25		10		45	

Time Fences

Consider the product structure shown in Figure 3.9. Item A is a master-scheduled item and is assembled from B, C, and D. Item D, in turn, is made from raw material E. The lead times to make or to buy the parts are shown in parentheses. The lead time to assemble A is two weeks. To purchase B and C, the respective lead times are six and five weeks. To make D takes eight weeks, and the purchase lead time for raw material E is 16 weeks. The longest cumulative lead time is thus 26 weeks (A + D + E = 2 + 8 + 16 = 26 weeks).

Figure 3.9 Product structure.

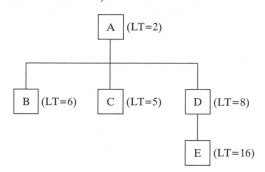

Since the cumulative lead time is 26 weeks, the MPS must have a planning horizon of at least 26 weeks.

Suppose that E is a long-lead-time electronic component and is used in the assembly of other boards as well as D. When E is received 16 weeks after ordering, a decision must be made to commit E to be made into a D or to use it in another board. In eight weeks, a decision must be made to commit D to the final assembly of A. The company would not have to commit the E to making D until ten weeks before delivery of the A. At each of these stages, the company commits itself to more cost and fewer alternatives. Therefore, the cost of making a change increases and the company's flexibility decreases as production gets closer to the delivery time.

Changes to the master production schedule will occur. For example:

- Customers cancel or change orders.
- Machines break down or new machines are added, changing capacity.
- Suppliers have problems and miss delivery dates.
- Processes create more scrap than expected.

A company wants to minimize the cost of manufacture and also be flexible enough to adapt to changing needs. Changes to production schedules can result in the following:

- Cost increases due to rerouting, rescheduling, extra setups, expediting, and buildup of work-in-process inventory.
- Decreased customer service. A change in quantity of delivery can disrupt the schedule of other orders.
- Loss of credibility for the MPS and the planning process.

Changes that are far off on the planning horizon can be made with little or no cost or disruption to manufacturing, but the nearer to delivery date, the more disruptive and costly changes will be. To help in the decision-making process, companies establish zones divided by time fences. Figure 3.10 shows how this might be applied to product A. The zones are as follows:

Figure 3.10 shows the concept. The time fences are as follows.

- *Frozen zone.* Capacity and materials are committed to specific orders. Since changes would result in excessive costs, reduced manufacturing efficiency, and poor customer service, senior management's approval is usually required to make changes. The extent of the frozen zone is defined by the **demand time**

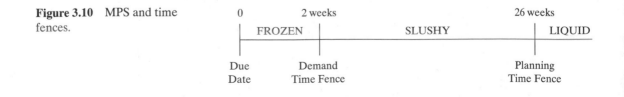

Figure 3.10 MPS and time fences.

fence. Within the demand time fence, demand is usually based on customer orders, not forecast.

- *Slushy zone.* Capacity and material are committed to less extent. This is an area for tradeoffs that must be negotiated between marketing and manufacturing. Materials have been ordered and capacity established; these are difficult to change. However, changes in priorities are easier to change. The extent of the slushy zone is defined by the **planning time fence.** Within this time fence the computer will not reschedule MPS orders.

- *Liquid zone.* Any change can be made to the MPS as long as it is within the limits set by the production plan. Changes are routine and are often made by the computer program.

Changes to the MPS will occur. They must be managed and decisions made with full knowledge of the costs involved.

SUMMARY

The master production schedule (MPS) is a plan for the production of individual end items. It must match demand for the product in total, but it is not a forecast of demand. The MPS must be realistic. It must be achievable and reflect a balance between required and available capacity.

The MPS is the meeting ground for sales and production. It provides a plan from which realistic delivery promises can be made to customers. If adjustments have to be made in deliveries or the booking of orders, they are done through the MPS.

Master production scheduling's major functions are as follows:

- To form the link between production planning and what manufacturing builds.
- To plan capacity requirements. The master production schedule determines the capacity required.
- To plan material requirements. The MPS drives the material requirements plan.
- To keep priorities valid. The MPS is a priority plan for manufacturing.
- To aid in making order promises. The MPS is a plan for what is to be produced and when. As such, it tells sales and manufacturing when goods will be available for delivery.
- To be a contract between marketing and manufacturing. It is an agreed-upon plan.

The MPS must be realistic and based on what production can and will do. If it is not, the results will be as follows:

- Overload or underload of plant resources.
- Unreliable schedules resulting in poor delivery performance.
- High levels of work-in-process (WIP) inventory.
- Poor customer service.
- Loss of credibility in the planning system.

QUESTIONS

1. What four functions does the master production schedule (MPS) perform in the production planning system?
2. What functions does the MPS perform between sales and production?
3. Does the MPS work with families of products or with individual items?
4. Where does the information come from to develop an MPS?
5. What are the three steps in making an MPS?
6. What is the purpose of a rough-cut capacity plan?
7. Where is the resource bill used?
8. At what level should master production scheduling take place?
   a. In a make-to-stock environment?
   b. In a make-to-order environment?
   c. In an assemble-to-order environment?
9. What is a final assembly schedule (FAS)? What is its purpose?
10. What is a planning horizon? What decides its minimum time? Why would it be longer?
11. How do the production plan and the MPS relate to sales and to the sales forecast?
12. What is the ATP (available to promise)? How is it calculated?
13. What is the purpose of time fences? Name and describe the three main divisions.

PROBLEMS

3.1 The Wicked Witch Whisk Company manufactures a line of broomsticks. The most popular is the 36-inch model, and the sales department has prepared a forecast for six weeks. The opening inventory is 20. As master scheduler, you must prepare an MPS. The brooms are manufactured in lots of 100.

Week		1	2	3	4	5	6
Forecast Sales		10	50	25	50	10	15
Projected Available	20						
MPS							

Answer. There should be scheduled receipts in weeks 2 and 4.

3.2 The Shades Sunglass Company assembles sunglasses from frames, which it makes, and lenses, which it purchases from an outside supplier. The sales department has prepared the following six-week forecast for Ebony, a popular model. The sunglasses are assembled in lots of 200, and the opening inventory is 400 pairs. Complete the projected available and the master production schedule.

Week		1	2	3	4	5	6
Forecast Sales		200	300	350	200	150	150
Projected Available	400						
MPS							

3.3 Amalgamated Mailbox Company manufactures a family of two mailboxes. The production plan and the MPS are developed on a quarterly basis. The forecast for the product group follows. The opening inventory is 270 units, and the company wants to reduce this to 150 units at the end of the year. Develop a level production plan.

Production Plan

Quarter		1	2	3	4	Total
Forecast Sales		220	300	200	200	
Projected Available	270					
Production Plan						

Answer. Quarterly production = 200 units

The forecast sales for each of the mailboxes in the family also follow. Develop an MPS for each item, bearing in mind that production is to be leveled as in the production plan. For each mailbox, the lot size is 200.

Mailbox A. Lot size: 200

Quarter		1	2	3	4	Total
Forecast Sales		120	180	100	120	
Projected Available	120					
MPS						

Answer. Scheduled receipts in weeks 2 and 3

Mailbox B.   Lot size: 200

Quarter		1	2	3	4	Total
Forecast Sales		100	120	100	80	
Projected Available	150					
MPS						

Answer.   Scheduled receipts in weeks 1 and 4

3.4   Worldwide Can-Openers, Inc., makes a family of two hand-operated can openers. The production plan is based on months. There are four weeks in this month. Opening inventory is 2000 dozen, and it is planned to increase that to 4000 dozen by the end of the month. The MPS is made using weekly periods. The forecast and projected available for the two models follow. The lot size for both models is 1000 dozen. Calculate the production plan and the MPS for each item.

Production Plan

Week		1	2	3	4	Total
Forecast		3000	3500	3500	4000	
Projected Available	2000					
Production Plan						

Model A

Week		1	2	3	4	Total
Forecast		2000	2000	2500	2000	
Projected Available	1500					
MPS						

Model B

Week		1	2	3	4	Total
Forecast		1000	1500	1000	2000	
Projected Available	500					
MPS						

3.5 In the example given on page 43, the MPS was unsatisfactory because there were poor inventory balances compared to the production plan. There was also a stockout for product C in periods 2 and 3. Revise the production plans for the three products to cut out or reduce these problems.

3.6 The Acme Widget Company makes widgets in two models, and the bottleneck operation is in work center 10. Following is the resource bill (in hours per part).

	Hours per Part	
Work Center	Model A	Model B
10	2.4	3.2

The master production schedule for the next five weeks is:

Week	1	2	3	4	5
Model A	70	50	50	60	45
Model B	20	40	55	30	45

a. Using the resource bill and the master production schedule, calculate the number of hours required in work center 10 for each of the five weeks. Use the following table to record the required capacity on the work center.

Week	1	2	3	4	5
Model A					
Model B					
Total Hours					

Answer. The total hours required are: week 1, 232; week 2, 248; week 3, 296; week 4, 240; and week 5, 252.

b. If the available capacity at work station 10 is 260 hours per week, suggest possible ways of meeting the demand in week 3.

3.7 Calculate the available to promise using the following data. There are 100 units on hand.

Week	1	2	3	4	5	6
Customer Orders	70	50	30	30	20	
MPS		100		100		100
ATP						

Answer.   ATP in week 1, 30; week 2, 20; week 4, 50; and week 6, 100.

3.8 Given the following data, calculate how many units are available to promise.

Week	1	2	3	4	5	6
Customer Orders	19		12	10		3
MPS	30		30	30		
ATP						

3.9 Using the scheduled receipts, calculate the ATP. There are zero units on hand.

Week	1	2	3	4	5	6	7	8	9	10
Customer Orders	10		2		60	20			10	
MPS	50				50				50	
ATP										

3.10 Using the scheduled receipts, calculate the ATP. There are 50 units on hand.

Week	1	2	3	4	5	6	7	8
Customer Orders	50	50	30	40	50	40	30	15
MPS		100		100		100		100
ATP								

3.11 Calculate the available to promise using the following data. There are 60 units on hand.

Week	1	2	3	4	5	6
Customer Orders	20	50	30	30	50	30
MPS		100			100	
ATP						

3.12 Given the data in problem 3.10, can an order for 20 for delivery in week 4 be accepted? Calculate the ATP using the following table.

Week	1	2	3	4	5	6	7	8
Customer Orders	50	50	30	60	50	40	30	15
MPS		100		100		100		100
ATP								

Answer. Yes. Ten can come from the ATP for week 4 and ten from the ATP for week 2.

3.13 Given the following data, can an order for 30 more units for delivery in week 5 be accepted? If not, what do you suggest can be done?

Week	1	2	3	4	5	6	7	8
Customer Orders	70	10	50	40	10	15	20	50
MPS	100		100			100		
ATP								

3.14 Given the following data, calculate the projected available and the planned MPS receipts. The order quantity is 200.

Week		1	2	3	4
Forecast		80	80	80	80
Customer Orders		100	90	50	40
Projected Available	160				
MPS					

Answer.   There is a planned MPS receipt in week 2.

3.15 Given the following data, calculate the projected available and the planned MPS receipts. The order quantity is 100.

Week		1	2	3	4
Forecast		50	50	50	50
Customer Orders		60	30	60	20
Projected Available	60				
MPS					

4

Material Requirements Planning

INTRODUCTION

Chapter 3 described the role of the master production schedule in showing the end items, or major components, that manufacturing intends to build. These items are made or assembled from components that must be available in the right quantities and at the right time to meet the master production schedule (MPS) requirements. If any component is missing, the product cannot be built and shipped on time. Material requirements planning (MRP) is the system used to avoid missing parts. It establishes a schedule (priority plan) showing the components required at each level of the assembly and, based on lead times, calculates the time when these components will be needed.

This chapter will describe bills of material (the major building block of material requirements planning), detail the MRP process, and explain how the material requirements plan is used. But first, some details about the environment in which MRP operates.

Nature of Demand

There are two types of demand: independent and dependent. Independent demand is not related to the demand for any other product. For example, if a company makes wooden tables, the demand for the tables is independent. Master production schedule items are independent demand items.

The demand for the sides, ends, legs, and tops depends on the demand for the tables, and these are dependent demand items.

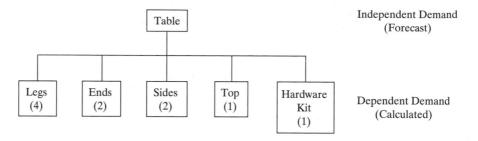

Figure 4.1   Product tree.

Figure 4.1 is a product tree that shows the relationship between independent and dependent demand items. The figures in parentheses show the required quantities of each component.

Since independent demand is not related to the demand for any other assemblies or products, it must be forecast. However, since dependent demand is directly related to the demand for higher level assemblies or products, it can be calculated. Material requirements planning is designed to do this calculation.

An item can have both a dependent and an independent demand. A service or replacement part has both. The manufacturer of vacuum cleaners uses flexible hose in the assembly of the units. In the assembly of the vacuums, the hose is a dependent demand item. However, the hose has a nasty habit of breaking, and the manufacturer must have replacement hoses available. Demand for replacement hoses is independent since demand for them does not depend directly upon the number of vacuums manufactured.

Objectives of MRP

Material requirements planning has two major objectives: determine requirements and keep priorities current.

Determine requirements.   The main objective of any manufacturing planning and control system is to have the right materials in the right quantities available at the right time to meet the demand for the firm's products. The material requirements plan's objective is to determine what components are needed to meet the master production schedule and, based on lead time, to calculate the periods when the components must be available. It must determine the following:

- What to order.
- How much to order.
- When to order.
- When to schedule delivery.

Keep priorities current.   The demand for, and supply of, components changes daily. Customers enter or change orders. Components get used up, suppliers are late with delivery, scrap occurs, orders are completed, and machines break down. In this ever-changing world, a material requirements plan must be able to reorganize priorities to keep plans current. It must be able to add and delete, expedite, delay, and change orders.

Linkages to Other MPC Functions

The master production schedule drives the material requirements plan. The MRP is a priority plan for the components needed to make the products in the MPS. The plan is valid only if capacity is available when needed to make the components, and the plan must be checked against available capacity. The process of doing so is called **capacity requirements planning** and is discussed in the next chapter.

Material requirements planning drives, or is input to, production activity control (PAC) and purchasing. MRP plans the release and receipt dates for orders. PAC and purchasing must plan and control the performance of the orders to meet the due dates.

Figure 4.2 shows a diagram of the production planning and control system with its inputs and outputs.

The Computer

If a company makes a few simple products, it might be possible to perform material requirements planning manually. However, most companies need to keep track of thousands of components in a world of changing demand, supply, and capacity.

In the days before computers, it was necessary to maintain extensive manual systems and to have large inventories and long lead times. These were needed as a

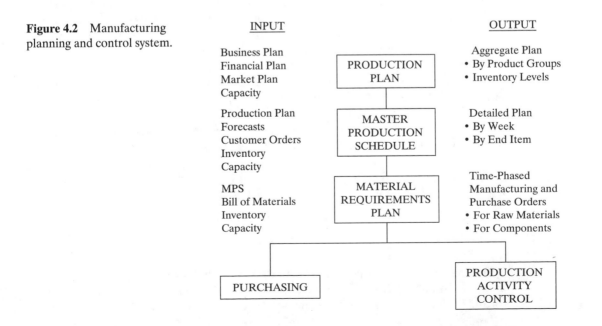

Figure 4.2 Manufacturing planning and control system.

INPUT

Business Plan
Financial Plan
Market Plan
Capacity

Production Plan
Forecasts
Customer Orders
Inventory
Capacity

MPS
Bill of Materials
Inventory
Capacity

PRODUCTION PLAN

MASTER PRODUCTION SCHEDULE

MATERIAL REQUIREMENTS PLAN

PURCHASING

PRODUCTION ACTIVITY CONTROL

OUTPUT

Aggregate Plan
• By Product Groups
• Inventory Levels

Detailed Plan
• By Week
• By End Item

Time-Phased Manufacturing and Purchase Orders
• For Raw Materials
• For Components

cushion due to the lack of accurate, up-to-date information and the inability to perform the necessary calculations quickly. Somehow, someone in the organization figured out what was required sooner, or very often, later than needed. "Get it early and get lots of it" was a good rule then.

Computers are incredibly fast, accurate, and ideally suited for the job at hand. With their ability to store and manipulate data and produce information rapidly, manufacturing now has a tool to use modern manufacturing planning and control systems properly. There are many application programs available that will perform the calculations needed in MRP systems. The computer software program that organizes and maintains the bills of material structures and their linkages is called a **bill of material processor.**

Inputs to the Material Requirements Planning System

There are three inputs to MRP systems:

1. Master production schedule.
2. Inventory records.
3. Bills of material.

Master production schedule.   The master production schedule is a statement of which end items are to be produced, the quantity of each, and the dates they are to be completed. It drives the MRP system by providing the initial input for the items needed.

Inventory records.   A major input to the MRP system is inventory. When a calculation is made to find out how many are needed, the quantities available must be considered.

There are two kinds of information needed. The first is called **planning factors** and includes information such as order quantities, lead times, safety stock, and scrap. This information does not change often; however, it is needed to plan what quantities to order and when to order for timely deliveries.

The second kind of information necessary is on the status of each item. The MRP system needs to know how much is available, how much is allocated, and how much is available for future demand. This type of information is dynamic and changes with every transaction that takes place.

These data are maintained in an **inventory record file,** also called a part master file or item master file. Each item has a record and all the records together form the file.

Bills of material.   The bill of material is one of the most important documents in a manufacturing company. It will be discussed next.

BILLS OF MATERIAL

Before we can make something, we must know what components are needed to make it. To bake a cake, we need a recipe. To mix chemicals together, we need a formula. To assemble a wheelbarrow, we need a parts list. Even though the names

Figure 4.3 Simplified bill of material.

| Description: TABLE | | |
| Part Number: 100 | | |
Part Number	Description	Quantity Required
203	Wooden Leg	4
411	Wooden Ends	2
622	Wooden Sides	2
023	Table Top	1
722	Hardware Kit	1

are different, recipes, formulas, and parts lists tell us what is needed to make the end product. All of these are bills of material.

The American Production and Inventory Control Society (APICS) defines a bill of material as "a listing of all the subassemblies, intermediates, parts, and raw materials that go into making the parent assembly showing the quantities of each required to make an assembly." Figure 4.3 shows a simplified bill of material. There are three important points:

1. The bill of material shows all the parts required to make *one* of the item.

2. Each part or item has only one part number. A specific number is unique to one part and is not assigned to any other part. Thus, if a particular number appears on two different bills of material, the part so identified is the same.

3. A part is defined by its form, fit, or function. If any of these change, then it is not the same part and it must have a different part number. For example, a part when painted becomes a different part and must have a different number. If the part could be painted in three different colors, then each must be identified with its unique number.

The bill of material shows the components that go into making the parent. It does not show the steps or process used to make the parent or the components. That information is recorded in a routing file. This file will be discussed in Chapter 6.

Bills of Material Structure

Bills of material structure refers to the overall design for the arrangement of bills of material files. Different departments in a company use bills of material for a variety of purposes. Although each user has individual preferences for the way the bill should be structured, there must be only one structure, and it should be designed to satisfy most needs. However, there can be several formats, or ways, to present the bill. Following are some important formats for bills.

Product tree. Figure 4.4 shows a product tree for the bill of material shown in Figure 4.3. The product tree is a convenient way to think about bills of material, but it is seldom used except for teaching and testing. In this text, it is used for that purpose.

Figure 4.4 Product tree.

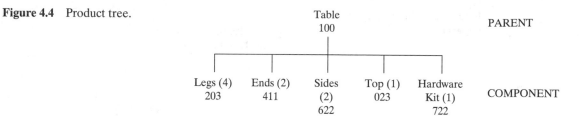

Parent–component relationship. The product tree and the bill of material shown in Figures 4.1 and 4.3 are called single-level structures. An assembly is considered a **parent,** and the items that comprise it are called its **component items.** Figure 4.4 shows the parent–component relationship of the table (P/N 100). Unique part numbers have also been assigned to each part. This makes identification of the part absolute.

Multilevel bill. Figure 4.5 shows the same product as the single-level bill shown in Figures 4.3 and 4.4. However, the single-level components have been expanded into their components.

Multilevel bills are formed as logical groupings of parts into subassemblies based on the way the product is assembled. For example, a frame, chassis, doors, windows, and engine are required to construct an automobile. Each of these forms a logical group of components and parts and, in turn, has its own bill of material.

It is the responsibility of manufacturing engineering to decide how the product is to be made: the operations to be performed, their sequence, and their grouping. The subassemblies created are the result of this. Manufacturing has decided to assemble the sides, ends, and leg supports (part of the hardware kit) of the table (P/N 100) in Figure 4.4 into a frame (P/N 300). The legs, leg bolts, and frame subassembly are to be assembled into the base (P/N 200). The top (P/N 023) is to be made from three boards glued together. Note that the original parts are all there, but they have been grouped into subassemblies and each subassembly has its own part number.

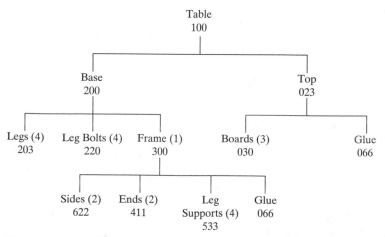

Figure 4.5 Multilevel bill.

One convention used with multilevel bills of material is that the last items on the tree (legs, leg bolts, ends, sides, glue, and boards) are all purchased items. Generally, a bill of material is not complete until all branches of the product structure tree end in a purchased part.

Each level in the bill of material is assigned a number starting from the top and working down. The top level, or end product level, is level zero, and its components are at level one.

Multiple bill.   Companies usually make more than one product, and the same components are often used in several products. This is particularly true with families of products. Using our example of a table, this company makes two models. They are similar except the tops are different. Figure 4.6 shows the two bills of material.

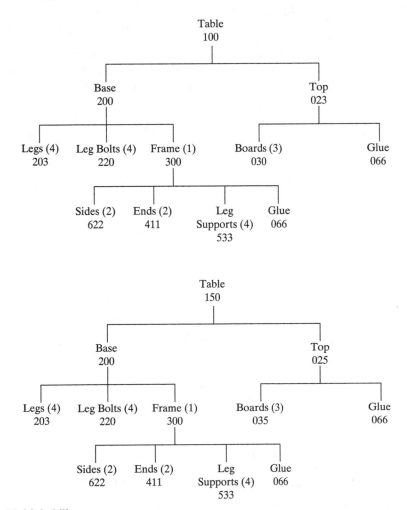

Figure 4.6   Multiple bills.

Because the boards used in the top are different, each top has a different part number. The balance of the components are common to both tables.

Single-level bill. A single-level bill of material contains only the parent and its immediate components, which is why it is called a single-level bill. The tables shown in Figure 4.6 have six single-level bills, and these are shown in Figure 4.7. Note that many components are common to both tables.

The computer stores information describing the product structure as a single-level bill. A series of single-level bills is needed to completely define a product. For example, the table needs four single-level bills, one each for the table, base, top, and frame. These can be chained together to form a multilevel, or indented, bill. Using this method, the information has to be stored only once. For example, the frame (P/N 300) might be used on other tables with different legs or tops.

There are several advantages to using single-level bills including the following:

- Duplication of records is avoided. For instance, base 200 is used in both table 100 and table 150. Rather than have two records of base 200—one in the bill for table 100 and one in the bill for table 150—only one record need be kept.

- The number of records and, in computer systems, the file size is reduced by avoiding duplication of records.

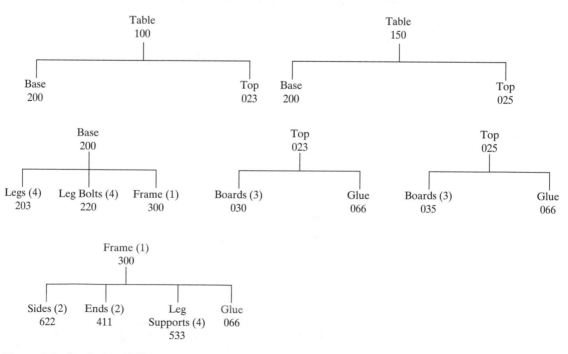

Figure 4.7 Single-level bills.

- Maintaining bills of material is simplified. For example, if there is a change in base 200, the change need be made in only one place.

EXAMPLE PROBLEM

Using the following product tree, construct the appropriate single-level trees. How many Ks are needed to make 100 Xs and 50 Ys?

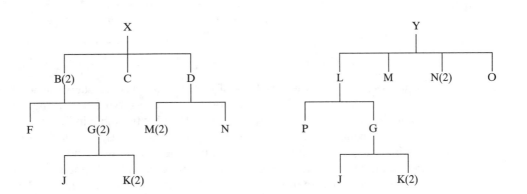

Answer

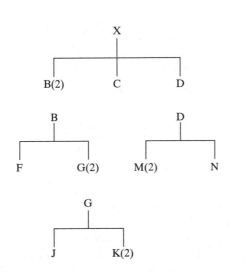

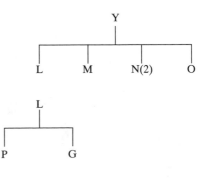

Each X requires two Bs

Each B requires two Gs 2×2 4 Gs

                                      100   Xs require 400   Gs

Each Y requires one L

Each L requires one G 1×1 1 G

                                       50   Ys require   50   Gs

                  Total Gs required   450

Each G requires two Ks

                  Total Ks required 2×450 =   900

Indented bill.   A multilevel bill of material can also be shown as an indented bill of material. This bill uses indentations as a way of identifying parents from components. Figure 4.8 shows an indented bill for the table in Figure 4.5.

The components of the parent table are listed flush left and their components are indented. The components of the base (legs, leg bolts, and frame) are indented immediately below their parent. The components of the frame are further indented immediately below their parent. Thus, the components are linked to their parents by indenting them as subentries and by listing them immediately below the parent.

Figure 4.8   Indented bill of material.

MANUFACTURING BILL OF MATERIAL TABLE P/N 100		
Part Number	Description	Quantity Required
200	Base	1
203	Legs	4
220	Leg Bolts	4
300	Frame	1
622	Sides	2
411	Ends	2
533	Leg Supports	4
066	Glue	
023	Top	1
030	Boards	3
066	Glue	

Summarized parts list.   The bill of material shown in Figure 4.3 is called a **summarized parts list.** It lists all the parts needed to make one complete assembly. The parts list is produced by the product design engineer and does not contain any information about the way the product is made or assembled.

Planning bill.   A major use of bills of material is to plan production. Planning bills are an artificial grouping of components for planning purposes. They are used to simplify forecasting, master production scheduling, and material requirements planning. They do not represent *buildable* products but an *average* product. Using the table example, suppose the company manufactured tables with three different leg styles, three different sides and ends, and three different tops. In total, they are making $3 \times 3 \times 3 = 27$ different tables, each with its own bill of material. For planning purposes, the 27 bills can be simplified by showing the percentage split for each type of component on one bill. Figure 4.9 shows how the product structure would look. The percentage usage of components is obtained from a forecast or past usage. Note the percentage for each category of component adds up to 100%.

Where-Used and Pegging

Where-used report.   Where-used reports give the same information as a bill of material, but the where-used report gives the parents for a component whereas the bill gives the components for a parent. A component may be used in making several parents. Wheels on an automobile, for example, might be used on several models of cars. A listing of all the parents in which a component is used is called a **where-used report.** This has several uses, such as in implementing an engineering change, or when materials are scarce, or in costing a product.

Pegging report.   A **pegging report** is similar to a where-used report. However, the pegging report shows only those parents for which there is an existing requirement whereas the where-used report shows all parents for a component. The pegging report shows the parents creating the demand for the components, the quantities needed, and when they are needed. Pegging keeps track of the origin of

Figure 4.9 Planning bill.

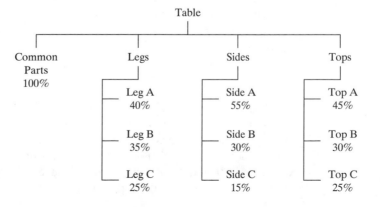

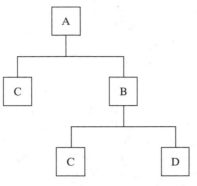

Pegged Requirements

Item Number	Week				
	1	2	3	4	5
C	50	125	25	50	150

Source of Requirements

A	50	25	25	50	50
B		100			100

Figure 4.10   Pegged requirements.

the demand. Figure 4.10 shows an example of a product tree in which part C is used twice and a pegging report.

Uses for Bills of Material

The bill of material is one of the most widely used documents in a manufacturing company. Some major uses are as follows:

- *Product definition.*   The bill specifies the components needed to make the product.
- *Engineering change control.*   Product design engineers sometimes change the design of a product and the components used. These changes must be recorded and controlled. The bill provides the method for doing so.
- *Service parts.*   Replacement parts needed to repair a broken component are determined from the bill of material.
- *Planning.*   Bills of material define what materials have to be scheduled to make the end product. They define what components have to be purchased or made to satisfy the master production schedule.
- *Order entry.*   When a product has a very large number of options (e.g., cars), the order-entry system very often configures the endproduct bill of materials. The bill can also be used to price the product.
- *Manufacturing.*   The bill provides a list of the parts needed to make or assemble a product.
- *Costing.*   Product cost is usually broken down into direct material, direct labor, and overhead. The bill provides not only a method of determining direct material but also a structure for recording direct labor and distributing overhead.

This list is not complete, but it shows the extensive use made of the bill of material in manufacturing. There is scarcely a department of the company that will not use the bill at some time. Maintaining bills of material and their accuracy is extremely important. Again, the computer is an excellent tool for centrally maintaining bills and for updating them.

MATERIAL REQUIREMENTS PLANNING PROCESS

Each component shown on the bill of material is planned for by the material requirements planning system. For convenience, it is assumed that each component will go into inventory and be accounted for. Whether the components actually go into a physical inventory or not is not important. However, it is important to realize that planning and control take place for each component on the bill. Raw material may go through several operations before it is processed and ready for assembly, or there may be several assembly operations between components and parent. These operations are planned and controlled by production activity control, not material requirements planning.

The purpose of material requirements planning is to determine the components needed, quantities, and due dates so items in the master production schedule are made on time. This section studies the basic MRP techniques for doing so. These techniques will be discussed under the following headings:

- Exploding and offsetting
- Gross and net requirements
- Releasing orders
- Low-level coding and netting

Exploding and Offsetting

Consider the product tree shown in Figure 4.11. It is similar to the ones used before but contains another necessary piece of information: lead times (LT).

Lead time. Lead time is the span of time needed to perform a processes. In manufacturing it includes time for order preparation, queuing, processing, moving, receiving and inspecting, and any expected delays. In this example, if B and C are available, it will take one week to assemble A. Thus, the lead time for A is one week. Similarly, if D and E are available, the time required to manufacture B is two weeks. The purchase lead times for D, E, and C are all one week.

Figure 4.11 Product tree with lead time.

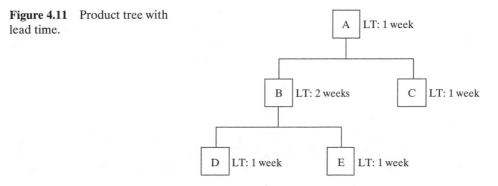

In this particular product tree, the usage quantities—the quantity of components needed to make one of a parent—are all one. To make an A requires one B and one C, and to make a B requires one D and one E.

Exploding the requirements. This is the process of multiplying the requirements by the usage quantity and recording the appropriate requirements throughout the product tree.

Offsetting. This is the process of placing the exploded requirements in their proper periods based on lead time. For example, if 50 units of A are required in week 5, the order to assemble the As must be released in week 4, and 50 Bs and 50 Cs must be available in week 4.

Planned orders. If it is planned to receive 50 of part A in week 5 and the lead time to assemble an A is one week, the order will have to be released and production started no later than week 4.

Thus, there should be a **planned order receipt** for 50 in week 5 and a **planned order release** for that number in week 4. If an order for 50 As is to be released in week 4, 50 Bs and 50 Cs must be available in that week. Thus, there must be planned order receipts for those components in week 4. Since the lead time to assemble a B is two weeks, there must be a planned order release for the Bs in week 2. Since the lead time to make a C is one week, there must be a planned order release for 50 in week 3. The planned order receipts and planned order releases for the Ds and Es are determined in the same manner. Figure 4.12 shows when orders must be released and received so the delivery date can be met.

Part Number		Week				
		1	2	3	4	5
A	Planned Order Receipt Planned Order Release				50	50
B	Planned Order Receipt Planned Order Release		50		50	
C	Planned Order Receipt Planned Order Release			50	50	
D	Planned Order Receipt Planned Order Release	50	50			
E	Planned Order Receipt Planned Order Release	50	50			

Figure 4.12 Exploding and offsetting.

▼

EXAMPLE PROBLEM

Using the product tree and lead times shown in Figure 4.11, complete the following table to determine the planned order receipts and releases. There are 50 As required in week 5 and 100 in week 6.

Answer

Part Number		Week					
		1	2	3	4	5	6
A	Planned Order Receipt					50	100
	Planned Order Release				50	100	
B	Planned Order Receipt				50	100	
	Planned Order Release		50	100			
C	Planned Order Receipt				50	100	
	Planned Order Release			50	100		
D	Planned Order Receipt		50	100			
	Planned Order Release	50	100				
E	Planned Order Receipt		50	100			
	Planned Order Release	50	100				

Gross and Net Requirements

The previous section assumed that no inventory was available for the As or any of the components. Often inventory is available and must be included when calculating quantities to be produced. If, for instance, there are 20 As in stock, only 30 need to be made. The requirements for component parts would be reduced accordingly. The calculation is as follows:

Gross requirements = 50

Inventory available = 20

Net requirements = gross requirements – available inventory

Net requirements = 50 – 20 = 30 units

84   Chapter 4

Since only 30 As need to be made, the gross requirement for Bs and Cs is only 30.

*The planned order release of the parent becomes the
gross requirement of the component.*

The time-phased inventory record shown in Figure 4.12 can now be modified to consider any inventory available. For example, suppose there are 10 Bs available as well as the 20 As. The requirements for the components D and E would change. Figure 4.13 shows the change in the MRP record.

Part Number		Week				
		1	2	3	4	5
A	Gross Requirements					50
	Projected Available 20	20	20	20	20	0
	Net Requirements					30
	Planned Order Receipt					30
	Planned Order Release				30	
B	Gross Requirements				30	
	Projected Available 10	10	10	10	0	
	Net Requirements				20	
	Planned Order Receipt				20	
	Planned Order Release		20			
C	Gross Requirements				30	
	Projected Available			0	0	
	Net Requirements				30	
	Planned Order Receipt				30	
	Planned Order Release			30		
D	Gross Requirements		20			
	Projected Available	0	0			
	Net Requirements		20			
	Planned Order Receipt		20			
	Planned Order Release	20				
E	Gross Requirements		20			
	Projected Available	0	0			
	Net Requirements		20			
	Planned Order Receipt		20			
	Planned Order Release	20				

Figure 4.13   Gross and net requirements.

EXAMPLE PROBLEM

Compare the following table. Lead time for the part is two weeks. The order quantity (lot size) is 100 units.

Week	1	2	3	4
Gross Requirements		50	45	20
Projected Available 75				
Net Requirements				
Planned Order Receipt				
Planned Order Release				

Answer

Week	1	2	3	4
Gross Requirements		50	45	20
Projected Available 75	75	25	80	60
Net Requirements			20	
Planned Order Receipt			100	
Planned Order Release	100			

Released Orders

So far we have looked at the process of planning when orders should be released so work is done in time to meet gross requirements. In many cases, requirements change daily. A computer-based material requirements planning system automatically recalculates the requirements for subassemblies and components and recreates planned order releases to meet the shifts in demand.

Planned order releases are just planned; they have not been released. It is the responsibility of the material planner to release planned orders, not the computer.

Since the objective of the MRP is to have material available when it is needed and not before, orders for material should not be released until the planned order release date arrives. Thus, an order is not normally released until the planned order is in the current week (week 1).

Releasing an order means that authorization is given to purchasing to buy the necessary material or to manufacturing to make the component.

Before a manufacturing order is released, component availability must be checked. The computer program checks the component inventory records to be sure that enough material is available and, if so, to allocate the necessary quantity to that work order. If the material is not available, the computer program will advise the planner of the shortage.

When the authorization to purchase or manufacture is released, the planned order receipt is canceled, and a scheduled receipt is created in its place. For the example shown in Figure 4.13, parts D and E have planned order releases of 20 scheduled for week 1. These orders will be released by the planner, and then the MRP records for parts D and E will appear as shown in Figure 4.14. Notice that scheduled receipts have been created, replacing the planned order releases.

When a manufacturing order is released the computer will *allocate* the required quantities of a parent's components to that order. This does not mean the components are withdrawn from inventory but that the projected *available* quantity is reduced. The allocated quantity of components is still in inventory but they are not available for other orders. They will stay in inventory until withdrawn for use.

Scheduled receipts. Scheduled receipts are orders placed on manufacturing or on a vendor and represent a commitment to make or buy. For an order in a factory, necessary materials are committed, and work-center capacity allocated to that order. For purchased parts, similar commitments are made to the vendor. The scheduled

Part Number		Week				
		1	2	3	4	5
D	Gross Requirements		20			
	Scheduled Receipts		20			
	Projected Available	0	0			
	Net Requirements		0			
	Planned Order Receipt					
	Planned Order Release					
E	Gross Requirements		20			
	Scheduled Receipts		20			
	Projected Available	0	0			
	Net Requirements		0			
	Planned Order Receipt					
	Planned Order Release					

Figure 4.14   Scheduled receipts.

receipts row shows the quantities ordered and when they are expected to be completed and available.

Open orders. Scheduled receipts on the MRP record are open orders on the factory or a vendor and are the responsibility of purchasing and of production activity control. When the goods are received into inventory and available for use, the order is closed out, and the scheduled receipt disappears to become part of the on-hand inventory.

Net requirements. The calculation for net requirements can now be modified to include scheduled receipts.

Net requirements = gross requirements – scheduled receipts – available inventory

▼ ━━

EXAMPLE PROBLEM

Complete the following table. Lead time for the item is two weeks, and the order quantity is 200. What action should be taken?

Week	1	2	3	4
Gross Requirements	50	250	100	50
Scheduled Receipts		200		
Projected Available 150				
Net Requirements				
Planned Order Receipt				
Planned Order Release				

Answer

Week	1	2	3	4
Gross Requirements	50	250	100	50
Scheduled Receipts		200		
Projected Available 150	100	50	150	100
Net Requirements			50	
Planned Order Receipt			200	
Planned Order Release	200			

The order for 200 units should be released.

Part Number		Week				
		1	2	3	4	5
	Gross Requirements					35
	Scheduled Receipts				20	
	Projected Available 10	10	10	10	30	
	Net Requirements					5
	Planned Order Receipt					5
	Planned Order Release				5	

Figure 4.15   Basic MRP record.

Basic MRP Record

Figure 4.15 shows a basic MRP record. There are several points that are important:

1. The current time is the beginning of the first period.
2. The top row shows periods, called **time buckets.** These are often a week but can be any length of time convenient to the company. Some companies are moving to daily time buckets.
3. The number of periods in the record is called the **planning horizon,** which shows the number of future periods for which plans are being made. It should be at least as long as the cumulative product lead time. Otherwise, the MRP system is not able to release planned orders of items at the lower level at the correct time.
4. An item is considered available at the beginning of the time bucket in which it is required.
5. The quantity shown in the projected on-hand row is the projected on-hand balance at the *end* of the period.
6. The immediate or most current period is called the **action bucket.** A quantity in the action bucket means that some action is needed now to avoid a future problem.

Capacity Requirements Planning

As occurred in the previous planning levels, the MRP priority plan must be checked against available capacity. At the MRP planning level, the process is called **capacity requirements planning (CRP).** The next chapter examines this problem in some detail. If the capacity is available, the plan can proceed. If not, either capacity has to be made available or the priority plans changed.

Low-Level Coding and Netting

A component may reside on more than one level in a bill of material. If this is the case, it is necessary to make sure that all gross requirements for that component

Figure 4.16 Multilevel product tree.

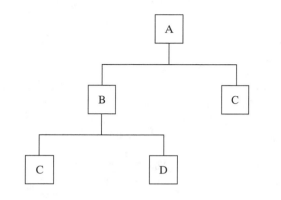

Level

A   0

B   C   1

C   D   2

have been recorded before netting takes place. Consider the product shown in Figure 4.16. Component C occurs twice in the product tree and at different levels. It would be a mistake to net the requirements for the Cs before calculating the gross requirements for those required for parent B.

The process of collecting the gross requirements and netting can be simplified by using low-level codes. The **low-level code** is the lowest level on which a part resides in all bills of material. Every part has only one low-level code. The low-level codes for the parts in the product tree shown in Figure 4.16 are:

Part	Low-Level Code
A	0
B	1
C	2
D	2

Low-level codes are determined by starting at the lowest level of a bill of material and, working up, recording the level against the part. If a part occurs on a higher level, its existence on the lower level has already been recorded.

Once the low-level codes are obtained, the net requirements for each part can be calculated using the following procedure. For the purpose of this exercise, there is a gross requirement for part A of 50 in week 5, all lead times are one week, and the following amounts are in inventory: A, 20 units; B, 10 units; and C, 10 units.

Procedure

1. Starting at level zero of the tree, determine if any of the parts on that level have a low-level code of zero. If so, those parts occur at no lower level, and all the gross requirements have been recorded. These parts can, therefore, be netted and exploded down to the next level, that is, into their components. If

Low-Level Code	Part Number		Week				
			1	2	3	4	5
0	A	Gross Requirements					50
		Scheduled Receipts					
		Projected Available 20	20	20	20	20	0
		Net Requirements					30
		Planned Order Receipt					30
		Planned Order Release				30	
1	B	Gross Requirements				30	
		Scheduled Receipts					
		Projected Available 10					
		Net Requirements					
		Planned Order Receipt					
		Planned Order Release					
2	C	Gross Requirements				30	
		Scheduled Receipts					
		Projected Available 10					
		Net Requirements					
		Planned Order Receipt					
		Planned Order Release					

Figure 4.17   Netting and exploding zero-level parts.

the low-level code is greater than zero, there are more gross requirements, and the part is not netted. In this example, A has a low-level code of zero so there is no further requirement for As; it can be netted and exploded into its components. Figure 4.17 shows the results.

2. The next step is to move down to level 1 on the product tree and to repeat the routine followed in step 1. Since B has a low-level code of one, all requirements for B are recorded, and it can be netted and exploded. The bill of material for B shows that it is made from a C and a D. Figure 4.18 shows the result of netting and exploding the Bs. Part C has a low-level code of two, which tells us there are further requirements for Cs and at this stage they are not netted.

3. Moving down to level 2 on the product tree, we find that part C has a low-level code of two. This tells us that all gross requirements for Cs are accounted for and that we can proceed and determine its net requirements. Notice there is a requirement for 30 Cs in week 4 to be used on the As and a requirement of 20

Low-Level Code	Part Number		Week				
			1	2	3	4	5
1	B	Gross Requirements Scheduled Receipts Projected Available 10 Net Requirements Planned Order Receipt Planned Order Release	10	10	10 20	30 0 20 20	
2	C	Gross Requirements Scheduled Receipts Projected Available 10 Net Requirements Planned Order Receipt Planned Order Release			20	30	
2	D	Gross Requirements Scheduled Receipts Projected Available Net Requirements Planned Order Receipt Planned Order Release			20		

Figure 4.18   Netting and exploding first-level parts.

Cs in week 3 to be used on the Bs. Looking at its bill of material, we see that it is a purchased part and no explosion is needed.

Figure 4.19 shows the completed material requirements plan. The process of level-by-level netting is now completed using the low-level codes of each part. The low-level codes are used to determine when a part is eligible for netting and exploding. In this way, each part is netted and exploded only once. There is no time-consuming re-netting and re-exploding each time a new requirement is met.

Multiple Bills of Material

Most companies make more than one product and often use the same components in many of their products. The material requirements planning system gathers the planned order releases from all the parents and creates a schedule of gross requirements for the components. Figure 4.20 illustrates what happens. Part F is a component of both C and B.

Low-Level Code	Part Number		Week 1	2	3	4	5
0	A	Gross Requirements					50
		Scheduled Receipts					
		Projected Available 20	20	20	20	20	0
		Net Requirements					30
		Planned Order Receipt					30
		Planned Order Release				30	
1	B	Gross Requirements				30	
		Scheduled Receipts					
		Projected Available 10	10	10	10	0	
		Net Requirements				20	
		Planned Order Receipt				20	
		Planned Order Release			20		
2	C	Gross Requirements			20	30	
		Scheduled Receipts					
		Projected Available 10	10	10	0	0	
		Net Requirements			10	30	
		Planned Order Receipt			10	30	
		Planned Order Release		10	30		
2	D	Gross Requirements			20		
		Scheduled Receipts					
		Projected Available	0	0	0		
		Net Requirements			20		
		Planned Order Receipt			20		
		Planned Order Release		20			

Figure 4.19   Completed material requirements plan.

The same procedure used for a single bill of material can be used when multiple products are being manufactured. All bills must be netted and exploded level by level as was done for a single bill.

Figure 4.21 shows the product trees for two products. Both are made from several components, but, for simplicity, only those components containing an F are shown in the product tree. Note both have F as a component but at different levels in their product tree. All lead times are one week. The quantities required are

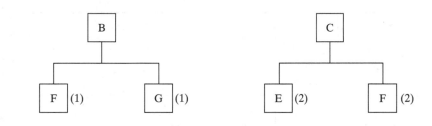

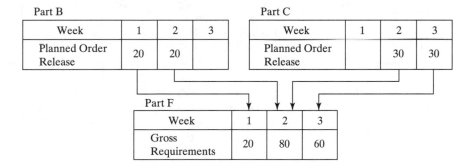

Figure 4.20   Multi-product MRP explosion.

Figure 4.21   Multi-product tree.

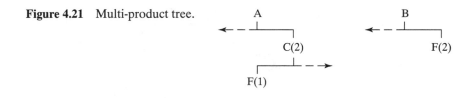

shown in parentheses; that is, two Cs are required to make an A, one F is required to make a C, and two Fs are needed to make a B. Figure 4.22 shows the completed material requirements plan that would result if 50 As were required in week 5 and 30 Bs in week 3.

USING THE MATERIAL REQUIREMENTS PLAN

The people who manage the material requirements planning system are planners. They are responsible for making detailed decisions that keep the flow of material moving into, through, and out of the factory. In many companies where there are

Low-Level Code	Part Number		Week 1	2	3	4	5
0	A	Gross Requirements					50
		Scheduled Receipts					
		Projected Available 20	20	20	20	20	0
		Net Requirements					30
		Planned Order Receipt					30
		Planned Order Release				30	
0	B	Gross Requirements			30		
		Scheduled Receipts					
		Projected Available 10	10	10	0		
		Net Requirements			20		
		Planned Order Receipt			20		
		Planned Order Release		20			
1	C	Gross Requirements				60	
		Scheduled Receipts					
		Projected Available 10	10	10	10	0	
		Net Requirements				50	
		Planned Order Receipt				50	
		Planned Order Release			50		
2	F	Gross Requirements		40	50		
		Scheduled Receipts					
		Projected Available		0	0		
		Net Requirements		40	50		
		Planned Order Receipt		40	50		
		Planned Order Release	40	50			

Figure 4.22   Partial material requirements plan.

thousands of parts to manage, planners are usually organized into logical groupings based on the similarity of parts or supply.

The basic responsibilities of a planner are to:

1. Launch (release) orders to purchasing or manufacturing.
2. Reschedule due dates of open (existing) orders as required.
3. Reconcile errors and try to find their cause.
4. Solve critical material shortages by expediting or replanning.

5. Coordinate with other planners, master production schedulers, production activity control, and purchasing to resolve problems.

The material planner works with three types of orders: planned, released, and firm.

Planned orders.   Planned orders are automatically scheduled and controlled by the computer. As gross requirements, projected available inventory, and scheduled receipts change, the computer recalculates the timing and quantities of planned order releases. The MRP program recommends to the planner the release of an order when the order enters the action bucket but does not release the order.

Released orders.   Releasing, or launching, a planned order is the responsibility of the planner. When released, the order becomes an open order to the factory or to purchasing and appears on the MRP record as a scheduled receipt. It is then under the control of the planner, who may expedite, delay, or even cancel the order.

Firm planned orders.   The computer-based MRP system automatically recalculates planned orders as the gross requirements change. At times, the planner may prefer to hold a planned order firm against changes in quantity and time despite what the computer calculates. This might be necessary because of future availability of material or capacity or special demands on the system. The planner can tell the computer that the order is not to be changed unless the planner advises the computer to do so. The order is "firmed" or frozen against the logic of the computer.

The MRP software nets, offsets, and explodes requirements and creates planned order releases. It keeps priorities current for all planned orders according to changes in gross requirements for the part. But it does not issue purchase or manufacturing orders or reschedule open orders. However, it does print action or exception messages suggesting that the planner should act and what kind of action might be appropriate.

Exception messages.   If the manufacturing process is under control and the material requirements planning system is working properly, the system will work according to plan. However, sometimes there are problems that need the attention of the planner. A good MRP system generates exception messages to advise the planner when some event needs attention. Following are some examples of situations that will generate exception messages.

- Components for which planned orders are in the action bucket and which should be considered for release.
- Open orders for which the timing or quantity of scheduled receipts does not satisfy the plan. Perhaps a scheduled receipt is timed to arrive too early or late, and its due date should be revised.
- Situations in which the standard lead times will result in late delivery of a zero-level part. This situation might call for expediting to reduce the standard lead times.

Transaction messages.   The planner must tell the MRP software of all actions taken that will influence the MRP records. For example, when the planner releases an order, or a scheduled receipt is received, or when any change to the data occurs, the MRP program must be told. Otherwise, the records will be inaccurate, and the plan will become unworkable.

Material requirements planners must manage the parts for which they are responsible. This means not only releasing orders to purchasing and the factory, rescheduling due dates of open orders, and reconciling differences and inconsistencies, but also finding ways to improve the system and removing the causes of potential error. If the right components are to be in the right place at the right time, the planner must manage the process.

Managing the Material Requirements Plan

The planner receives feedback from many sources such as:

- Suppliers' actions through purchasing.
- Changes to open orders in the factory such as early or late completions or differing quantities.
- Management action such as changing the master production schedule.

The planner must evaluate this feedback and take corrective action if necessary. The planner must consider three important factors in managing the material requirements plan.

Priority.   Priority refers to maintaining the correct due dates by constantly evaluating the true due-date need for released orders and, if necessary, expediting or de-expediting.

Consider the following MRP record. The order quantity is 300 units and the lead time is three weeks.

Week	1	2	3	4	5
Gross Requirements	100	50	100	150	200
Scheduled Receipts			300		
Projected Available 150	50	0	200	50	150
Net Requirements					150
Planned Order Receipt					300
Planned Order Release		300			

What will happen if the gross requirements in week 2 are changed from 50 units to 150? The MRP record will look like the following.

Week	1	2	3	4	5
Gross Requirements	100	150	100	150	200
Scheduled Receipts			300		
Projected Available 150	50	−100	100	250	50
Net Requirements				50	
Planned Order Receipt				300	
Planned Order Release	300				

Note that there is a shortage of 100 units in week 2 and that the planned order release originally in week 2 is now in week 1. What can the planner do? One solution is to expedite the scheduled receipt of 300 units from week 3 to week 2. If this is not possible, the extra 100 units wanted in week 2 must be rescheduled into week 3. Also, there is now a planned order release in week 1, and this order should be released.

Bottom-up replanning.   Action to correct for changed conditions should occur as low in the product structure as possible. Suppose the part in the previous example is a component of another part. The first alternative is to expedite the scheduled receipt of 300 into week 2. If this can be done, there is no need to make any changes to the parent. If the 300 units cannot be expedited, the planned order release and net requirement of the parent must be changed.

Reducing system nervousness.   Sometimes requirements change rapidly and by small amounts, causing the material requirements plan to change back and forth. The planner must judge whether the changes are important enough to react to and whether an order should be released. One method of reducing system nervousness is firm planned orders.

▼

EXAMPLE PROBLEM

As the MRP planner, you arrive at work Monday morning, look at the MRP record for part 2876 as shown below.

Order quantity = 30 units
Lead time = 2 weeks

Week		1	2	3	4	5	6
Gross Requirements		35	10	15	30	15	20
Scheduled Receipts		30					
Projected Available	20	15	5	20	20	5	15
Net Requirements				10	10		15
Planned Order Receipt				30	30		30
Planned Order Release		30	30		30		

The computer draws attention to the need to release the planned order for 30 in week 1. Either you release this order, or there will be a shortage in week 3. During the first week, the following transactions take place:

a. Only 25 units of the scheduled receipt are received into inventory. The balance is scrapped.

b. The gross requirement for week 3 is changed to 10.

c. The gross requirement for week 4 is increased to 50.

d. The requirement for week 7 is 15.

e. An inventory count reveals there are ten more in inventory than the record shows.

f. The 35 gross requirement for week 1 is issued from inventory.

g. The planned order release for 30 in week 1 is released and becomes a scheduled receipt in week 3.

As these transactions occur during the first week, you must enter these changes in the computer record. At the beginning of the next week, the MRP record appears as follows:

Order quantity = 30 units
Lead time = 2 weeks

Week		2	3	4	5	6	7
Gross Requirements		10	10	50	15	20	15
Scheduled Receipts			30				
Projected Available	20	10	30	10	25	5	20
Net Requirements				20	5		10
Planned Order Receipt				30	30		30
Planned Order Release		30	30		30		

The opening on-hand balance for week 2 is 20 (20 + 25 + 10 – 35 = 20). The planned order release originally set in week 4 has shifted to week 3. Another planned order has been created for release in week 5. More importantly, the scheduled receipt in week 3 will not be needed until week 4. You should reschedule this to week 4. The planned order in week 2 should be released and become a scheduled receipt in week 4.

QUESTIONS

1. What is a material requirements plan?
2. What is the difference between dependent and independent demand?
3. Should an MRP be used with dependent or independent demand items?

4. What are the objectives of the MRP?

5. What is the relationship between the MPS and the MRP?

6. Why is a computer necessary in an MRP system?

7. What are the major inputs to the MRP system?

8. What data are found in a part master file or an item master file?

9. What is a bill of material? What are two important points about bills of material?

10. To what does "bill of material structure" refer? Why is it important?

11. Describe the parent–component relationship.

12. Describe the following types of bills of material:

    a. Product tree.

    b. Multilevel bill.

    c. Single-level bill.

    d. Indented bill.

    e. Summarized parts list.

    f. Planning bill.

13. Describe the parent–component relationship.

14. Why do MRP computer programs store single-level bills?

15. Describe each of the seven uses of a bill of material described in the text.

16. What are where-used and pegging reports? Give some of their uses.

17. Describe the process of offsetting and exploding.

18. What is a planned order? How is it created?

19. From where does the gross requirement of a component come?

20. Who is responsible for releasing an order? Describe what happens to the inventory records and to PAC and purchasing.

21. What is a scheduled receipt? From where does it originate?

22. What is an open order? How does it get closed?

23. What is the meaning of the term *low-level code?* What is the low-level code of an MPS part?

24. What are the responsibilities of a material requirements planner?

25. Describe the differences between planned orders, released orders, and firm planned orders. Who controls each?

26. What are exception messages? What is their purpose?

27. What is a transaction message? Why is it important?

28. What are the three important factors in managing the material requirements plan? Why is each important?

PROBLEMS

4.1 Using the following product tree, construct the appropriate single-level trees. How many Cs are needed to make 50 Xs and 100 Ys?

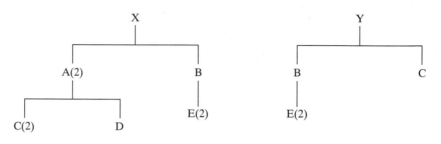

Answer. 300 Cs

4.2 Given the following parents and components, construct a product tree. Figures in parentheses show the quantities per item. How many Gs are needed to make one A?

Parent	A	B	C	E
Component	B(2)	E(2)	G(2)	G(4)
	C(4)	F(1)		F(3)
	D(3)			H(2)

4.3 Using the following product tree, determine the planned order receipts and planned order releases if 100 As are to be produced in week 5. All lead times are one week except for component E which has a lead time of two weeks.

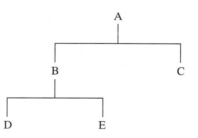

	Week	1	2	3	4	5
Part A Lead Time: 1 week	Planned Order Receipt Planned Order Release					
Part B Lead Time: 1 week	Planned Order Receipt Planned Order Release					
Part C Lead Time: 1 week	Planned Order Receipt Planned Order Release					
Part D Lead Time: 1 week	Planned Order Receipt Planned Order Release					
Part E Lead Time: 2 weeks	Planned Order Receipt Planned Order Release					

4.4 Complete the following table. Lead time for the part is two weeks, and the order quantity is 30. What action should be taken?

Week	1	2	3	4
Gross Requirements	10	15	10	20
Projected Available 30				
Net Requirements				
Planned Order Receipt				
Planned Order Release				

Answer. An order for 30 should be released in week 1.

4.5 Given the following product tree, explode, offset, and determine the gross and net requirements. All lead times are one week, and the quantities required are shown in parentheses. The master production schedule calls for production of 100 As in week 5. There are 20 Bs available.

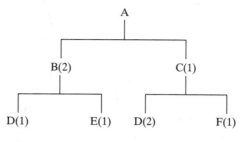

	Week	1	2	3	4	5
Part A	Gross Requirements Projected Available Net Requirements Planned Order Receipt Planned Order Release					
Part B	Gross Requirements Projected Available Net Requirements Planned Order Receipt Planned Order Release					
Part C	Gross Requirements Projected Available Net Requirements Planned Order Receipt Planned Order Release					
Part D	Gross Requirements Projected Available Net Requirements Planned Order Receipt Planned Order Release					
Part E	Gross Requirements Projected Available Net Requirements Planned Order Receipt Planned Order Release					
Part F	Gross Requirements Projected Available Net Requirements Planned Order Receipt Planned Order Release					

Answer. Planned order releases are:

Part A:	100 in week 4
Part B:	180 in week 3
Part C:	100 in week 3
Part D:	380 in week 2
Part E:	180 in week 2
Part F:	100 in week 2

4.6 Complete the following table. Lead time for the part is two weeks. The lot size is 100. What is the projected available at the end of week 3? When is it planned to release an order?

Week	1	2	3	4
Gross Requirements	20	65	40	25
Scheduled Receipts		100		
Projected Available 30				
Net Requirements				
Planned Order Receipt				
Planned Order Release				

Answer.   Projected available at the end of week 3 is 5.

An order release is planned for the beginning of week 2.

4.7 Complete the following table. Lead time for the part is three weeks. The lot size is 50. What is the projected available at the end of week 3? When is it planned to release an order?

Week	1	2	3	4
Gross Requirements	30	25	10	10
Scheduled Receipts	50			
Projected Available 20				
Net Requirements				
Planned Order Receipt				
Planned Order Release				

4.8   Given the following partial product tree, explode, offset, and determine the gross and net requirements for components H, I, J, and K. There are other components, but they are not connected to this problem. The quantities required are shown in parentheses. The master production schedule calls for production of 50 Hs in week 3 and 80 in week 5. There is a scheduled receipt of 100 Is in week 2. There are 400 Js and 400 Ks available.

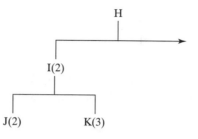

	Week	1	2	3	4	5
Part H Lead Time: 1 week	Gross Requirements Scheduled Receipts Projected Available Net Requirements Planned Order Receipt Planned Order Release					
Part I Lead Time: 2 weeks	Gross Requirements Scheduled Receipts Projected Available Net Requirements Planned Order Receipt Planned Order Release					
Part J Lead Time: 1 week	Gross Requirements Scheduled Receipts Projected Available 400 Net Requirements Planned Order Receipt Planned Order Release					
Part K Lead Time: 1 week	Gross Requirements Scheduled Receipts Projected Available 400 Net Requirements Planned Order Receipt Planned Order Release					

Answer.   There is a planned order release for part K of 80 in week 1.

4.9 MPS parent X has planned order releases of 30 in weeks 2 and 4. Given the following product tree, complete the MRP records for parts Y and Z. Quantities required are shown in brackets.

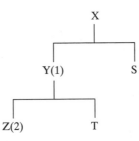

Part Y Lead Time: 2 weeks Lot Size: 50	Week			
	1	2	3	4
Gross Requirements Scheduled Receipts Projected Available 40 Net Requirements Planned Order Receipt Planned Order Release				

Part Z Lead Time: 1 week Lot Size: 100	Week			
	1	2	3	4
Gross Requirements Scheduled Receipts Projected Available 20 Net Requirements Planned Order Receipt Planned Order Release				

4.10 Given the following product tree, explode, offset, and determine the gross and net requirements. The quantities required are shown in parentheses. The master production schedule calls for production of 100 As in week 5. There is a scheduled receipt of 100 Bs in week 1. There are 200 Fs available. All order quantities are lot-for-lot.

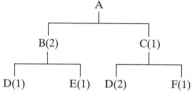

	Week	1	2	3	4	5
Part A Lead Time: 1 week	Gross Requirements Scheduled Receipts Projected Available Net Requirements Planned Order Receipt Planned Order Release					
Part B Lead Time: 1 week	Gross Requirements Scheduled Receipts Projected Available Net Requirements Planned Order Receipt Planned Order Release					
Part C Lead Time: 1 week	Gross Requirements Scheduled Receipts Projected Available Net Requirements Planned Order Receipt Planned Order Release					
Part D Lead Time: 1 week	Gross Requirements Scheduled Receipts Projected Available Net Requirements Planned Order Receipt Planned Order Release					
Part E Lead Time: 1 week	Gross Requirements Scheduled Receipts Projected Available Net Requirements Planned Order Receipt Planned Order Release					
Part F Lead Time: 1 week	Gross Requirements Scheduled Receipts Projected Available Net Requirements Planned Order Receipt Planned Order Release					

4.11 Given the following product tree, complete the MRP records for parts X, Y, and Z. Note that parts X and Y have specified order quantities.

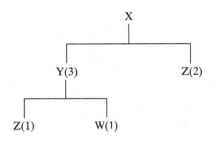

Week		1	2	3	4	5
Part X	Gross Requirements	15	5	15	10	15
	Scheduled Receipts	20				
Lead Time: 1 week	Projected Available 10					
	Net Requirements					
Lot Size: 20	Planned Order Receipt					
	Planned Order Release					
Part Y	Gross Requirements					
	Scheduled Receipts		50			
Lead Time: 2 weeks	Projected Available 30					
	Net Requirements					
Lot Size: 50	Planned Order Receipt					
	Planned Order Release					
Part Z	Gross Requirements					
	Scheduled Receipts		90			
Lead Time: 2 weeks	Projected Available					
	Net Requirements					
Lot Size: lot-for-lot	Planned Order Receipt					
	Planned Order Release					
Part W	Gross Requirements					
	Scheduled Receipts					
Lead Time: 1 week	Projected Available					
	Net Requirements					
Lot Size: 400	Planned Order Receipt					
	Planned Order Release					

4.12 Given the following product tree, determine the low-level codes for all the components.

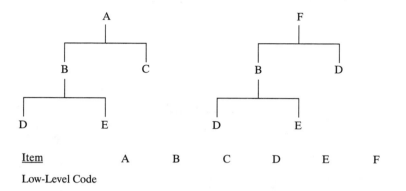

Item	A	B	C	D	E	F
Low-Level Code						

4.13 Given the following product tree, determine the low-level codes for all the components.

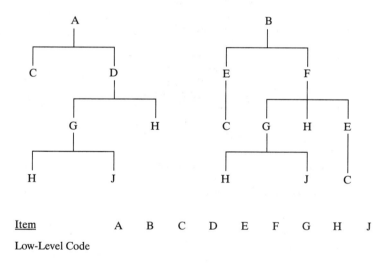

Item	A	B	C	D	E	F	G	H	J
Low-Level Code									

4.14 Given the following product tree, develop a material requirements plan for the components. Quantities per are shown in parentheses. The following worksheet shows the present active orders, the available balances, and the lead times.

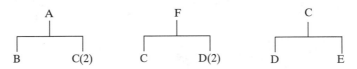

Low-Level Code		Week	1	2	3	4	5
0	Part A Lead Time: 1 week Lot-for-lot	Gross Requirements Scheduled Receipts Projected Available Net Requirements Planned Order Receipt Planned Order Release			60		70
0	Part F Lead Time: 1 week Lot-for-lot	Gross Requirements Scheduled Receipts Projected Available Net Requirements Planned Order Receipt Planned Order Release				100	
	Part B Lead Time: 2 weeks Lot Size: 300	Gross Requirements Scheduled Receipts Projected Available 200 Net Requirements Planned Order Receipt Planned Order Release					
	Part C Lead Time: 2 weeks Lot Size: Lot-for- lot	Gross Requirements Scheduled Receipts Projected Available Net Requirements Planned Order Receipt Planned Order Release		120			
	Part D Lead Time: 2 weeks Lot Size: 300	Gross Requirements Scheduled Receipts Projected Available Net Requirements Planned Order Receipt Planned Order Release	300				
	Part E Lead Time: 3 weeks Lot Size: 500	Gross Requirements Scheduled Receipts Projected Available 400 Net Requirements Planned Order Receipt Planned Order Release					

Answer. The low-level-code for part D is 2. There is a planned order release of 300 for part D in week 1. There are no planned order releases for part E.

4.15 Given the following product tree, explode, offset, and determine the gross and net requirements. All lead times are one week, and the quantities required are shown in parentheses. The master production schedule calls for production of 100 As in week 4 and 50 in week 5. There are 300 Bs scheduled to be received in week 1 and 200 Ds in week 3. There are also 20 As available.

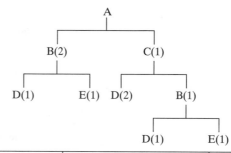

Low-Level Code		Week	1	2	3	4	5
	Part A Lead Time: 1 week Lot Size: lot-for-lot	Gross Requirements Scheduled Receipts Projected Available 20 Net Requirements Planned Order Receipt Planned Order Release					
	Part B Lead Time: 1 week Lot Size: lot-for-lot	Gross Requirements Scheduled Receipts Projected Available Net Requirements Planned Order Receipt Planned Order Release					
	Part C Lead Time: 1 week Lot Size: lot-for-lot	Gross Requirements Scheduled Receipts Projected Available Net Requirements Planned Order Receipt Planned Order Release					
	Part D Lead Time: 1 week Lot Size: lot-for-lot	Gross Requirements Scheduled Receipts Projected Available Net Requirements Planned Order Receipt Planned Order Release					
	Part E Lead Time: 1 week Lot Size: lot-for-lot	Gross Requirements Scheduled Receipts Projected Available Net Requirements Planned Order Receipt Planned Order Release					

4.16 Given the following product tree, determine the low-level codes and the gross and net quantities for each part. There is a requirement for 100 As in week 4 and 50 Bs in week 5. There is a scheduled receipt of 100 Cs in week 2. Quantities required of each are also shown.

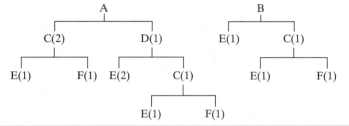

Low-Level Code		Week	1	2	3	4	5
	Part A Lead Time: 1 week Lot Size: lot-for-lot	Gross Requirements Scheduled Receipts Projected Available Net Requirements Planned Order Receipt Planned Order Release					
	Part B Lead Time: 1 week Lot Size: lot-for-lot	Gross Requirements Scheduled Receipts Projected Available Net Requirements Planned Order Receipt Planned Order Release					
	Part C Lead Time: 1 week Lot Size: lot-for-lot	Gross Requirements Scheduled Receipts Projected Available Net Requirements Planned Order Receipt Planned Order Release					
	Part D Lead Time: 1 week Lot Size: lot-for-lot	Gross Requirements Scheduled Receipts Projected Available Net Requirements Planned Order Receipt Planned Order Release					
	Part E Lead Time: 1 week Lot Size: 500	Gross Requirements Scheduled Receipts Projected Available Net Requirements Planned Order Receipt Planned Order Release					
	Part F Lead Time: 1 week Lot Size: lot-for-lot	Gross Requirements Scheduled Receipts Projected Available Net Requirements Planned Order Receipt Planned Order Release					

4.17 Complete the following MRP record. The lead time is four weeks, and the lot size is 200. What will happen if the gross requirements in week 3 are increased to 150 units? As a planner, what actions can you take?

Initial MRP

Week	1	2	3	4	5
Gross Requirements	50	125	100	60	40
Scheduled Receipts		200		200	
Projected Available 100					
Net Requirements					
Planned Order Receipt					
Planned Order Release					

Revised MRP

Week	1	2	3	4	5
Gross Requirements					
Scheduled Receipts					
Projected Available 100					
Net Requirements					
Planned Order Receipt					
Planned Order Release					

4.18 It is Monday morning, and you have just arrived at work. Complete the following MRP record as it would appear Monday morning. Lead time is two weeks, and the lot size is 100.

Initial MRP

Week	1	2	3	4	5
Gross Requirements	70	40	80	50	40
Scheduled Receipts	100				
Projected Available 50					
Net Requirements					
Planned Order Receipt					
Planned Order Release					

During the week, the following events occur. Enter them in the MRP record.

a. The planned order for 100 in week 1 is released.

b. Thirty of the scheduled receipt for week 1 are scrapped.

c. An order for 20 is received for delivery in week 3.

d. An order for 40 is received for delivery in week 6.

e. The gross requirements of 70 in week 1 is issued.

MRP record at the end of week 1

Week	2	3	4	5	6
Gross Requirements Scheduled Receipts Projected Available Net Requirements Planned Order Receipt Planned Order Release					

5

Capacity Management

INTRODUCTION

So far we have been concerned with planning priority, that is, determining what is to be produced and when. The system is hierarchical, moving from long planning horizons and few details (production plan) through medium time spans (master production schedule) to a high level of detail and short time spans (material requirements plan). At each level, manufacturing develops priority plans to satisfy demand. However, without the resources to achieve the priority plan, the plan will be unworkable. Capacity management is concerned with supplying the necessary resources. This chapter looks more closely at the question of capacity: what it is, how much is available, how much is required, and how to balance priority and capacity.

DEFINITION OF CAPACITY

Capacity is the amount of work that can be done in a specified time period. In the eighth edition of the APICS dictionary, capacity is defined as "the capability of a worker, machine, work center, plan, or organization to produce output per period of time." Capacity is a *rate* of doing work, not the *quantity* of work done.

Two kinds of capacity are important: capacity available and capacity required. **Capacity available** is the capacity of a system or resource to produce a quantity of output in a given time period.

Capacity required is the capacity of a system or resource needed to produce a desired output in a given time period. A term closely related to capacity required is **load.** This is the amount of released and planned work assigned to a facility for a particular time period. It is the sum of all the required capacities.

These three terms—capacity required, load, and capacity available—are important in capacity management and will be discussed in subsequent sections of this chapter. Capacity is often pictured as a funnel as shown in Figure 5.1. Capacity available is the rate at which work can be withdrawn from the system. Load is the amount of work in the system.

Capacity management is responsible for determining the capacity needed to achieve the priority plans as well as providing, monitoring, and controlling that capacity so the priority plan can be met. The eighth edition of the APICS dictionary defines capacity management as "the function of establishing, measuring, monitoring, and adjusting limits or levels of capacity in order to execute all manufacturing schedules." As with all management processes, it consists of planning and control functions.

Capacity planning is the process of determining the resources required to meet the priority plan and the methods needed to make that capacity available. It takes place at each level of the priority planning process. Production planning, master production scheduling, and material requirements planning determine priorities: what is wanted and when. These priority plans cannot be implemented, however, unless the firm has sufficient capacity to fulfill the demand. Capacity planning, thus, links the various production priority schedules to manufacturing resources.

Capacity control is the process of monitoring production output, comparing it with capacity plans, and taking corrective action when needed. Capacity control will be examined in Chapter 6.

CAPACITY PLANNING

Capacity planning involves calculating the capacity needed to achieve the priority plan and finding ways of making that capacity available. If the capacity requirement cannot be met, the priority plans have to be changed.

Figure 5.1   Capacity versus load.

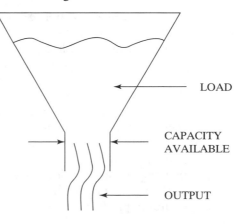

LOAD

CAPACITY
AVAILABLE

OUTPUT

Priority plans are usually stated in units of product or some standard unit of output. Capacity can sometimes be stated in the same units, for example, tons of steel or yards of cloth. If there is no common unit, capacity must be stated as the hours available. The priority plan must then be translated into hours of work required and compared to the hours available. The process of capacity planning is as follows:

1. Determine the capacity available at each work center in each time period.
2. Determine the load at each work center in each time period.
   - Translate the priority plan into the hours of work required at each work center in each time period.
   - Sum up the capacities required for each item on each work center to determine the load on each work center in each time period.
3. Resolve differences between available capacity and required capacity. If possible, available capacity should be adjusted to match the load. Otherwise, the priority plans must be changed to match the available capacity.

This process occurs at each level in the priority planning process, varying only in the level of detail and time spans involved.

Planning Levels

Resource planning involves long-range capacity resource requirements and is directly linked to production planning. Typically, it involves translating monthly, quarterly, or annual product priorities from the production plan into some total measure of capacity, such as gross labor hours. Resource planning involves changes in manpower, capital equipment, product design, or other facilities that take a long time to acquire and eliminate. If a resource plan cannot be devised to meet the production plan, the production plan has to be changed. The two plans set the limits and levels for production. If they are realistic, the master production schedule should work. (See Chapter 2, p. 36.)

Rough-cut capacity planning takes capacity planning to the next level of detail. The master production schedule is the primary information source. The purpose of rough-cut capacity planning is to check the feasibility of the MPS, provide warnings of any bottlenecks, ensure utilization of work centers, and advise vendors of capacity requirements. (See Chapter 3, p. 50.)

Capacity requirements planning is directly linked to the material requirements plan. Since this type of planning focuses on component parts, greater detail is involved than in rough-cut capacity planning. It is concerned with individual orders at individual work centers and calculates work center loads and labor requirements for each time period at each work center.

Figure 5.2 shows the relationship between the different levels of priority planning and capacity planning. Notice that, although the upper levels of priority planning are input to lower levels, the various capacity plans relate only to their level in the priority plan, not to subsequent capacity planning levels. Resource planning relates to production planning but is not an input to rough-cut capacity planning.

Figure 5.2 Planning levels.

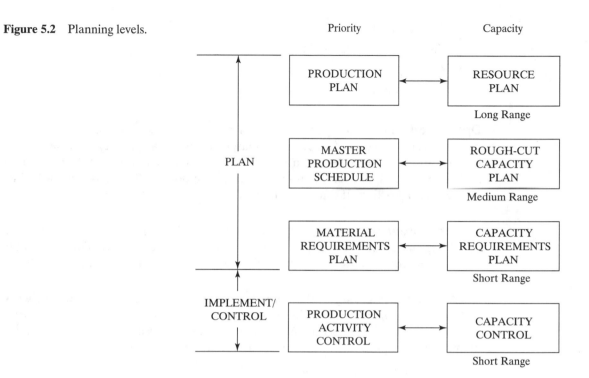

After the plans are completed, production activity control and purchasing must be authorized to process, or implement, shop orders and purchase orders. Capacity must still be considered. Work center capacity control will be covered in the next chapter.

CAPACITY REQUIREMENTS PLANNING (CRP)

The capacity requirements plan (CRP) occurs at the level of the material requirements plan. It is the process of determining in detail the amount of labor and machine resources needed to achieve the required production. Planned orders from the MRP and open shop orders (scheduled receipts) are converted into demand for time in each work center in each time period. This process takes into consideration the lead times for operations and offsets the operations at work centers accordingly. In considering open shop orders, it accounts for work already done on a shop order. Capacity planning is the most detailed, complete, and accurate of the capacity planning techniques. This accuracy is most important in the immediate time periods. Because of the detail, a great amount of data and computation are required.

Inputs

The inputs needed for a CRP are open shop orders, planned order releases, routings, time standards, lead times, and work center capacities. This information can be obtained from the following:

- Open order file.
- Material requirements plan.
- Routing file.
- Work center file.

Open order file.   An open shop order appears as a scheduled receipt on the material requirements plan. It is a released order for a quantity of a part to be manufactured and completed on a specific date. It shows all relevant information such as quantities, due dates, and operations. The open order file is a record of all the active shop orders. It can be maintained manually or as a computer file.

Planned order releases.   Planned orders are determined by the computer's MRP logic based upon the gross requirements for a particular part. They are inputs to the CRP process in assessing the total capacity required in future time periods.

Routing file.   A routing is the path that work follows from work center to work center as it is completed. Routing is specified on a route sheet or, in a computer-based system, in a route file. A routing file should exist for every component manufactured and contain the following information:

- Operations to be performed.
- Sequence of operations.
- Work centers to be used.
- Possible alternate work centers.
- Tooling needed at each operation.
- Standard times: setup times and run times per piece.

Figure 5.3 shows an example of a routing file.

Part Name: Gear shaft		Part Number: SG 123		
Drawing Number: D123X				
Operation No.	Work Center	S/U Time (standard hours)	Run Time/piece (standard hours)	Operation
10	12	1.50	0.20	Turn shaft
20	14	0.50	0.25	Mill slot
30	17	0.30	0.05	Drill 2 holes
40	03	0.45	0.10	Grind
50	Stores			Inventory

Figure 5.3   Routing file.

Work center file.   A work center is composed of a number of machines or workers capable of doing the same work. The machinery will normally be similar so there are no differences in the kind of work the machines can do or the capacity of each. Several sewing machines of similar capacity could be considered a work center. A work center file contains information on the capacity and move, wait, and queue times associated with the center.

The **move time** is the time normally taken to move material from one workstation to another. The **wait time** is the time a job is at a work center after completion and before being moved. The **queue time** is the time a job waits at a work center before being handled. **Lead time** is the sum of queue, setup, run, wait, and move times.

Shop calendar.   Another piece of information needed is the number of working days available. The Gregorian calendar (which is the one we use every day) has some serious drawbacks for manufacturing planning and control. The months do not have the same number of days, holidays are spread unevenly throughout the year, and the calendar does not work on a decimal base. Suppose the lead time for an item is 35 working days and on December 13 we are asked if we can deliver by January 22. This is about six weeks away, but with the Gregorian calendar, some calculations have to be made to decide if there is enough time to make the delivery. Holidays occur in that period, and the plant will be shut down for inventory the first week in January. How many working days do we really have?

Because of these problems, it is desirable to develop a shop calendar. This can be set up in different ways, but the example shown in Figure 5.4 is typical.

MONTH	WEEK	MON.	TUES.	WED.	THURS.	FRI.	SAT.	SUN.
JULY	27	2 / 123	3 / 124	④ /	5 / 125	6 / 126	⑦ /	⑧ /
	28	9 / 127	10 / 128	11 / 129	12 / 130	13 / 131	⑭ /	⑮ /
	29	16 / 132	17 / 133	18 / 134	19 / 135	20 / 136	㉑ /	㉒ /
	30	23 / 137	24 / 138	25 / 139	26 / 140	27 / 141	㉘ /	㉙ /
	31	30 / 142	31 / 143	1 / 144	2 / 145	3 / 146	④ /	⑤ /

JULY 2 ⟶ [2 / 123] ⟵ WORK DAY − 123   ◯ DEFINES NON-WORK DAYS

Figure 5.4   Planning calendar. *(Source: The American Production and Inventory Control Society, Inc., Material Requirements Planning Training Aid, 5–21. Reprinted with permission.)*

CAPACITY AVAILABLE

Capacity available is the capacity of a system or resource to produce a quantity of output in a given time period. It is affected by the following:

- *Product specifications.* If the product specifications change, the work content (work required to make the product) will change, thus affecting the number of units that can be produced.
- *Product mix.* Each product has its own work content measured in the time it takes to make the product. If the mix of products being produced changes, the total work content (time) for the mix will change.
- *Plant and equipment.* This relates to the methods used to make the product. If the method is changed—for example, a faster machine is used—the output will change. Similarly, if more machines are added to the work center, the capacity will change.
- *Work effort.* This relates to the speed or pace at which the work is done. If the workforce changes pace, perhaps producing more in a given time, the capacity will be altered.

Product specification and product mix depend on the design of the product and the mix of products made. If these vary considerably, it is difficult to use units of product to measure capacity. So what units should be used to measure capacity?

Measuring Capacity

Units of output. If the variety of products produced at a work center or in a plant is not large, it is often possible to use a unit common to all products. Paper mills measure capacity in tons of paper, breweries in barrels of beer, and automobile manufacturers in numbers of cars. However, if a variety of products is made, a good common unit may not exist. In this case, the unit common to all products is time.

Standard time. The work content of a product is expressed as the time required to make the product using a given method of manufacture. Using time-study techniques, the standard time for a job can be determined—that is, the time it would take a qualified operator working at a normal pace to do the job. It provides a yardstick for measuring work content and a unit for stating capacity. It is also used in loading and scheduling.

Levels of Capacity

Capacity needs to be measured on at least three levels:

- Machine or individual worker.
- Work center.
- Plant, which can be considered as a group of different work centers.

Determining Capacity Available

There are two ways of determining the capacity available: measurement and calculation. **Demonstrated (measured) capacity** is figured from historical data. **Calculated** or **rated capacity** is based on available time, utilization, and efficiency.

Rated capacity.   Rated, or calculated, capacity is the product of available time, utilization, and efficiency.

Available time.   The available time is the number of hours a work center can be used. For example, a work center working one eight-hour shift for five days a week is available 40 hours a week. The available time depends on the number of machines, the number of workers, and the hours of operation.

▼

EXAMPLE PROBLEM

A work center has three machines and is operated for eight hours a day five days a week. What is the available time?

Answer

$$\text{Available time} = 3 \times 8 \times 5 = 120 \text{ hours per week}$$

Utilization.   The available time is the maximum hours we can expect from the work center. However, it is unlikely this will be attained all the time. Downtime can occur due to machine breakdown, absenteeism, lack of material, and all those problems that cause unavoidable delays. The percentage of time that the work center is active compared to the available time is called **work center utilization:**

$$\text{Utilization} = \frac{\text{hours actually worked}}{\text{available hours}} \times 100\%$$

▼

EXAMPLE PROBLEM

A work center is available 120 hours but actually produced goods for 100 hours. What is the utilization of the work center?

Answer

$$\text{Utilization} = \frac{100}{120} \times 100\% = 83.3\%$$

Utilization can be determined from historical records or by a work sampling study.

Efficiency.   It is possible for a work center to utilize 100 hours a week but not produce 100 standard hours of work. The workers might be working at a faster or slower pace than the standard working pace, causing the efficiency of the work center to be more or less than 100%:

$$\text{Efficiency} = \frac{\text{standard hours of work produced}}{\text{hours actually worked}} \times 100\%$$

▼ ──

EXAMPLE PROBLEM

A work center is utilized for 100 hours a week and produces 120 standard hours of work in that time. What is the efficiency of the work center?

Answer

$$\text{Efficiency} = \frac{120}{100} \times 100\% = 120\%$$

Rated capacity.   Rated capacity is calculated by taking into account the work center utilization and efficiency:

$$\text{Rated capacity} = \text{available time} \times \text{utilization} \times \text{efficiency}$$

▼ ──

EXAMPLE PROBLEM

A work center consists of four machines and is operated eight hours per day for five days a week. Historically, the utilization has been 85% and the efficiency 110%. What is the rated capacity?

Answer

Available time $= 4 \times 8 \times 5 = 160$ hours per week

Rated capacity $= 160 \times 0.85 \times 1.10 = 149.6$ standard hours

We expect to get 149.6 standard hours of work from that work center in an average week.

Demonstrated Capacity

One way to find out the capacity of a work center is to examine the previous production records and to use that information as the available capacity of the work center.

▼ ━━━

EXAMPLE PROBLEM

Over the previous four weeks, a work center produced 120, 130, 150, and 140 standard hours of work. What is the demonstrated capacity of the work center?

Answer

$$\text{Demonstrated capacity} \; = \; \frac{120 + 130 + 150 + 140}{4} \; = \; 135 \text{ standard hours}$$

Notice that demonstrated capacity is average, not maximum, output. It also depends on the utilization and efficiency of the work center, although these are not included in the calculation.

Efficiency and utilization can be obtained from historical data if a record is maintained of the hours available, hours actually worked, and the standard hours produced by a work center.

▼ ━━━

EXAMPLE PROBLEM

Over a four-week period, a work center produced 540 standard hours of work, was available for work 640 hours, and actually worked 480 hours. Calculate the utilization and the efficiency of the work center.

Answer

$$\text{Utilization} \; = \; \frac{\text{hours actually worked}}{\text{available hours}} \; \times \; 100 \; = \; \frac{480}{640} \; \times \; 100\% \; = \; 75\%$$

$$\text{Efficiency} = \frac{\text{standard hours of work produced}}{\text{hours actually worked}} \times 100 \; = \frac{540}{480} \times 100\% = 112.5\%$$

CAPACITY REQUIRED (LOAD)

Capacity requirements are generated by the priority planning system and involve translating priorities, given in units of product or some common unit, into hours of work required at each work center in each time period. This translation takes place at each of the priority planning levels from production planning to master production scheduling to material requirements planning. Figure 5.2 illustrates this relationship.

The level of detail, the planning horizon, and the techniques used vary with each planning level. In this text, we will study the material requirements planning/capacity requirements planning level.

Determining the capacity required is a two-step process. First, determine the time needed for each order at each work center; then, sum up the capacity required for individual orders to obtain the load.

Time Needed for Each Order

The time needed for each order is the sum of the setup time and the run time. The run time is equal to the run time per piece multiplied by the number of pieces in the order.

▼ ────────────────────────────────

EXAMPLE PROBLEM

A work center is to process 150 units of gear shaft SG 123 on work order 333. The setup time is 1.5 hours, and the run time is 0.2 hours per piece. What is the standard time needed to run the order?

Answer

$$
\begin{aligned}
\text{Total standard time} &= \text{setup time} + \text{run time} \\
&= 1.5 + (150 \times 0.2) \\
&= 31.5 \text{ standard hours}
\end{aligned}
$$

▼ ────────────────────────────────

EXAMPLE PROBLEM

In the previous problem, how much actual time will be needed to run the order if the work center has an efficiency of 120% and a utilization of 80%?

Answer

$$
\text{Capacity required} = (\text{actual time})(\text{efficiency})(\text{utilization})
$$

$$
\text{Actual time} = \frac{\text{capacity required}}{(\text{efficiency})(\text{utilization})}
$$

$$
= \frac{31.5}{(1.2)(0.8)}
$$

$$
= 32.8 \text{ hours}
$$

Load

The load on a work center is the sum of the required times for all the planned and actual orders to be run on the work center in a specified period. The steps in calculating load are as follows:

1. Determine the standard hours of operation time for each planned and released order for each work center by time period.

2. Add all the standard hours together for each work center in each period. The result is the total required capacity (load) on that work center for each time period of the plan.

▼ ━━

EXAMPLE PROBLEM

A work center has the following open orders and planned orders for week 20. Calculate the total standard time required (load) on this work center in week 20. Order 222 is already in progress, and there are 100 remaining to run.

	Order Quantity	Setup Time (hours)	Run Time (hours/piece)	Total Time (hours)
Released Orders				
222	100	0	0.2	
333	150	1.5	0.2	
Planned Orders				
444	200	3	0.25	
555	300	2.5	0.15	
Total Time				

Answer

Released Orders	222	Total time $= 0 + (100 \times 0.2)$	$=$	20.0	standard hours
	333	Total time $= 1.5 + (150 \times 0.2)$	$=$	31.5	standard hours
Planned Orders	444	Total time $= 3 + (200 \times 0.25)$	$=$	53.0	standard hours
	555	Total time $= 2.5 + (300 \times 0.15)$	$=$	47.5	standard hours
Total Time			$=$	152.0	standard hours

In week 20, there is a load (requirement) for 152 standard hours.

━━

The load must now be compared to the available capacity. One way of doing this is with a work center load report.

Work Center Load Report

The work center load report shows future capacity requirements based on released and planned orders for each time period of the plan.

The load of 152 hours calculated in the previous example is for week 20. Similarly, loads for other weeks can be calculated and recorded on a load report such as is shown in Figure 5.5. Figure 5.6 shows the same data in graphical form. Note that the report shows released and planned load, total load, rated capacity and (over)/under capacity. The term overcapacity means that the work center is overloaded and the term undercapacity means the work center is underloaded. This type of display gives information used to adjust available capacity or to adjust the

Week	20	21	22	23	24	Total
Released Load	51.5	45	30	30	25	181.5
Planned Load	100.5	120	100	90	100	510.5
Total Load	152	165	130	120	125	692
Rated Capacity	140	140	140	140	140	700
(Over)/Under capacity	(12)	(25)	10	20	15	8

Figure 5.5   Work center load report.

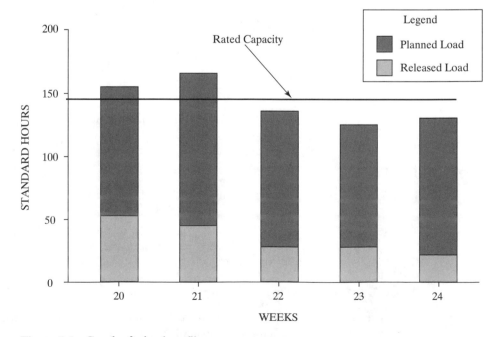

Figure 5.6   Graph of a load profile.

load by changing the priority plan. In this example, weeks 1 and 2 are overloaded, the balance are underloaded, and the cumulative load is less than the available. For the planner, this shows there is enough total capacity over the planning horizon, and available capacity or priority can be juggled to meet the plan.

SCHEDULING ORDERS

So far we have assumed that we know when an order should be run on one work center. Most orders are processed across a number of work centers, and it is necessary to calculate when orders must be started and completed on each work center so the final due date can be met. This process is called **scheduling.** In the eighth edition of the APICS dictionary, scheduling is defined as "a timetable for planned occurrences."

Back scheduling.   The usual process is to start with the due date and, using the lead times, to work back to find the start date for each operation. This process is called **back scheduling.** To schedule, we need to know for each order:

- The quantity and due date.
- Sequence of operations and work centers needed.
- Setup and run times for each operation.
- Queue, wait, and move times.
- Work center capacity available (rated or demonstrated).

The information needed is obtained from the following:

- *Order file.*   Quantities and due dates.
- *Route file.*   Sequence of operations, work centers needed, setup time, and run time.
- *Work center file.*   Queue, move, and wait times and work center capacity.

The process is as follows:

1. For each work order, calculate the capacity required (time) at each work center.
2. Starting with the due date, schedule back to get the completion and start dates for each operation.

EXAMPLE PROBLEM

Suppose there is an order for 150 of gear shaft SG 123. The due date is day 135. The route sheet, shown in Figure 5.3, gives information about the operations to be performed and the setup and run times. The work center file, shown in Figure 5.7, gives lead time data for each work center. Calculate the start and finish dates for each operation. Use the following scheduling rules.

Figure 5.7 Lead time data from work center file.

Work Center	Queue Time (days)	Wait Time (days)	Move Time (days)
12	4	1	1
14	3	1	1
17	5	1	1
03	8	1	1

a. Operation times are rounded up to the nearest eight hours and expressed as days on a one-shift-basis. That is, if an operation takes 6.5 standard hours, round it up to eight hours, which represents one day.

b. Assume an order starts at the beginning of the day and finishes at the end of a day. For example, if an order starts on day 1 and is finished on day 5, it has taken five days to complete. If move time is one day, the order will be available to the next workstation at the start of day 7.

Answer

The calculations for the operation time at each work center are as follows:

$$\text{Setup time} + \text{run time} = \text{total time (standard hours)}$$

Operation 10: Work center 12: $1.5 + 0.20 \times 150$ = 31.5 standard hours = 4 days

Operation 20: Work center 14: $0.50 + 0.25 \times 150$ = 38.0 standard hours = 5 days

Operation 30: Work center 17: $0.30 + 0.05 \times 150$ = 7.8 standard hours = 1 day

Operation 40: Work center 03: $0.45 + 0.10 \times 150$ = 15.45 standard hours = 2 days

The next step is to schedule back from the due date (day 135) to get the completion and start dates for each operation. To do so, we need to know not only the operation times just calculated, but also the queue, wait, and move times. These are in the work center file. Suppose the information shown in Figure 5.7 is obtained from these files.

The process starts with the last operation. The goods are to be in the stores on day 135. It takes one day to move them, so the order must be completed on operation 40 on day 133. Subtracting the wait, queue, and operation times (11 days), the order must be started on day 123. With a move time of one day, it must be completed on operation 30 on day 121. Using this process, the start and completion date can be calculated for all operations. Figure 5.8 shows the resulting schedule and Figure 5.9 shows the same thing graphically.

Operation Number	Work Center	Arrival Date	Queue (days)	Operation (days)	Wait (days)	Finish Date
10	12	95	4	4	1	103
20	14	105	3	5	1	113
30	17	115	5	1	1	121
40	3	123	8	2	1	133
50	Stores	135				

Figure 5.8

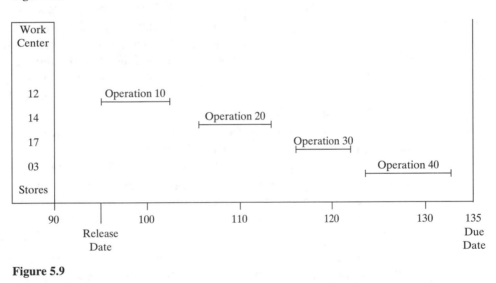

Figure 5.9

MAKING THE PLAN

So far we have discussed the data needed for a capacity requirements plan, where the data come from, and the scheduling and loading of shop orders through the various work centers. The next step is to compare the load to available capacity to see if there are imbalances and if so, to find possible solutions.

There are two ways of balancing capacity available and load: alter the load, or change the capacity available. Altering the load means shifting orders ahead or back so the load is leveled. If orders are processed on other work stations, the schedule and load on the other work stations have to be changed as well. It may also mean that other components should be rescheduled and the master production schedule changed.

Figure 5.10   Simple bill of material.

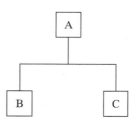

Consider the bill of material shown in Figure 5.10. If component B is to be rescheduled to a later date, then the priority for component C is changed, as is the master production schedule for A. For these reasons, changing the load may not be the preferred course of action. In the short run, capacity can be adjusted. Some ways that this may be done are as follows:

- Schedule overtime or undertime. This will provide temporary and quick relief for cases where the load/capacity imbalance is not too large.

- Adjust the level of the workforce by hiring or laying off workers. The ability to do so will depend on the availability of the skills required and the training needed. The higher the skill level and the longer the training needed, the more difficult it becomes to change quickly the level of the workforce.

- Shift workforce from underloaded to overloaded work centers. This requires a flexible cross-trained workforce.

- Use alternate routings to shift some load to another work center. Often the other work center is not as efficient as the original. Nevertheless, the important thing is to meet the schedule, and this is a valid way of doing so.

- Subcontract work when more capacity is needed or bring in previously subcontracted work to increase required capacity. It may be more costly to subcontract rather than make the item in-house, but again it is important to maintain the schedule.

The result of capacity requirements planning should be a detailed workable plan that meets the priority objectives and provides the capacity to do so. Ideally, it will satisfy the material requirements plan and allow for adequate utilization of the workforce, machinery, and equipment.

SUMMARY

Capacity management occurs at all levels of the planning process. It is directly related to the priority plan, and the level of detail and time spans will be similar to the related priority plan.

Capacity planning is concerned with translating the priority plan into the hours of capacity required in manufacturing to make the items in the priority plan and with methods of making that capacity available. Capacity available depends upon the number of workers and machines, their utilization, and efficiency.

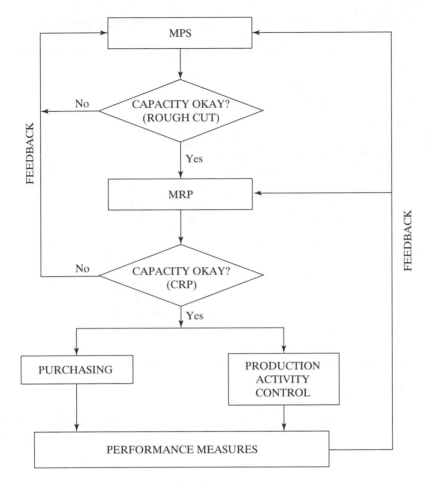

Figure 5.11   MRP and CRP closed-loop system.

Capacity requirements planning occurs at the material requirements planning level. It takes planned orders from the MRP and open shop orders from the open order file and converts them to a load on each work center. It considers lead times and actual order quantities. It is the most detailed of the capacity planning techniques.

Material requirements planning and capacity requirements planning should form part of a closed-loop system which not only includes planning and control functions but also provides feedback so planning can always be current. Figure 5.11 illustrates the concept.

QUESTIONS

1. What are the responsibilities of capacity management?
2. What is capacity planning?
3. Describe the three steps of capacity planning.

4. Relate the three levels of priority planning to capacity planning. Describe each level in terms of the detail and the time horizons used.

5. What is capacity requirements planning? At what level of the priority planning process does it occur?

6. What are the inputs to the CRP process? Where is this information obtained?

7. Describe each of the following and the information they contain.
   a. Open order file.
   b. Route file.
   c. Work center file.

8. What is a shop calendar? Why is it needed?

9. In which file would you find the following information?
   a. A scheduled receipt.
   b. A planned receipt.
   c. Efficiency and utilization.
   d. Sequence of operations on a part.

10. Define "capacity available." What are the four factors that affect it?

11. Why is standard time usually used to measure capacity?

12. What are rated capacity, utilization, and efficiency? How are they related?

13. What is measured or demonstrated capacity? How is it different from rated capacity?

14. What is load?

15. What is a work center load report? What information does it contain?

16. What is a schedule?

17. Describe the process of back scheduling.

18. What are the two ways of balancing capacity available and load? Which is preferred? Why?

19. What are some of the ways capacity available can be altered in the short run?

20. Why is feedback necessary in a control system?

PROBLEMS

5.1 A work center consists of three machines each working a 16-hour day for five days a week. What is the weekly available time?

   Answer.   240 hours per week

5.2 The work center in problem 5.1 is utilized 75% of the time. What are the hours per week actually worked?

   Answer.   180 hours per week

5.3 If the efficiency of the work center in problem 5.1 is 115%, what is the rated capacity of the work center?

   Answer.   207 standard hours per week

5.4 A work center consisting of seven machines is operated 16 hours a day for a five-day week. Utilization is 80%, and efficiency is 110%. What is the rated weekly capacity in standard hours?

   Answer.   492.8 standard hours per week

5.5   A work center consists of three machines working eight hours a day for a five-day week. If the utilization is 75% and the efficiency is 125%, what is the rated capacity of the work center?

5.6   Over a period of four weeks, a work center produced 50, 45, 35, and 55 standard hours of work. What is the demonstrated capacity of the work center?

  Answer. 46.25 standard hours of work per week

5.7   In a 12-week period, a work center produces 1140 standard hours of work. What is the measured capacity of the work center?

5.8   In one week, a work center produces 85 standard hours of work. The hours scheduled are 80, and 75 hours are actually worked. Calculate the utilization and efficiency of the work center.

  Answer. Utilization is 93.75%; efficiency is 113.33%.

5.9   A work center consisting of three machines operates 40 hours a week. In a four-week period, it actually worked 360 hours and produced 468 standard hours of work. Calculate the utilization and efficiency of the work center. What is the demonstrated weekly capacity of the work center?

5.10 A firm wishes to determine the efficiency and utilization of a work center composed of three machines each working 16 hours per day for five days a week. A study undertaken by the materials management department found that over the past year the work center was available for work 12,000 hours, work was actually being done for 10,440 hours, and work was performed 11,480 standard hours. Calculate the utilization, efficiency, and demonstrated weekly capacity. Assume a 50-week year.

5.11 How many standard hours are needed to run an order of 200 pieces if the setup time is 1.3 hours and the run time 0.3 hours per piece? How many actual hours are needed at the work center if the efficiency is 130% and the utilization is 70%?

  Answer. 61.3 standard hours. 67.4 actual hours.

5.12 How many standard hours are needed to run an order of 500 pieces if the setup time is 2.0 hours and the run time 0.15 hours per piece? How many actual hours are needed at the work center if the efficiency is 120% and the utilization is 75%?

5.13 A work center has the following open and planned orders for week 4. Calculate the total standard time required (load).

		Order Quantity	Setup time (hours)	Run time (hours /piece)	Total time (hours)
Released Orders	120	250	1.00	0.10	
	340	100	2.50	0.30	
Planned Orders	560	300	3.00	0.25	
	780	500	2.00	0.15	
Total Time (standard hours)					

  Answer. Total time = 213.5 standard hours

5.14 A work center has the following open and planned orders for week 4. Calculate the total standard time required (load).

		Order Quantity	Setup time (hours)	Run time (hours /piece)	Total time (hours)
Released Orders	125	150	0.25	0.10	
	345	50	0.40	0.05	
Planned Orders	565	75	1.00	0.20	
	785	35	0.50	0.15	
Total Time (standard hours)					

5.15 Using the information in the following route file, open order file, and MRP planned orders, calculate the load on the work center.

Routing:	Part 123:	Setup time	=	2 standard hours
		Run time per piece	=	3 standard hours per piece
	Part 456:	Set up time	=	3 standard hours
		Run time per piece	=	1 standard hour per piece

Open Orders for parts

Week	1	2	3
123	12	8	5
456	15	5	5

Planned Orders for parts

1	2	3
0	5	10
0	10	15

Load report

Week	1	2	3
Released Load 123			
456			
Planned Load 123			
456			
Total Load			

5.16 Complete the following load report and suggest possible courses of action.

Week	18	19	20	21	Total
Released Load	145	155	90	60	450
Planned Load	0	0	80	80	160
Total Load					
Rated Capacity	150	150	150	150	600
(Over)/Under					

5.17 Back schedule the following shop order. All times are given in days. Move time between operations is one day, and wait time is one day. Due date is day 150. Assume orders start at the beginning of a day and finish at the end of a day.

Operation Number	Work Center	Operation Time (days)	Queue Time (days)	Arrival Date	Finish Date
10	111	2	3		
20	130	4	5		
30	155	1	2		
	Stores			150	

Answer. The order must arrive at work center 111 on day 127.

5.18 Back schedule the following shop order. All times are given in days. Move time between operations is one day, and wait time is one day. Due date is day 200. Assume orders start at the beginning of a day and finish at the end of a day.

Operation Number	Work Center	Operation Time (days)	Queue Time (days)	Arrival Date	Finish Date
10	110	4	3		
20	120	2	4		
30	130	3	2		
	Stores			200	

6

Production Activity Control

INTRODUCTION

The time comes when plans must be put into action. Production activity control (PAC) is responsible for executing the master production schedule and the material requirements plan. At the same time, it must make good use of labor and machines, minimize work-in-process inventory, and maintain customer service.

The material requirements plan authorizes PAC:

- To release work orders to the shop for manufacturing
- To take control of work orders and make sure they are completed on time
- To be responsible for the immediate detailed planning of the flow of orders through manufacturing, carrying out the plan, and controlling the work as it progresses to completion
- To manage day-to-day activity and provide the necessary support.

Figure 6.1 shows the relationship between the planning system and PAC.

The activities of the PAC system can be classified into planning, implementation, and control functions.

Planning

The flow of work through each of the work centers must be planned to meet delivery dates, which means production activity control must do the following:

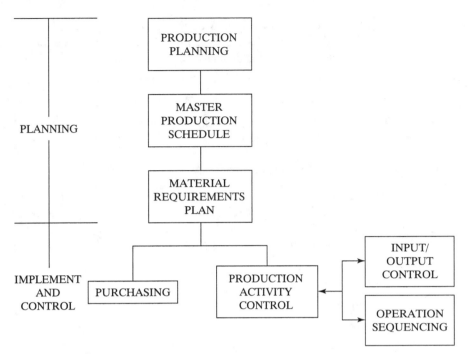

Figure 6.1 Priority planning and production activity control.

- Ensure that the required materials, tooling, personnel, and information are available to manufacture the components when needed.
- Schedule start and completion dates for each shop order at each work center so the scheduled completion date of the order can be met. This will involve the planner in developing a load profile for the work centers.

Implementation

Once the plans are made, production activity control must put them into action by advising the shop floor what must be done. Usually instructions are given by issuing a shop order. Production activity control will:

- Gather the information needed by the shop floor to make the product.
- Release orders to the shop floor as authorized by the material requirements plan. This is called **dispatching.**

Control

Once plans are made and shop orders released, the process must be monitored to learn what is actually happening. The results are compared to the plan to decide whether corrective action is necessary. Production activity control will do the following:

- Rank the shop orders in desired priority sequence by work center and establish a dispatch list based on this information.

- Track the actual performance of work orders and compare it to planned schedules. Where necessary, PAC must take corrective action by replanning, rescheduling, or adjusting capacity to meet final delivery requirements.
- Monitor and control work-in-process, lead times, and work center queues.
- Report work center efficiency, operation times, order quantities, and scrap.

The functions of planning, implementing, and controlling are shown schematically in Figure 6.2.

Manufacturing Systems

The particular type of production control system used varies from company to company, but all should perform the preceding functions. However, the relative importance of these functions will depend on the type of manufacturing process. As discussed in Chapter 1, manufacturing processes can be conveniently broken down into three categories:

1. Flow manufacturing
2. Intermittent manufacturing
3. Project manufacturing

Flow manufacturing.   Flow manufacturing is concerned with the production of high-volume standard products. If the units are discrete (e.g., cars and appliances), the process is usually called **repetitive manufacturing,** and if the goods are made in a continuous flow (e.g., gasoline), **continuous manufacturing.** There are four major characteristics to flow manufacturing:

1. Routings are fixed, and work centers are arranged according to the routing. The time taken to perform work at one work center is almost the same as at any other work center in the line.

Figure 6.2   Schematic of a production control system.

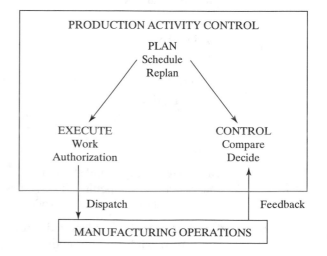

2. Work centers are dedicated to producing a limited range of similar products. Machinery and tooling are especially designed to make the specific products.

3. Material flows from one workstation to another using some form of mechanical transfer. There is little buildup in work-in-process inventory, and throughput times are low.

4. Capacity is fixed by the line.

Production activity control concentrates on planning the flow of work and making sure that the right material is fed to the line as stated in the planned schedule. Since work flows from one workstation to another automatically, implementation and control are relatively simple.

Intermittent manufacturing.   Intermittent manufacturing is characterized by many variations in product design, process requirements, and order quantities. This kind of manufacturing is characterized by the following:

1. Flow of work through the shop is varied and depends on the design of a particular product. As orders are processed, they will take more time at one workstation than at another. Thus, the work flow is not balanced.

2. Machinery and workers must be flexible enough to do the variety of work. Machinery and work centers are usually grouped according to the function they perform (e.g., all lathes in one department).

3. Throughput times are generally long. Scheduling work to arrive just when needed is difficult, the time taken by an order at each work center varies, and work queues before work centers, causing long delays in processing. Work-in-process inventory is often large.

4. The capacity required depends on the particular mix of products being built and is difficult to predict.

Production activity control in intermittent manufacturing is complex. Because of the number of products made, the variety of routings, and scheduling problems, PAC is a major activity in this type of manufacturing. Planning and control are typically exercised using shop orders for each batch being produced. Our discussion of PAC assumes this kind of environment.

Project manufacturing.   Project manufacturing usually involves the creation of one or a small number of units. Large ship-building is an example. Because the design of a product is often carried out or modified as the project develops, there is close coordination between manufacturing, marketing, purchasing, and engineering.

DATA REQUIREMENTS

To plan the processing of materials through manufacturing, PAC must have the following information:

- What and how much to produce.
- When parts are needed so the completion date can be met.
- What operations are required to make the product and how long the operations will take.
- What the available capacities of the various work centers are.

Production activity control must have a data or information system from which to work. Usually the data needed to answer these questions are organized into databases. The files contained in the databases are of two types: planning and control.

Planning Files

Four planning files are needed: item master file, product structure file, routing file, and work center master file.

Item master file.   There is one record in this file for each part number. The file contains, in one place, all of the pertinent data related to the part. For PAC, this includes the following:

- Part number, a unique number assigned to a component.
- Part description.
- Manufacturing lead time, the normal time needed to make this part.
- Quantity on hand.
- Quantity available.
- Allocated quantity, quantities assigned to specific work orders but not yet withdrawn from inventory.
- On-order quantities, the balance due on all outstanding orders.
- Lot-size quantity, the quantity normally ordered at one time.

Product structure file (bill of material file).   This file contains a list of the single-level components and quantities needed to assemble a parent. It forms a basis for a "pick list" to be used by storeroom personnel to collect the parts required to make the assembly.

Routing file.   There is a record in this file for each part manufactured. The routing consists of a series of operations required to make the item. For each product, this file contains a step-by-step set of instructions describing how the product is made. It gives details of the following:

- The operations required to make the product and the sequence in which those operations are performed.
- A brief description of each operation.
- Equipment, tools, and accessories needed for each operation.

- Setup times, the standard time required for setting up the equipment for each operation.
- Run times, the standard time to process one unit through each operation.
- Lead times for each operation.

Work center master file.   The purpose of this file is to collect all of the relevant data on a work center. For each work center, it gives details on the following:

- Work center number.
- Capacity.
- Number of shifts worked per week.
- Number of machine hours per shift.
- Number of labor hours per shift.
- Efficiency.
- Utilization.
- Queue time, the average time that a job waits at the work center before work is begun.
- Alternate work centers, work centers that may be used as alternatives.

Control Files

Control in intermittent manufacturing is exercised through shop orders and control files that contain data on these orders. There are generally two kinds of files: the shop order master file and the shop order detail file.

Shop order master file.   Each active manufacturing order has a record in this file. The purpose is to provide summarized data on each shop order such as the following information:

- Shop order number, a unique number identifying the shop order.
- Order quantity.
- Quantity completed.
- Quantity scrapped.
- Quantity of material issued to the order.
- Due date, the date the order is expected to be finished.
- Priority, a value used to rank the order in relation to others.
- Balance due, the quantity not yet completed.
- Cost information.

Shop order detail file.   Each shop order has a detail file that contains a record for each operation needed to make the item. Each record contains the following information:

- Operation number.
- Setup hours, planned and actual.
- Run hours, planned and actual.
- Quantity reported complete at that operation.
- Quantity reported scrapped at that operation.
- Due date or lead time remaining.

ORDER PREPARATION

Once authorization to process an order has been received, production activity control is responsible for planning and preparing its release to the shop floor. The order should be reviewed to be sure that the necessary tooling, material, and capacity are available. If they are not, the order cannot be completed and should not be released.

Tooling is not generally considered in the material requirements planning (MRP) program, so at this stage, material availability must be checked. If MRP software is used, it will have checked the availability of material and allocated it to a shop order so no further checking is necessary. If MRP software is not used, production activity control must manually check material availability.

If a capacity requirements planning system has been used, necessary capacity should be available. However, at this stage, there may be some differences between planned capacity and what is actually available. When capacity requirements planning is not used, it is necessary to determine if capacity is available.

Checking capacity availability is a two-step process. First, the order must be scheduled to see when the capacity is needed, and second, the load on work centers must be checked in that period. Scheduling and loading are discussed in the next two sections.

SCHEDULING

The objective of scheduling is to meet delivery dates and to make the best use of manufacturing resources. It involves establishing start and finish dates for each operation required to complete an item. To develop a reliable schedule, the planner must have information on routing, required and available capacity, competing jobs, and manufacturing lead times (MLT) at each work center involved.

Manufacturing Lead Time

Manufacturing lead time is the time normally required to produce an item in a typical lot quantity. Typically, MLT consists of five elements:

1. Queue time, amount of time the job is waiting at a work center before operation begins.

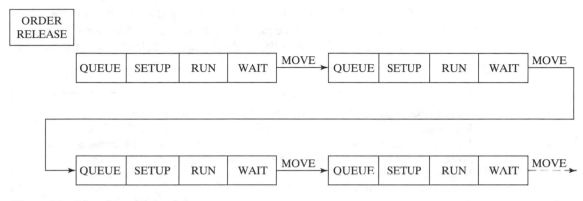

Figure 6.3   Manufacturing lead time.

2. Setup time, time required to prepare the work center for operation.
3. Run time, time needed to run the order through the operation.
4. Wait time, amount of time the job is at the work center before being moved to the next work center.
5. Move time, transit time between work centers.

The total manufacturing lead time will be the sum of order preparation and release plus the MLTs for each operation. Figure 6.3 shows the elements making up manufacturing lead time. Setup time and run time are straightforward, and determining them is the responsibility of the industrial engineering department. Queue, wait, and move times are under the control of manufacturing and PAC.

The largest of the five elements is queue time. Typically, in an intermittent manufacturing operation, it accounts for 85%–95% of the total lead time. Production activity control is responsible for managing the queue by regulating the flow of work into and out of work centers. If the number of orders waiting to be worked on (load) is reduced, so is the queue time, the lead time, and work-in-process. Increasing capacity also reduces queue. Production activity control must manage both the input of orders to the production process and the available capacity to control queue and work-in-process.

A term that is closely related to manufacturing lead time is **cycle time.** The eighth edition of the APICS dictionary defines cycle time as the length of time from when material enters a production facility until it exits. A synonym is **throughput time.**

EXAMPLE PROBLEM

An order for 100 of a product is processed on work centers A and B. The setup time on A is 30 minutes, and run time is ten minutes per piece. The setup time on B is 50 minutes, and the run time is five minutes per piece. Wait time between the two operations is four hours. The move time between A and B is ten minutes. Wait time after operation B is four hours, and the move time into stores is 15 minutes. There is no queue at either workstation. Calculate the total manufacturing lead time for the order.

Answer

Work Center A operation time $= 30 + (100 \times 10) =$	1030	minutes
Wait time $=$	240	minutes
Move time from A to B $=$	10	minutes
Work Center B operation time $= 50 + (100 \times 5) =$	550	minutes
Wait time $=$	240	minutes
Move time from B to stores $=$	15	minutes
Total manufacturing lead time $=$	2085	minutes
$=$		34 hours, 45 minutes

Scheduling Techniques

There are many techniques to schedule shop orders through a plant, but all of them require an understanding of forward and backward scheduling as well as finite and infinite loading.

Forward scheduling assumes that material procurement and operation scheduling for a component start when the order is received, whatever the due date, and that operations are scheduled forward from this date. The first line in Figure 6.4 illustrates this method. The result is completion before the due date, which usually results in a buildup of inventory. This method is used to decide the earliest delivery date for a product.

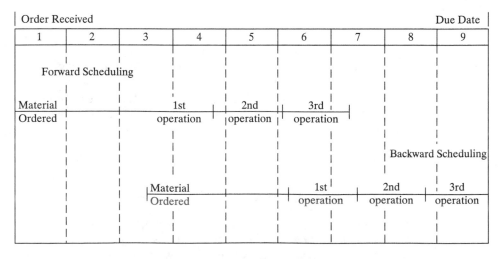

Figure 6.4   Forward and backward scheduling: infinite loading.

Forward scheduling is used to calculate how long it will take to complete a task. The technique is used for purposes such as developing promise dates for customers or figuring out whether an order behind schedule can be caught up.

Backward scheduling is illustrated by the second line in Figure 6.4. The last operation on the routing is scheduled first and is scheduled for completion at the due date. Previous operations are scheduled back from the last operation. This schedules items to be available as needed and is the same logic as used in the MRP system. Work-in-process inventory is reduced, but because there is little slack time in the system, customer service may suffer.

Backward scheduling is used to determine when an order must be started. Backward scheduling is common in industry because it reduces inventory.

Infinite loading is also illustrated in Figure 6.4. The assumption is made that the workstations on which operations 1, 2, and 3 are done have capacity available when required. It does not consider the existence of other shop orders competing for capacity at these work centers. It assumes infinite capacity will be available. Figure 6.5 shows a load profile for infinite capacity. Notice the over and under load.

Finite loading assumes there is a defined limit to available capacity at any workstation. If there is not enough capacity available at a workstation because of other shop orders, the order has to be scheduled in a different time period. Figure 6.6 illustrates the condition.

In the forward-scheduling example shown in Figure 6.6, the first and second operations cannot be performed at their respective workstations when they should be because the required capacity is not available at the time required. These operations must be rescheduled to a later time period. Similarly, in the example of scheduling back, the second and first operations cannot be performed when they should be and must be rescheduled to an earlier time period. Figure 6.7 shows a load profile for finite loading. Notice the load is smoothed so there is no overload condition.

Chapter 5 gives an example of backward scheduling as it relates to capacity requirements planning. The same process is used in PAC.

Figure 6.5 Infinite load profile.

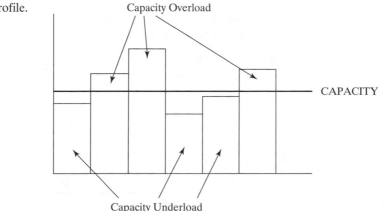

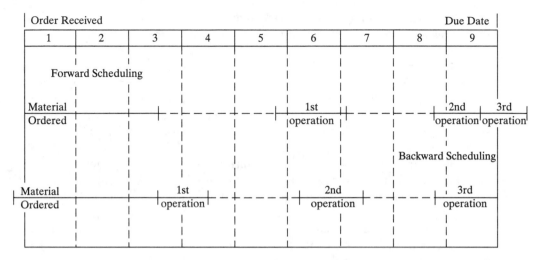

Figure 6.6   Forward and backward scheduling: finite loading.

Figure 6.7   Finite load profile.

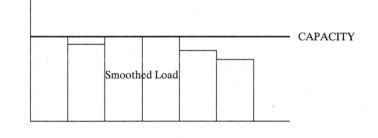

EXAMPLE PROBLEM

A company has an order for 50 brand X to be delivered on day 100. Draw a backward schedule based on the following:

a. Only one machine is assigned to each operation

b. The factory works one 8-hour shift five days a week

c. The parts move in one lot of 50.

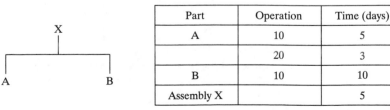

Part	Operation	Time (days)
A	10	5
	20	3
B	10	10
Assembly X		5

Answer

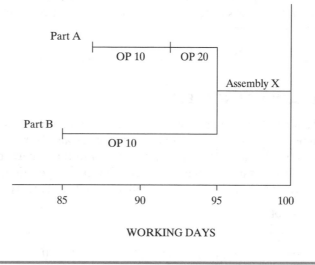

Operation Overlapping

In operation overlapping, the next operation is allowed to begin before the entire lot is completed on the previous operation. This reduces the total manufacturing lead times because the second operation starts before the first operation finishes all the parts in the order. Figure 6.8 shows schematically how it works and the potential reduction in lead time.

An order is divided into at least two lots. When the first lot is completed on operation A, it is transferred to operation B. In Figure 6.8, it is assumed operation B cannot be set up until the first lot is received, but this is not always the case. While operation A continues with the second lot, operation B starts on the first lot. When operation A finishes the second lot, it is transferred to operation B. If the lots are sized properly, there will be no idle time at operation B. The manufacturing lead time is reduced by the overlap time and the elimination of queue time.

Operation overlapping is a method of expediting an order, but there are some costs involved. First, move costs are increased, especially if the overlapped operations are not close together. Second, it may increase the queue and lead time for other orders. Third, it does not increase capacity but potentially reduces it if the second operation is idle waiting for parts from the first operation.

Figure 6.8 Operation overlapping.

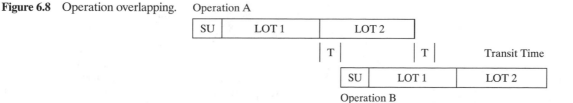

The problem is deciding the size of the sublot. If the run time per piece on operation B is shorter than that on A, the first batch must be large enough to avoid idle time on operation B.

▼

EXAMPLE PROBLEM

Refer to the data given in the example problem in the section on manufacturing lead time. It is decided to overlap operations A and B by splitting the lot of 100 into two lots of 70 and 30. Wait time between A and B and between B and stores is eliminated. The move times remain the same. Setup on operation B cannot start until the first batch arrives. Calculate the manufacturing lead time. How much time has been saved?

Answer

Operation time for A for lot of 70	$= 30 + (70 \times 10) =$	730 minutes
Move time between A and B	$=$	10 minutes
Operation time for B for lot of 100	$= 50 + (100 \times 5) =$	550 minutes
Move time from B to stores	$=$	15 minutes
Total manufacturing lead time	$=$	1305 minutes
	$=$	21 hours, 45 minutes
Time saved $= 2085 - 1305 = 780$ minutes	$=$	13 hours

Operation Splitting

Operation splitting is a second method of reducing manufacturing lead time. The order is split into two or more lots and run on two or more machines simultaneously. If the lot is split in two, the run-time component of lead time is effectively cut in half, although an additional setup is incurred. Figure 6.9 shows a schematic of operation splitting.

Figure 6.9   Operation splitting.

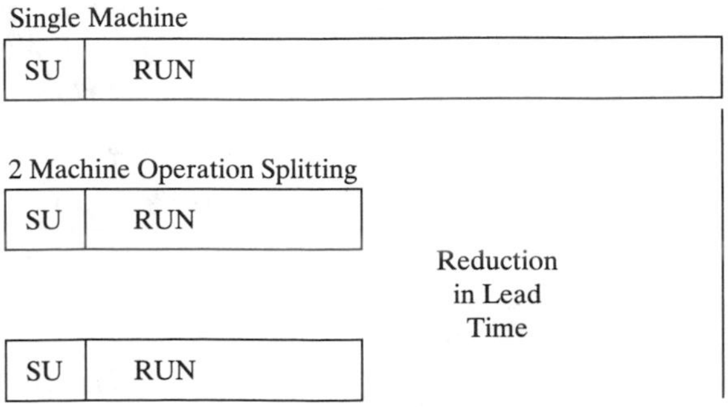

Single Machine

SU	RUN

2 Machine Operation Splitting

SU	RUN

Reduction
in Lead
Time

SU	RUN

Operation splitting is practical when:

- Setup time is low compared to run time.
- A suitable work center is idle.
- It is possible for an operator to run more than one machine at a time.

The last condition often exists when a machine cycles through its operation automatically, leaving the operator time to set up another machine. The time needed to unload and load must be shorter than the run time per piece. For example, if the unload/load time was two minutes and the run time was three minutes, the operator would have time to unload and load the first machine while the second was running.

EXAMPLE PROBLEM

A component made on a particular work center has a setup time of 100 minutes and a run time of three minutes per piece. An order for 500 is to be processed on two machines simultaneously. The machines can be set up at the same time. Calculate the elapsed operation time.

Answer

$$\text{Elapsed operation time} = 100 + 3 \times 250 = 850 \text{ minutes}$$
$$= 14 \text{ hours and 10 minutes}$$

LOAD LEVELING

Load profiles were discussed in Chapter 5 in the section on capacity requirements planning. The load profile for a work center is constructed by calculating the standard hours of operation for each order in each time period and adding them together by time period. Figure 6.10 is an example of a load report.

This report tells PAC what the load is on the work center. There is a capacity shortage in week 20 of 30 hours. This means there was no point in releasing all of the planned orders that week. Perhaps some could be released in week 18 or 19, and perhaps some overtime could be worked to help reduce the capacity crunch.

SCHEDULING BOTTLENECKS

In intermittent manufacturing, it is almost impossible to balance the available capacities of the various workstations with the demand for their capacity. As a result, some workstations are overloaded and some underloaded. The overloaded workstations are called **bottlenecks** and, by definition, are those workstations where

Work Center: 10   Available Time: 120 hours/week
Description: Lathes   Efficiency: 115%
Number of Machines: 3   Utilization 80%
Rated Capacity:
   110 standard hours/week

Week	18	19	20	21	22	23	Total
Released Load Planned Load	105	100	80 60	30 80	0 130	0 80	315 350
Total Load	105	100	140	110	130	80	665
Rated Capacity	110	110	110	110	110	110	660
(Over)/Under Capacity	5	10	(30)	0	(20)	30	(5)

Figure 6.10 Work center load report.

the required capacity is greater than the available capacity. In the eighth edition of their dictionary, APICS defines a bottleneck as "a facility, function, department, or resource whose capacity is equal to or less than the demand placed upon it."

Throughput. **Throughput** is the total volume of production passing through a facility. Bottlenecks control the throughput of all products processed by them. If work centers feeding bottlenecks produce more than the bottleneck can process, excess work-in-process inventory is built up. Therefore, work should be scheduled through the bottleneck at the rate it can process the work. Work centers fed by bottlenecks have their throughput controlled by the bottleneck, and their schedules should be determined by that of the bottleneck.

EXAMPLE PROBLEM

Suppose a manufacturer makes wagons composed of a box body, a handle assembly, and two wheel assemblies. Demand for the wagons is 500 a week. The wheel assembly capacity is 1200 sets a week, the handle assembly capacity is 450 a week, and final assembly can produce 550 wagons a week.

   a. What is the capacity of the factory?
   b. What limits the throughput of the factory?

    c. How many wheel assemblies should be made each week?

    d. What is the utilization of the wheel assembly operation?

    e. What happens if the wheel assembly utilization is increased to 100%?

Answer

a. 450 units a week.

b. Throughput is limited by the capacity of the handle assembly operation.

c. 900 wheel assemblies should be made each week. This matches the capacity of the handle assembly operation.

d. Utilization of the wheel assembly operation is $900 \div 1200 = 75\%$

e. Excess inventory builds up.

Some bottleneck principles.   Since bottlenecks control the throughput of a facility, some important principles should be noted:

1. *Utilization of a non-bottleneck resource is not determined by its potential, but by another constraint in the system.* In the previous example problem, the utilization of the wheel assembly operation was determined by the handle assembly operation.

2. *Using a non-bottleneck 100% of the time does not produce 100% utilization.* If the wheel assembly operation was utilized 100% of the time, it would produce 1200 sets of wheels a week, 300 sets more than needed. Because of the buildup of inventory, this operation would eventually have to stop.

3. *The capacity of the system depends on the capacity of the bottleneck.* If the handle assembly operation breaks down, the throughput of the factory is reduced.

4. *Time saved at a non-bottleneck saves the system nothing.* Suppose, in a flash of brilliance, the industrial engineering department increased the capacity of the wheel assembly operation to 1500 units a week. This extra capacity could not be utilized, and nothing would be gained.

5. *Capacity and priority must be considered together.* Suppose the wagon manufacturer made wagons with two styles of handles. During setup, nothing is produced, which reduces the capacity of the system. Since handle assembly is the bottleneck, every setup in this operation reduces the throughput of the system. Ideally, the company would run one style of handle for six months, then switch over to the second style. However, customers wanting the second style of handle might not be willing to wait six months. A compromise is needed whereby runs are as long as possible but priority (demand) is satisfied.

6. *Loads can, and should, be split.* Suppose the handle assembly operation (the bottleneck) produces one style of handle for two weeks, then switches to the

second style. The batch size is 900 handles. Rather than waiting until the 900 are produced before moving them to the final assembly area, the manufacturer can move a day's production (90) at a time. The process batch size and the transfer batch size are different. Thus, delivery to the final assembly is matched to usage, and work-in-process inventory is reduced.

Managing bottlenecks.   Since bottlenecks are so important to the throughput of a system, scheduling and controlling them is extremely important. The following must be done:

1. *Establish a time buffer before each bottleneck.*   A time buffer is an inventory (queue) place before each bottleneck. Because it is of the utmost importance to keep the bottleneck working, it must never be starved for material, and it can be starved only if the flow from feeding workstations is disrupted. The time buffer should be only as long as the time of any expected delay caused by feeding workstations. In this way, the time buffer ensures that the bottleneck will not be shut down for lack of work and this queue will be held at a predetermined minimum quantity.

2. *Control the rate of material feeding the bottleneck.*   A bottleneck must be fed at a rate equal to its capacity so the time buffer remains constant. The first operation in the sequence of operations is called a *gate operation*. This operation controls the work feeding the bottleneck and must operate at a rate equal to the output of the bottleneck so the time buffer queue is maintained.

3. *Do everything to provide the needed bottleneck capacity.*   Anything that increases the capacity of the bottleneck increases the capacity of the system. Better utilization, fewer setups, improved methods to reduce setup and run time are some methods for increasing capacity.

4. *Adjust loads.*   This is similar to item 3 but puts emphasis on reducing the load on a bottleneck by using such things as using alternate work centers and subcontracting. These may be more costly than using the bottleneck, but utilization of non-bottlenecks and throughput of the total system is increased, resulting in more company sales and increased profits.

5. *Change the schedule.*   Do this as a final resort, but it is better to be honest about delivery promises.

Once the bottleneck is scheduled according to its available capacity and the market demand it must satisfy, the non-bottleneck resources can be scheduled. When a work order is completed at the bottleneck, it can be scheduled on subsequent operations.

Feeding operations have to protect the time buffer by scheduling backward in time from the bottleneck. If the time buffer is set at four days, the operation immediately preceding the bottleneck is scheduled to complete the required parts four days before they are scheduled to run on the bottleneck. Each preceding operation can be back-scheduled in the same way so the parts are available as required for the next operation.

Any disturbances in the feeding operations are absorbed by the time buffer, and throughput is not affected. Also, work-in-process inventory is reduced. Since the queue is limited to the time buffer, lead times are reduced.

▼ ──

EXAMPLE PROBLEM

Parent X requires one each of component Y and Z. Both Y and Z are processed on work center 20 which has an available capacity of 40 hours. The setup time for component Y is one hour and the run time 0.3 hours per piece. For component Z, setup time is two hours and the run time is 0.20 hours per piece. Calculate the number of Ys and Zs that can be produced.

Answer

Available capacity for Ys and Zs $= 40$ hours

Let $x = $ number of Ys and Zs to produce

$\text{Time}_Y + \text{Time}_Z = 40$ hours

$1 + 0.3x + 2 + 0.2x = 40$ hours

$0.5x = 37$ hours

$x = 74$

Therefore, work center 20 can produce 74 Ys and 74 Zs.

IMPLEMENTATION

Orders that have tooling, material, and capacity have a good chance of being completed on time and can be released to the shop floor. Other orders that do not have all of the necessary elements should not be released because they only cause excess work-in-process inventory and may interrupt work on orders that can be completed. The process for releasing an order is shown in Figure 6.11.

Implementation is arrived at by issuing a shop order to manufacturing authorizing them to proceed with making the item. A shop packet is usually compiled which contains the shop order and whatever other information is needed by manufacturing. It may include any of the following:

- Shop order showing the shop order number, the part number, name, description, and quantity.
- Engineering drawings.
- Bills of material.
- Route sheets showing the operations to be performed, equipment and accessories needed, materials to use, and the setup and run times.
- Material issue tickets that authorize manufacturing to get the required material from stores. These are also used for charging the material against the shop order.

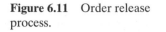

Figure 6.11 Order release process.

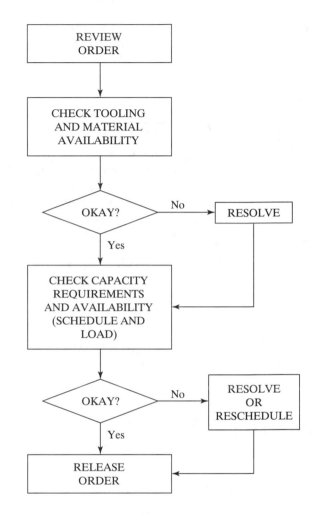

- Tool requisitions authorizing manufacturing to withdraw necessary tooling from the tool crib.
- Job tickets for each operation to be performed. As well as authorizing the individual operations to be performed, they also can function as part of a reporting system. The worker can log on and off the job using the job ticket, and it then becomes a record of that operation.
- Move tickets that authorize and direct the movement of work between operations.

CONTROL

Once work orders have been issued to manufacturing, their progress has to be controlled. To control progress, performance has to be measured and compared to what is planned. If what is actually happening (what is measured) varies significantly from what was planned, either the plans have to be changed or corrective action must be taken to bring performance back to plan.

The objectives of production activity control are to meet delivery dates and to make the best use of company resources. To meet delivery dates, a company must control the progress of orders on the shop floor, which means controlling the lead time for orders. As discussed earlier in this chapter, the largest component of lead time is queue. If queue can be controlled, delivery dates can be met. Chapter 1 discussed some characteristics of intermittent operations in which many different products and order quantities have many different routings, each requiring different capacities. In this environment, it is almost impossible to balance the load over all the workstations. Queue exists because of this erratic input and output.

To control queue and meet delivery commitments, production activity control must:

- Control the work going into and coming out of a work center. This is generally called **input/output control.**

- Set the correct priority of orders to run at each work center.

Input/Output Control

Production activity control must balance the flow of work to and from different work centers. This is to ensure queue, work-in-process, and lead times are controlled. The input/output control system is a method of managing queues and work-in-process lead times by monitoring and controlling the input to, and output from, a facility. It is designed to balance the input rate in hours with the output rate so these will be controlled.

The input rate is controlled by the release of orders to the shop floor. If the rate of input is increased, queue, work-in-process, and lead times increase. The output rate is controlled by increasing or decreasing the capacity of a work center. Capacity change is a problem for manufacturing, but it can be attained by overtime or undertime, shifting workers, and so forth. Figure 6.12 shows the idea graphically.

Figure 6.12   Input/output control.

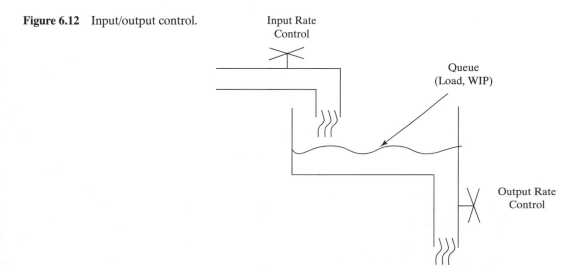

Input/output report.   To control input and output, a plan must be devised, along with a method for comparing what actually occurs against what was planned. This information is shown on an input/output report. Figure 6.13 is an example of such a report. The values are in standard hours.

Cumulative variance is the difference between the total planned for a given period and the actual total for that period. It is calculated as follows:

Cumulative variance = previous cumulative variance + actual − planned

Cumulative input variance week $2 = -4 + 32 - 32 = -4$

Backlog is the same as queue and expresses the work to be done in hours. It is calculated as follows:

Planned backlog for period 1 = previous backlog + planned input − planned output

$$= 32 + 38 - 40$$

$$= 30 \text{ hours}$$

The report shows the plan was to maintain a level output in each period and to reduce the queue and lead time by ten hours, but input and output were lower than expected.

Work Center: 201
Capacity per Period: 40 standard hours

Period	1	2	3	4	5	Total
Planned Input	38	32	36	40	44	190
Actual Input	34	32	32	42	40	180
Cumulative Variance	–4	–4	–8	–6	–10	–10

	1	2	3	4	5	Total
Planned Output	40	40	40	40	40	200
Actual Output	32	36	44	44	36	192
Cumulative Variance	–8	–12	–8	–4	–8	–8

		1	2	3	4	5	
Planned Backlog	32	30	22	18	18	22	
Actual Backlog	32	34	30	18	16	20	

Figure 6.13   Input/output report.

Planned and actual inputs monitor the flow of work coming to the work center. Planned and actual outputs monitor the performance of the work center. Planned and actual backlogs monitor the queue and lead time performance.

▼

EXAMPLE PROBLEM

Complete the following input/output report for weeks 1 and 2.

Week		1	2
Planned Input		45	40
Actual Input		42	46
Cumulative Variance			
Planned Output		40	40
Actual Output		42	44
Cumulative Variance			
Planned Backlog	30		
Actual Backlog	30		

Answer

Cumulative input variance week 1 $= 42 - 45 = -3$
Cumulative input variance week 2 $= -3 + 46 - 40 = 3$
Cumulative output variance week 1 $= 42 - 40 = 2$
Cumulative output variance week 2 $= 2 + 44 - 40 = 6$
Planned backlog week 1 $= 30 + 45 - 40 = 35$
Planned backlog week 2 $= 35 + 40 - 40 = 35$
Actual backlog week 1 $= 30 + 42 - 42 = 30$
Actual backlog week 2 $= 30 + 46 - 44 = 32$

Operation Sequencing

The eighth edition of the APICS dictionary defines operation sequencing as a technique for short-term planning of actual jobs to be run in each work center based on capacity and priorities. Priority, in this case, is the sequence in which jobs at a work center should be worked on.

The material requirements plan establishes proper need dates and quantities. Over time, these dates and quantities change for a variety of reasons. Customers may require different delivery quantities or dates. Deliveries of component parts, either from vendors or internally, may not be met. Scrap, shortages, and overages may occur, and so on. Control of priorities is exercised through dispatching.

Dispatching.   Dispatching is the function of selecting and sequencing available jobs to be run at individual work centers. The **dispatch list** is the instrument of priority control. It is a listing by operation of all the jobs available to be run at a work center with the job listed in priority sequence. It normally includes the following information and is updated and published at least daily:

- Plant, department, and work center.
- Part number, shop order number, operation number, and operation description of jobs at the work center.
- Standard hours.
- Priority information.
- Jobs coming to the work center.

Figure 6.14 is an example of a daily dispatch list.

DISPATCH LIST

Work Center: 10
Rated Capacity: 16 standard hours per day
Shop Date: 250

Order Number	Part Number	Order Quantity	Setup Hours	Run Hours	Total Hours	Quantity Completed	Load Remaining	Operation Dates Start	Finish
123	6554	100	1.5	15	16.5	75	3.75	249	250
121	7345	50	0.5	30	30.5	10	24	249	251
142	2687	500	0.2	75	75.2	0	75	250	259
		Total Available Load in Standard Hours					102.75		
Jobs Coming									
145	7745	200	0.7	20	20.7	0	20.7	251	253
135	2832	20	1.2	1.0	2.7	0	2.7	253	254
		Total Future Load in Standard Hours					23.4		

Figure 6.14   Dispatch list (based on 2 machines working one 8 hour shift per day).

Dispatching rules. The ranking of jobs for the dispatch list is created through the application of priority rules. There are many rules, some attempting to reduce work-in-process inventory, others attempting to minimize the number of late orders or maximize the output of the work center. None is perfect or will satisfy all objectives. Some commonly used rules are:

- *First come, first served (FCFS).* Jobs are performed in the sequence in which they are received. This rule ignores due dates and processing time.
- *Earliest job due date (EDD).* Jobs are performed according to their due dates. Due dates are considered, but processing time is not.
- *Earliest operation due date (ODD).* Jobs are performed according to their operation due dates. Due dates and processing time are taken into account. As well, the operation due date is easily understood on the shop floor.
- *Shortest process time (SPT).* Jobs are sequenced according to their process time. This rule ignores due dates, but it maximizes the number of jobs processed. Orders with long process times tend to be delayed.

Figure 6.15 illustrates how these sequencing rules work. Notice that each rule produces a different sequence.

One other rule that should be mentioned is called **critical ratio (CR).** This is an index of the relative priority of an order to other orders at a work center. It is based on the ratio of time remaining to work remaining and is usually expressed as:

$$CR = \frac{due\ date\ -\ present\ date}{lead\ time\ remaining} = \frac{actual\ time\ remaining}{lead\ time\ remaining}$$

Lead time remaining includes all elements of manufacturing lead time and expresses the amount of time the job normally takes to completion.

If the actual time remaining is less than the lead time remaining, it implies there is not sufficient time to complete the job and the job is behind schedule. Similarly, if

Job	Process Time (days)	Arrival Date	Due Date	Operation Due Date	Sequencing Rule			
					FCFS	EDD	ODD	SPT
A	4	223	245	233	2	4	1	3
B	1	224	242	239	3	2	2	1
C	5	231	240	240	4	1	3	4
D	2	219	243	242	1	3	4	2

Figure 6.15 Application of sequencing rules.

lead time remaining and actual time remaining are the same, the job is on schedule. If the actual time remaining is greater than the lead time remaining, the job is ahead of schedule. If the actual time remaining is less than one, the job is late already. The following table summarizes these facts and relates them to the critical ratio:

CR less than 1 (actual time less than lead time).	Order is behind schedule.
CR equal to 1 (actual time equal to lead time).	Order is on schedule.
CR greater than 1 (actual time greater than lead time).	Order is ahead of schedule.
CR zero or less (today's date greater than due date).	Order is already late.

Thus, orders are listed in order of their critical ratio with the lowest one first.

Critical ratio considers due dates and process time. However, it is not easily understood.

EXAMPLE PROBLEM

Today's date is 175. Orders A, B, and C have the following due dates and lead time remaining. Calculate the actual time remaining and the critical ratio for each.

Order	Due Date	Lead Time Remaining (days)
A	185	20
B	195	20
C	205	20

Answer

Order A has a due date of 185, and today is day 175. There are 10 actual days remaining. Since the lead time remaining is 20 days,

$$\text{Critical ratio} = \frac{10}{20} = 0.5$$

Similarly, the actual time remaining and the critical ratios are calculated for orders B and C. The following table gives the results:

Order	Due Date	Lead Time Remaining (days)	Actual Time Remaining (days)	CR
A	185	20	10	0.5
B	195	20	20	1.0
C	205	20	30	1.5

Order A has less actual time remaining than lead time remaining, so the CR is less than 1. It is, therefore, behind schedule. Order B has a CR of 1 and is exactly on schedule. Order C has a CR of 1.5—greater than 1—and is ahead of schedule.

Dispatching rules should be simple to use and easy to understand. As shown in the preceding example, each rule produces a different sequence and has its own advantages and disadvantages. Whichever rule is selected should be consistent with the objectives of the planning system.

PRODUCTION REPORTING

Production reporting provides feedback of what is actually happening on the plant floor. It allows PAC to maintain valid records of on-hand and on-order balances, job status, shortages, scrap, material shortages, and so on. Production activity control needs this information to establish proper priorities and to answer questions regarding deliveries, shortages, and the status of orders. Manufacturing management needs this information to make decisions about plant operation. Payroll needs this information to calculate employees' pay.

Data must be collected, sorted, and reported. The particular data collected depend upon the needs of the various departments. The methods of data collection vary. Sometimes the operator reports the start and completion of an operation, order, movement, and so on, using an on-line system directly reporting events as they occur via data terminals. In other cases, the operator, supervisor, or timekeeper reports this information on an operation reporting form included in the shop packet. Information about inventory withdrawals and receipts must be reported as well.

Once the data are collected, they must be sorted, and appropriate reports produced. Types of information needed for the various reports include:

- Order status.
- Weekly input/output by department or work center.
- Exception reports on such things as scrap, rework, and late shop orders.
- Inventory status.
- Performance summaries on order status, work center and department efficiencies, and so on.

SUMMARY

Production activity control is concerned with converting the material requirements plan into action, reporting the results achieved, and when required, revising the plans and actions to meet the required results. Order release, dispatching, and progress reporting are the three primary functions. To accomplish the plans, PAC must establish detailed schedules for each order, set priorities for work to be done at each work center, and keep them current. Production activity control is also responsible for managing the queue and lead times. If PAC is performing its functions, the company should be able to meet delivery dates, utilize labor and equipment efficiently, and keep inventory levels down.

QUESTIONS

1. What is the responsibility of production activity control?
2. What are the major functions of planning, implementation, and control?
3. What are the major characteristics of flow, intermittent, and project manufacturing?
4. Why is production activity control more complex in intermittent manufacturing?
5. To plan the flow of materials through manufacturing, what four things must production activity control know? Where will information on each be obtained?
6. What are the four planning files used in production activity control? What information does each contain?
7. What are the two control files? What are their purposes?
8. What should production activity control check before releasing a shop order?
9. What is manufacturing lead time? Name and describe each of its elements.
10. Describe forward and backward scheduling. Why is backward scheduling preferred?
11. Describe infinite and finite loading.
12. What is operation overlapping? What is its purpose?
13. What is operation splitting? What is its purpose?
14. What information does a load report contain? Why is it useful to production activity control?
15. What is a bottleneck operation?
16. What is the definition of throughput?
17. What are the six bottleneck principles discussed in the text?
18. What are the five things discussed in the text that are important in managing bottlenecks?
19. What is a shop order? What kind of information does it usually contain?
20. What two things must be done to control queue and meet delivery commitments?
21. What is an input/output control system designed to do? How is input controlled? How is output controlled?
22. What is dispatching? What is a dispatch list?
23. Describe each of the following dispatching rules giving their advantages and disadvantages.
    a. First come, first served.
    b. Earliest due date.
    c. Earliest operation due date.
    d. Shortest processing time.
    e. Critical ratio.

24. If the time remaining to complete a job is 10 days and the lead time remaining is 12 days, what is the critical ratio? Is the order ahead of schedule, on schedule, or behind schedule?

25. What is the purpose of production reporting? Why is it needed?

PROBLEMS

6.1 Shop order 7777 is for 700 of part 8900. From the routing file, it is found that operation 20 is done on work center 300. The setup time is 3.1 hours, and run time is 0.133 hours per piece. What is the required capacity on work center 300 for shop order 7777?

Answer. 96.2 standard hours

6.2 An order for 100 of a product is processed on work centers A and B. The setup time on A is 50 minutes, and run time is 5 minutes per piece. The setup time on B is 60 minutes, and the run time 5 minutes per piece. Wait time between the two operations is 6 hours. The move time between A and B is 40 minutes. Wait time after operation B is 6 hours, and the move time into stores is 3 hours. Queue at work center A is 30 hours and at B is 25 hours. Calculate the total manufacturing lead time for the order.

Answer. 89 hours and 10 minutes

6.3 In problem 6.2, what percent of the time is the order actually running?

Answer. 18.7%

6.4 An order for 50 of a product is processed on work centers A and B. The setup time on A is 60 minutes, and run time is 5 minutes per piece. The setup time on B is 30 minutes, and the run time is 6 minutes per piece. Wait time between the two operations is 10 hours. The move time between A and B is 60 minutes. Wait time after operation B is 8 hours, and the move time into stores is 2 hours. Queue at work center A is 40 hours and at B is 35 hours. Calculate the total manufacturing lead time for the order.

6.5 In problem 6.4, what percent of time is the order actually running?

6.6 Amalgamated Skyhooks, Inc., has an order for 200 Model SKY3 Skyhooks for delivery on day 200. The Skyhook consists of three parts. Components B and C form subassembly A. Subassembly A and component D form the final assembly. Following are the work centers and times for each operation. Using a piece of graph paper, draw a backward schedule based on the following. When must component C be started to meet the delivery date?

a. Only one machine is assigned to each operation.

b. The factory works one 8-hour shift five days a week.

c. All parts move in one lot of 200.

Part	Operation	Standard Time (days)
D	10	5
	20	7
B	10	5
	20	7
C	10	10
	20	5
Subassembly A		5
Final Assembly SKY3		5

Answer. Day 175

6.7 International Door Slammers has an order to deliver 500 door slammers on day 130. Draw up a backward schedule under the following conditions:

   a. Only one machine is assigned to each operation.

   b. Schedule one 8-hour shift per day for five days per week.

   c. All parts are to move in one lot of 500 pieces.

   d. Allow eight hours between operations for queue and move times.

   A slammer consists of three parts. Purchased components C and D form subassembly A. Subassembly A and component B form the final assembly. Part B is machined in three operations. No special tooling is required except for part B, operation 20. It takes 24 hours to make the tooling. Material is available for all parts.

   Standard times for the lot of 500 are as follows:

Part	Operation	Standard Time (days)
B	10	10
	20	6
	30	6
Subassembly A		16
Final Assembly		10

6.8 An order for 100 of a product is processed on operation A and operation B. The setup time on A is 50 minutes, and the run time per piece is 9 minutes. The setup time on B is 30 minutes, and the run time is 6 minutes per piece. It takes 20 minutes to move a lot between A and B. Since this is a rush order, it is given top priority (president's edict) and is run as soon as it arrives at either workstation.

   It is decided to overlap the two operations and to split the lot of 100 into two lots of 60 and 40. When the first lot is finished on operation A, it is moved to operation B where it is set up and run. Meanwhile, operation A completes the balance of the 100 units (40) and sends the units over to operation B. These 40 units should arrive as operation B is completing the first batch of 60; thus, operation B can continue without interruption until all 100 are completed.

   a. Calculate the total manufacturing lead time for operation A and for B without overlapping.

   b. Calculate the manufacturing lead time if the operations are overlapped. How much time is saved?

   Answer. **a.** Total manufacturing lead time = 1600 minutes

   b. Total manufacturing lead time = 1240 minutes

   Saving in lead time = 360 minutes

6.9 An order for 200 bell ringers is processed on work centers 10 and 20. The setup and run times are as follows. It is decided to overlap the lot on the two work centers and to split the lot into two lots of 75 and 125. Move time between operations is 30 minutes. Work center 20 cannot be set up until the first lot arrives. Calculate the saving in manufacturing lead time.

Setup on A = 50 minutes   Run time on A = 5 minutes per piece

Setup on B = 100 minutes   Run time on B = 7 minutes per piece

6.10 An order for 100 of a product is processed on operation A. The setup time is 50 minutes, and the run time per piece is 9 minutes. Since this is a rush order, it is to be split into two lots of 50 each and run on two machines in the work center. The machines can be set up simultaneously.

a. Calculate the manufacturing lead time if the 100 units are run on one machine.

b. Calculate the manufacturing lead time when run on two machines simultaneously.

c. Calculate the reduction in lead time.

Answer.   **a.** 950 minutes

           b. 500 minutes

           c. 450 minutes

6.11 What would be the reduction in MLT if the second machine could not be set up until the setup was completed on the first machine?

6.12 An order for 100 of a product is run on work center 40. The setup time is 3 hours, and the run time is 3 minutes per piece. Since the order is a rush and there are two machines in the work center, it is decided to split the order and run it on both machines. Calculate the manufacturing lead time before and after splitting.

6.13 In problem 6.12, what would be the manufacturing lead time if the second machine could not be set up until the setup on the first machine was completed? Would there be any reduction in manufacturing lead time?

6.14 Parent W requires one of component B and two of component C. Both B and C are run on work center 10. Setup time for B is 2 hours, and run time is 0.1 hours per piece. For component C, setup time is 2 hours, and run time is 0.15 per piece. If the rated capacity of the work center is 80 hours, how many Ws should be produced in a week?

Answer.   190 Ws

6.15 Parent S requires three of component T and two of component U. Both T and U are run on work center 30. Setup time for T is 7 hours, and run time is 0.1 hours per piece. For component U, setup time is 8 hours, and run time is 0.2 hours per piece. If the rated capacity of the work center is 120 hours, how many Ts and Us should be produced in a week?

Answer.   450 Ts and 300 Us a week

6.16 Complete the following input/output report. What are the planned and actual backlogs at the end of period 4?

Period	1	2	3	4	Total
Planned Input	36	38	35	40	
Actual Input	34	32	32	42	
Cumulative Variance					

	1	2	3	4	Total
Planned Output	40	40	40	40	
Actual Output	38	36	40	38	
Cumulative Variance					

		1	2	3	4	
Planned Backlog	32					
Actual Backlog	32					

Answer. Planned backlog = 21 units. Actual backlog = 20 units.

6.17 Complete the following input/output report. What is the actual backlog at the end of period 5?

Period	1	2	3	4	5	Total
Planned Input	78	78	78	78	78	
Actual Input	82	80	74	82	84	
Cumulative Variance						

		1	2	3	4	5	
Planned Output		80	80	80	80	80	
Actual Output		87	83	74	80	84	
Cumulative Variance							

Planned Backlog	45						
Actual Backlog	45						

Answer. 39 units

6.18 Complete the following table to determine the run sequence for each of the sequencing rules.

Job	Process Time (days)	Arrival Date	Due Date	Operation Due Date	Sequencing Rule			
					FCFS	EDD	ODD	SPT
A	5	123	142	132				
B	2	124	144	131				
C	3	131	140	129				

6.19 Jobs X, Y, and Z are in queue at work center 10 before being completed on work center 20. The following information pertains to the jobs and the work centers. For this problem, there is no move time. Today is day 1. If the jobs are scheduled by the earliest due date, can they be completed on time?

Job	Process Time (days)		Due Date
	Work Center 10	Work Center 20	
A	7	3	12
B	5	2	24
C	9	4	18

Job	Work Center 10		Work Center 20	
	Start Day	Stop Day	Start Day	Stop Day
A				
C				
B				

6.20 Calculate the critical ratios for the following orders and establish in what order they should be run. Today's date is 75.

Order	Due Date	Lead Time Remaining (days)	Actual Time Remaining (days)	CR
A	87	12		
B	95	25		
C	100	20		

7

Purchasing

INTRODUCTION

Purchasing is the "process of buying." Many assume purchasing is solely the responsibility of the purchasing department. However, the function is much broader and, if it is carried out effectively, all departments in the company are involved. Obtaining the right material, in the right quantities, with the right delivery (time and place), from the right source, and at the right price are all purchasing functions.

Choosing the right material requires input from the marketing, engineering, manufacturing, and purchasing departments. Quantities and delivery of finished goods are established by the needs of the marketplace. However, manufacturing planning and control (MPC) must decide when to order which raw materials so that marketplace demands can be satisfied. Purchasing is then responsible for placing the orders and for ensuring that the goods arrive on time.

The purchasing department has the major responsibility for locating suitable sources of supply and for negotiating prices. Input from other departments is required in finding and evaluating sources of supply and to help the purchasing department in price negotiation. Purchasing, in its broad sense, is everyone's business.

Purchasing and Profit Leverage

On the average, manufacturing firms spend about 50% of their sales dollar in the purchase of raw materials, components, and supplies. This gives the purchasing function tremendous potential to increase profits. As a simple example, suppose a firm spends 50% of its revenue on purchased goods and shows a net profit before taxes of 10%. For every $100 of sales, they receive $10 of profit and spend $50 on purchases. Other expenses are $40. For the moment, assume that all costs vary with sales. These figures are shown in the following as a simplified income statement:

INCOME STATEMENT

Sales		$100
Cost of Goods Sold		
Purchases	$50	
Other Expenses	40	90
Profit Before Tax		$10

To increase profits by $1, a 10% increase in profits, sales must be increased to $110. Purchases and other expenses increase to $55 and $44. The following modified income statement shows these figures:

INCOME STATEMENT (SALES INCREASE)

Sales		$110
Cost of Goods Sold		
Purchases	$55	
Other Expenses	44	99
Profit Before Tax		$11

However, if the firm can reduce the cost of purchases from $50 to $49, a 2% reduction, it would gain the same 10% increase in profits. In this particular example, a 2% reduction in purchase cost has the same impact on profit as a 10% increase in sales.

INCOME STATEMENT (REDUCED PURCHASE COST)

Sales		$100
Cost of Goods Sold		
Purchases	$49	
Other Expenses	40	89
Profit Before Tax		$11

Purchasing Objectives

Purchasing is responsible for establishing the flow of materials into the firm, following-up with the supplier, and expediting delivery. Missed deliveries can create havoc for manufacturing and sales, but purchasing can reduce problems for both areas, further adding to the profit.

The objectives of purchasing can be divided into four categories:

- Obtaining goods and services of the required quantity and quality.
- Obtaining goods and services at the lowest cost.
- Ensuring the best possible service and prompt delivery by the supplier.

- Developing and maintaining good supplier relations and developing potential suppliers.

  To satisfy these objectives, some basic functions must be performed:

- Determining purchasing specifications: right quality, right quantity, and right delivery (time and place).
- Selecting supplier (right source).
- Negotiating terms and conditions of purchase (right price).
- Issuing and administration of purchase orders.

Purchasing Cycle

The purchasing cycle consists of the following steps:

1. Receiving and analyzing purchase requisitions.
2. Selecting suppliers. Finding potential suppliers, issuing requests for quotations, receiving and analyzing quotations, and selecting the right supplier.
3. Determining the right price.
4. Issuing purchase orders.
5. Following-up to assure delivery dates are met.
6. Receiving and accepting goods.
7. Approving supplier's invoice for payment.

Receiving and analyzing purchase requisition. Purchase requisitions start with the department or person who will be the ultimate user. In the material requirements planning environment, the planner releases a planned order authorizing the purchasing department to go ahead and process a purchase order. At a minimum, the purchase requisition contains the following information:

- Identity of originator, signed approval, and account to which cost is assigned.
- Material specification.
- Quantity and unit of measure.
- Required delivery date and place.
- Any other supplemental information needed.

Selecting suppliers. Identifying and selecting suppliers are important responsibilities of the purchasing department. For routine items or those that have not been purchased before, a list of approved suppliers is kept. If the item has not been purchased before or there is no acceptable supplier on file, a search must be made. If the order is of small value or for standard items, a supplier can probably be found in a catalogue, trade journal, or directory.

Requesting quotations.   For major items, it is usually desirable to issue a request for quotation. This is a written inquiry that is sent to enough suppliers to be sure competitive and reliable quotations are received. *It is not a sales order.* After the suppliers have completed and returned the quotations to the buyer, the quotations are analyzed for price, compliance to specification, terms and conditions of sale, delivery, and payment terms. For items where specifications can be accurately written, the choice is probably made on price, delivery, and terms of sale. For items where specifications cannot be accurately written, the items quoted will vary. The quotations must be evaluated for technical suitability. The final choice is a compromise between technical factors and price. Usually both the issuing and purchasing departments are involved in the decision.

Determining the right price.   This is the responsibility of the purchasing department and is closely tied to the selection of suppliers. The purchasing department is also responsible for price negotiation and will try to obtain the best price from the supplier. Price negotiation will be discussed in a later section of the chapter.

Issuing a purchase order.   A purchase order is a legal offer to purchase. Once accepted by the supplier, it becomes a legal contract for delivery of the goods according to the terms and conditions specified in the purchase agreement. The purchase order is prepared from the purchase requisition or the quotations and from any other additional information needed. A copy is sent to the supplier; copies are retained by purchasing and are also sent to other departments such as accounting, the originating department, and receiving.

Following-up and delivery.   The supplier is responsible for delivering the items ordered on time. The purchasing department is responsible for ensuring that suppliers do deliver on time. If there is doubt that delivery dates can be met, purchasing must find out in time to take corrective action. This might involve expediting transportation, alternate sources of supply, working with the supplier to solve its problems, or rescheduling production.

   The purchasing department is also responsible for working with the supplier on any changes in delivery requirements. Demand for items changes with time, and it may be necessary to expedite certain items or push delivery back on some others. The buyer must keep the supplier informed of the true requirements so that the supplier is able to provide what is wanted and when.

Receiving and accepting goods.   When the goods are received, the receiving department inspects the goods to be sure the correct ones have been sent, are in the right quantity, and have not been damaged in transit. Using their copy of the purchase order and the bill of lading supplied by the carrier, the receiving department then accepts the goods and writes up a receiving report noting any variance. If further inspection is required, such as by quality control, the goods are sent to quality control or held there for inspection. If the goods are received damaged, the receiving department will advise the purchasing department and hold the goods for further action. Provided the goods are in order and require no further inspection, they will be sent to the originating department or to inventory.

A copy of the receiving report is then sent to the purchasing department noting any variance or discrepancy from the purchase order. If the order is considered complete, the receiving department closes out its copy of the purchase order and advises the purchasing department. If it is not, the purchase order is held open awaiting completion. If the goods have also been inspected by the quality control department, they, too, will advise the purchasing department whether the goods have been accepted or not.

Approving supplier's invoice for payment. When the supplier's invoice is received, there are three pieces of information that should agree: the purchase order, the receiving report, and the invoice. The items and the quantities should be the same on all; the prices, and extensions to prices, should be the same on the purchase order and the invoice. All discounts and terms of the original purchase order must be checked against the invoice. It is the job of the purchasing department to verify these and to resolve any differences. Once approved, the invoice is sent to accounts payable for payment.

ESTABLISHING SPECIFICATIONS

The first concern of purchasing—what to buy—is not necessarily a simple decision. For example, someone deciding to buy a car should consider how the car will be used, how often, how much one is willing to pay, and so on. Only then can an individual specify the type of car needed to make the "best buy." This section looks at the problems that organizations face when developing specifications of products and the types of specifications that may be used.

In purchasing an item or a service from a supplier, several factors are included in the package bought. These must be considered when specifications are being developed and can be divided into three broad categories.

- Quantity requirements
- Price requirements
- Functional requirements

Quantity Requirements

Market demand first determines the quantities needed. The quantity is important because it will be a factor in the way the product is designed, specified, and manufactured. For example, if the demand was for only one item, it would be designed to be made at least cost, or a suitable standard item would be selected. However, if the demand were for several thousand, the item would be designed to take advantage of economies of scale, thus satisfying the functional needs at a better price.

Price Requirements

The price specification represents the economic value that the buyer puts on the item—the amount the individual is willing to pay. If the item is to be sold at a low price, the manufacturer will not want to pay a high price for a component part. The economic value placed on the item must relate to the use of the item and its anticipated selling price.

Functional Requirements

Functional specifications are concerned with the end use of the item and what the item is expected to do. By their very nature, functional specifications are the most important of all categories and govern the others.

In a sense, functional specifications are the most difficult to define. To be successful, they must satisfy the real need or purpose of an item. In many cases, the real need has both practical and aesthetic elements to it. A coat is meant to keep one warm, but under what circumstances does it do so and what other functions is it expected to perform? How cold must it get before one needs a coat? On what occasions will it be worn? Is it for working or dress wear? What color and style should it be? What emotional needs is it expected to fill? In the same way, we can ask what practical and aesthetic needs a door handle or side-view mirror on a car is expected to satisfy.

Functional specifications and quality.

Functional specifications are intimately tied to the quality of a product or service. Everyone knows, or thinks he or she knows, what quality is, but there are several misconceptions about what it is and what it is not. Ask someone what is meant by quality, and you will get replies such as, "The best there is," "Perfection," "Degree of excellence," and "Very good." All sound great but do not mean very much.

There are many definitions of quality, but they all center on the idea of user satisfaction. On this basis, it can be said that an item has the required quality if it satisfies the needs of the user.

There are four phases to providing user satisfaction:

1. Quality and product planning.
2. Quality and product design.
3. Quality and manufacturing.
4. Quality and use.

Product planning is involved with decisions about which products and services a company is to market. It must decide the market segment to be served, the product features and quality level expected by that market, the price, and the expected sales volume. The basic quality level is thus specified by senior management according to their understanding of the needs and wants of the marketplace. The success of the product depends on how well they do this.

The result of the firm's market studies is a general specification of the product outlining the expected performance, appearance, price, and sales volume of the product. It is then the job of the product designer to build into the design of the product the quality level described in the general specification. If this is not properly done, the product may not be successful in the marketplace.

For manufactured products, it is the responsibility of manufacturing, as a minimum, to meet the specifications laid down by the product designer. If the item is bought, it is purchasing's responsibility to make sure the supplier can provide the required quality level. For purchasing and manufacturing, quality means conforming to specifications or requirements.

To the final user, quality is related to his or her expectation of how the product should perform. Customers do not care why a product or service is defective. They expect satisfaction. If the product is what the customer wants, well designed, well made, and well serviced, the quality is satisfactory.

Functional specifications should define the quality level needed. They should describe all those characteristics of a product determined by its final use.

Function, quantity, service, and price are interrelated. It is difficult to specify one without consideration of the others. Indeed, the final specification is a compromise of them all, and the successful specification is the best combination of the lot. However, functional specifications ultimately are the ones that drive the others. If the product does not perform adequately for the price, it will not sell.

FUNCTIONAL SPECIFICATION DESCRIPTION

Functional specification can be described in the following ways or by a combination of them:

1. By brand.
2. By specification of physical and chemical characteristics, material and method of manufacture, and performance.
3. By engineering drawings.
4. Miscellaneous.

Description by Brand

Description by brand is most often used in wholesale or retail businesses but is also used extensively in manufacturing. This is particularly true under the following circumstances:

- Items are patented, or the process is secret.
- The supplier has special expertise that the buyer does not have.
- The quantity bought is so small that it is not worth the buyer's effort to develop specifications.
- The supplier, through advertising or direct sales effort, has created a preference on the part of the buyer's customers or staff.

When buying by brand, the customer is relying on the reputation and integrity of the supplier. The assumption is that the supplier wishes to maintain the brand's reputation and will maintain and guarantee the quality of the product so repeat purchases will give the buyer the same satisfaction.

Most of the objections to purchasing by brand center on cost. Branded items, as a group, usually have price levels that are higher than nonbranded items. It may be less costly to develop specifications for generic products than to rely on brands. The other major disadvantage to specifying by brand is that it restricts the number of potential suppliers and reduces competition. Consequently, the usual practice, when specifying by brand, is to ask for the item by brand name or equivalent. In theory, this allows for competition.

Description by Specification

There are several ways of describing a product, but they usually include one or more of the following. Whatever method is used, description by specification depends on the buyer describing in detail exactly what is wanted:

- *Physical and chemical characteristics.*   The buyer must define the physical and chemical properties of the materials wanted. Petroleum products, pharmaceuticals, and paints are often specified in this way.
- *Material and method of manufacture.*   Sometimes the method of manufacture determines the performance and use of a product. For example, hot- and cold-rolled steels are made differently and have different characteristics.
- *Performance.*   This method is used where the buyer is primarily concerned with what the item is required to do and is prepared to have the supplier decide how performance is to be attained. For example, a water pump might be specified as having to deliver so many gallons per minute. Performance specifications are relatively easy to prepare and take advantage of the supplier's special knowledge.

        Whatever the method of specification, there are several characteristics to description by specification:

- To be useful, specifications must be carefully designed. If they are too loosely drawn, they may not provide a satisfactory product. If they are too detailed and elaborate, they are costly to develop, are difficult to inspect, and may discourage possible suppliers.
- Specifications must allow for multiple sources and for competitive bidding.
- If performance specifications are used, the buyer is assured that if the product does not give the desired results, the seller is responsible. They provide a standard for measuring and checking the materials supplied.
- Not all items lend themselves to specification. For example, it may not be easy to specify color schemes or the appearance of an item.
- An item described by specification may be no more suitable, and a great deal more expensive, than a supplier's standard product.

- If the specifications are set by the buyer, they may be expensive to develop. They will be used only where there is sufficient volume of purchases to warrant the cost or where it is not possible to describe what is wanted in any other way.

Sources of specifications.   There are two major sources of specifications:

1. Buyer specifications.
2. Standard specifications.

Buyer specifications.   Buyer-developed specifications are usually expensive and time consuming to develop. Companies usually do not use this method unless there is no suitable standard specification available or unless the volume of work makes it economical to do so.

Standard specifications.   Standard specifications have been developed as a result of much study and effort by governmental and nongovernmental agencies. They usually apply to raw or semifinished products, component parts, or the composition of material. In many cases, they have become *de facto* standards used by consumers and by industry. When we buy motor oil for a car and ask for SAE 10W30, we are specifying a standard grade of motor oil established by the Society of Automotive Engineers. Most of the electrical products we buy are manufactured to Underwriters Laboratory (UL) standards. Steel and structural steel members are manufactured to standards set by the American Society of Mechanical Engineers.

There are several advantages to using standard specifications. First, they are widely known and accepted and, because of this, are readily available from most suppliers. Second, because they are widely accepted, manufactured, and sold, they are lower in price than nonstandard items. Finally, because they have been developed with input from a broad range of producers and users, they are usually adaptable to the needs of many purchasers.

Market grades are a type of standard specification usually set by the government and used for commodities and foodstuffs. When we buy eggs, we buy them by market grade—small, medium, or large.

Engineering Drawings

Engineering drawings describe in detail the exact configuration of the parts and the assembly. They also give information on such things as finishes, tolerances, and material to be used. These drawings are a major method of specifying what is wanted and are widely used because often there is no other way to describe the configuration of parts or the way they are to fit together. They are produced by the engineering design department and are expensive to produce, but give an exact description of the part required.

Miscellaneous

There are a variety of other methods of specification including the famous phrase, "Gimme one just like the last one." Sometimes samples are used, for example, where colors or patterns are to be specified. Often a variety of methods can be used, and the buyer must select the best one.

The method of description is communication with the supplier. How well it is done will affect the success of the purchase and sometimes the price paid.

SELECTING SUPPLIERS

The objective of purchasing is to get all the right things together: quality, quantity, delivery, and price. Once the decision is made about what to buy, the selection of the right supplier is the next most important purchasing decision. A good supplier is one that has the technology to make the product to the required quality, has the capacity to make the quantities needed, and can run the business well enough to make a profit and still sell a product competitively.

Sourcing

There are three types of sourcing: sole, multiple, and single.

1. *Sole sourcing* implies that only one supplier is available because of patents, technical specifications, raw material, location, and so forth.
2. *Multiple sourcing* is the use of more than one supplier for an item. The potential advantages of multiple sourcing are that competition will result in lower price and better service and that there will be a continuity of supply. In practice there is a tendency toward an adversarial relationship between supplier and customer.
3. *Single sourcing* is a planned decision by the organization to select one supplier for an item when several sources are available. It is intended to produce a long-term partnership. This aspect is discussed at more length in the section in Chapter 16 on supplier partnerships.

Factors in Selecting Suppliers

The previous section discussed the importance of function, quantity, service, and price specifications. These are what the supplier is expected to provide and are the basis for selection and evaluation. Considering this, there are several factors in selecting a supplier.

Technical ability.   Does the supplier have the technical ability to make or supply the product wanted? Does the supplier have a program of product development and improvement? Can the supplier assist in improving the products? These questions are important since, often, the buyer will depend upon the supplier to provide product improvements that will enhance or reduce the cost of the buyer's products. Sometimes the supplier can suggest changes in product specification that will improve the product and reduce cost.

Manufacturing capability. Manufacturing must be able to meet the specifications for the product consistently while producing as few defects as possible. This means that the supplier's manufacturing facilities must be able to supply the quality and quantity of the products wanted. The supplier must have a good quality control program, competent and capable manufacturing personnel, and good manufacturing planning and control systems to ensure timely delivery. These are important in ensuring that the supplier can supply the quality and quantity wanted.

Reliability. In selecting a supplier, it is desirable to pick one that is reputable, stable, and financially strong. If the relationship is to continue, there must be an atmosphere of mutual trust and assurance that the supplier is financially strong enough to stay in business.

After-sales service. If the product is of a technical nature or likely to need replacement parts or technical support, the supplier must have a good after-sales service. This should include a good service organization and inventory of service parts.

Supplier location. Sometimes it is desirable that the supplier be located near the buyer, or at least maintain an inventory locally. A close location helps shorten delivery times and means emergency shortages can be delivered quickly.

Other considerations. Sometimes other factors such as credit terms, reciprocal business, and willingness of the supplier to hold inventory for the buyer should be considered.

Price. The supplier should be able to provide competitive prices. This does not necessarily mean the lowest price. It is one that considers the ability of the supplier to provide the necessary goods in the quantity and quality wanted, at the time wanted, as well as any other services needed.

In a modern business environment, the type of relationship between the supplier and the buyer is crucial to both. Ideally, the relationship will be on-going with a mutual dependency. The supplier can rely on future business, and the buyer will have an assured supply of quality product, technical support, and product improvement. Communications between buyer and supplier must be open and full so both parties understand the problems of the other and can work together to solve problems to their mutual advantage. Thus, supplier selection and supplier relations are of the utmost importance.

Identifying Suppliers

One major responsibility of the purchasing department is to continue to research all available sources of supply. Some aids for identifying sources of supply follow:

- Salespersons of the supplier company.
- Catalogues.
- Trade magazines.
- Trade directories.
- Information obtained by the salespeople of the buyer firm.

Final Selection of Supplier

Some factors in evaluating potential suppliers are quantitative, and a dollar value can be put on them. Price is the obvious example. Other factors are qualitative and demand some judgment to determine them. These are usually set out in a descriptive fashion. The supplier's technical competence might be an example.

The challenge is finding some method of combining these two major factors that will enable a buyer to pick the best supplier. One method is the **ranking method,** described next.

1. Select those factors that must be considered in evaluating potential suppliers.
2. Assign a weight to each factor. This weight determines the importance of the factor in relation to the other factors. Usually a scale of one to ten is used. If one factor is assigned a weight of five and another factor a weight of ten, the second factor is considered twice as important as the first.
3. Rate the suppliers for each factor. This rating is not associated with the weight. Rather, suppliers are rated on their ability to meet the requirements of each factor. Again, usually a scale of one to ten is used.
4. Rank the suppliers. For each supplier, the weight of each factor is multiplied by the supplier rating for that factor. For example, if a factor had a weight of 8 and a supplier was rated 3 for that factor, the ranking value for that factor would be 24. The supplier rankings are then added to produce a total ranking. The suppliers can then be listed by total ranking and the supplier with the highest ranking chosen.

Figure 7.1 shows an example of this method of selecting suppliers. Theoretically, supplier B, with the biggest total of 223, will be selected.

The ranking method is an attempt to quantify those things that are not quantified by nature. It attempts to put figures on subjective judgment. It is not a perfect method, but it forces the buying company to consider the relative importance of the various factors.

Factor	Weight	Rating of Suppliers				Ranking of Suppliers			
Suppliers		A	B	C	D	A	B	C	D
Function	10	8	10	6	6	80	100	60	60
Cost	8	3	5	9	10	24	40	72	80
Service	8	9	4	5	7	72	32	40	56
Technical Assistance	5	7	9	4	2	35	45	20	10
Credit Terms	2	4	3	6	8	8	6	12	16
Total (rank of suppliers)						219	223	204	222

Figure 7.1 Supplier rating.

PRICE DETERMINATION

Price is not the only factor in making purchasing decisions. However, all other things being equal, it is the most important. In the average manufacturing company, purchases account for about 50% of the cost of goods sold, and any savings made in purchase cost has a direct influence on profits.

However, remember that "you only get what you pay for." The trick is to know what you want and not pay more than necessary. When a purchase is made, the buyer receives a package of function, quantity, service, and price characteristics that are suited to the individual's needs. The idea of "best buy" is the mixture that serves the purpose best.

Basis for Pricing

The term "fair price" is sometimes used to describe what should be paid for an item. But what is a fair price? One answer is that it is the lowest price at which the item can be bought. However, there are other considerations, especially for repeat purchases where the buyer and seller want to establish a good working relationship. One definition of a fair price is one that is competitive, gives the seller a profit, and allows the buyer ultimately to sell at a profit. Sellers who charge too little to cover their costs will not stay in business. To survive, they may attempt to cut costs by reducing quality and service. In the end, both the buyer and seller must be satisfied.

Since we want to pay a fair price and no more, it is good to develop some basis for establishing what is a fair price.

Prices have an upper and a lower limit. The market decides the upper limit. What buyers are willing to pay is based on their perception of demand, supply, and their needs. The seller sets the lower limit. It is determined by the costs of manufacturing and selling the product and profit expectation. If buyers are to arrive at a fair price, they must develop an understanding of market demand and supply, competitive prices, and the methods of arriving at a cost.

One widely used method of analyzing costs is to break them down into fixed and variable costs. **Fixed costs** are costs incurred no matter the volume of sales. Examples are equipment depreciation, taxes, insurance, and administrative overhead. **Variable costs** are those directly associated with the amount produced or sold. Examples are direct labor, direct material, and commissions of the sales force.

$$\text{Total cost} = \text{fixed cost} + (\text{variable cost per unit})(\text{number of units})$$

$$\text{Unit (average) cost} = \frac{\text{total cost}}{\text{number of units}}$$

$$= \frac{\text{fixed cost}}{\text{number of units}} + \text{variable cost per unit}$$

Figure 7.2 shows the relationship of fixed and variable costs to sales volume and how revenue will behave. The sum of the fixed and variable costs is labeled total cost on the graph. The third line represents the sales revenue. Where this line intercepts the total cost line, revenue equals total cost, and profit is zero. This is called the **break-even point.** When the volume is less than the break-even point, a loss is incurred; when the volume is greater, a profit is realized. The break-even point occurs where the revenue equals the total cost.

Figure 7.2   Break-even analysis.

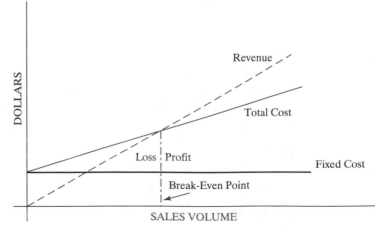

$$\text{Revenue} = \text{total cost}$$

$$(\text{Price per unit})(\text{number of units}) = \text{fixed cost} + (\text{variable cost per unit})(\text{number of units})$$

EXAMPLE PROBLEM

To make a particular component requires an overhead (fixed) cost of $5000 and a variable unit cost of $6.50 per unit. What is the total cost and the average cost of producing a lot of 1000? If the selling price is $15 per unit, what is the break-even point?

Answer

$$\text{Total cost} = \$5000 + (6.5 \times \$1000) = \$11,500$$
$$\text{Average cost} = \$11,500 \div 1000 = \$11.50 \text{ per unit}$$
$$\text{Break-even point: Let } X = \text{ number of units sold}$$
$$\$15X = \$5000 + \$6.5X$$
$$\$8.5X = \$5000$$
$$X = 588.2 \text{ units}$$

Break-even occurs when 588.2 units are made and sold.

Price Negotiation

Prices can be negotiated if the buyer has the knowledge and the clout to do so. A small retailer probably has little of the latter, but a large buyer may have much. Through negotiation, the buyer and seller try to resolve conditions of purchase to the mutual benefit of both parties. Skill and careful planning are required for the negotiation to be successful. It also takes a great deal of time and effort, so the potential profit must justify the expense.

One important factor in the approach to negotiation is the type of product. There are four categories:

1. *Commodities.* Commodities are materials such as copper, coal, wheat, meat, and metals. Price is set by market supply and demand and can fluctuate widely. Negotiation is concerned with contracts for future prices.

2. *Standard products.* These items are provided by many suppliers. Since the items are standard and the choice of suppliers large, prices are determined on the basis of listed catalog prices. There is not much room for negotiation except for large purchases.

3. *Items of small value.* These are items such as maintenance or cleaning supplies and represent purchases of such small value that price negotiation is of little purpose. The prime objective should be to keep the cost of ordering low. Firms will negotiate a contract with a supplier that can supply many items and set up a simple ordering system that reduces the cost of ordering.

4. *Made-to-order items.* This category includes items made to specification or on which quotations from several sources are received. These can generally be negotiated.

IMPACT OF MATERIAL REQUIREMENTS PLANNING ON PURCHASING

The material in this chapter has described the traditional role and responsibilities of purchasing. This section will study the effect material requirements planning (MRP) has on the purchasing function and the changing role of purchasing.

Purchasing can be separated into two types of activities: procurement, and supplier scheduling and follow-up. Much of what has been covered in this chapter is in the area of procurement. Procurement includes the functions of establishing specifications, selecting suppliers, price determination, and negotiation. Supplier scheduling and follow-up is concerned with the release of orders to suppliers, working with suppliers to schedule delivery, and follow-up. The goals of supplier scheduling are the same as those of production activity control: to execute the master production schedule and the material requirements plan, ensure good use of resources, minimize work-in-process inventory, and to maintain the desired level of customer service.

Planner/buyer concept.   In a traditional system, the material requirements planner releases an order either to production activity control or to purchasing. Purchasing issues purchase orders based on the material requirements plan. Production activity control prepares shop orders, schedules components into the work flow, and controls material progress through the plant. When plans change, as they invariably do, the production planner must advise the buyer of the change, and the buyer must advise the supplier. The production planner is in closer, more continuing contact with MRP and their frequently changing schedules than is the buyer. To improve the effectiveness of the planner/buyer activity, many companies have combined the two functions of buying and planning.

The planner's job and the buyer's job are combined into a single job done by one person. Planner/buyers do the material planning for the items under their control, communicate the schedules to their suppliers, follow up, resolve problems, and work with other planners and the master scheduler when delivery problems arise. The planner/buyer handles fewer components than either a planner or a buyer, but has the responsibilities of both. The planner/buyer is responsible for:

- Determining material requirements.
- Developing schedules.
- Issuing shop orders.
- Issuing material releases to suppliers.
- Establishing delivery priorities.
- Controlling orders in the factory and to suppliers.
- Handling all the activities associated with the buying and production planning functions.
- Maintaining close contact with supplier personnel.

Because the role of production planning and buying are combined, there is a smoother flow of information and material between the supplier and the factory. The planner/buyer has a keener knowledge of factory needs than the buyer does and can better coordinate the material flow with suppliers. At the same time, the planner/buyer is better able to match material requirements with supplier's manufacturing capabilities and constraints.

Contract buying. Usually a material requirements planning system generates frequent orders for relatively small quantities. This is particularly true for components that are ordered lot-for-lot. It is costly, inefficient, and sometimes impossible to issue a new purchase order for every weekly requirement. The alternative is to develop a long-term contract with a supplier and to authorize releases against the contract. Often the supplier is given a copy of the material requirements plan so they are aware of future demands. The buyer then issues a release against the schedule. This approach is efficient and cost effective but requires close coordination and communication with the supplier. Again, contract buying can be managed very well by a planner/buyer.

Supplier responsiveness and reliability. Because material requirements often change, suppliers must be able to react quickly to change. They must be highly flexible and reliable so they can react quickly to changes in schedules.

Contract buying assures suppliers a given amount of business and commits them to allocating that amount of their capacity to the customer. Suppliers are more responsive to customer needs and can react quickly to changes in schedules. Because customers know the capacity will be available when needed, they can delay ordering until they are more sure of their requirements.

Close relationship with suppliers. Contract buying and the need for supplier flexibility and reliability mean the buyer-supplier relationship must be close and cooperative. There must be excellent two-way communication, cooperation, and teamwork. Both parties have to understand their own and the other's operations and problems.

The planner/buyer and the supplier counterpart (often the supplier's production planner) must work on a weekly basis to ensure both parties are aware of any changes in material requirements or material availability.

Electronic data interchange (EDI). EDI enables customers and suppliers to electronically exchange transaction information such as purchase orders, invoices, and material requirements planning information. This eliminates time-consuming paper work and facilitates easy communication between planners/buyer and suppliers.

QUESTIONS

1. What are four objectives of purchasing?

2. List the seven steps in the purchasing cycle.

3. Describe the purposes, similarities, and differences among purchase requisitions, purchase orders, and requests for quotation.

4. What are the responsibilities of the purchasing department in follow-up?

5. Describe the duties of the receiving department upon receipt of goods.

6. Besides functional specifications, what three other specifications must be determined? Why is each important?

7. What is quality?

8. Name and describe the four phases of quality. How do they interrelate? Who is responsible for quality?

9. Describe the advantages and disadvantages of the following ways of describing functional requirements. Give examples of when each is used.

    a. By brand.

    b. By specification of physical and chemical characteristics, material and method of manufacture, and performance.

10. What are the advantages of using standard specifications?

11. Why is it important to select the right supplier and to maintain a relationship with him or her?

12. Name and describe the three types of sourcing.

13. Describe the six factors that should be used in selecting a supplier.

14. What is the concept of "best buy"?

15. Type of product is a factor that influences the approach to negotiation. Name the four categories of products and state what room there is for negotiation.

PROBLEMS

7.1 If purchases were 40% of sales and other expenses were 50% of sales, what would be the increase in profit if, through better purchasing, the cost of purchases was reduced to 38% of sales?

7.2 If suppliers were to be rated on the following basis, what would be the ranking of the two suppliers listed?

Factor	Weight	Rating of Suppliers		Ranking of Suppliers	
		Supplier A	Supplier B	Supplier A	Supplier B
Function	8	9	9		
Cost	5	8	5		
Technical Assistance	7	5	7		
Credit Terms	2	8	4		

7.3 A company is negotiating with a potential supplier for the purchase of 10,000 widgets. The company estimates that the supplier's variable costs are $5 per unit and that the fixed costs, depreciation, overhead, etc., are $5000. The supplier quotes a price of $10 per unit. Calculate the estimated average cost per unit. Do you think $10 is too much to pay? Could the purchasing department negotiate a better price?

8

Forecasting

INTRODUCTION

Forecasting is a prelude to planning. Before making plans, an estimate must be made of what conditions will exist over some future period. How estimates are made, and with what accuracy, is another matter, but little can be done without some form of estimation.

Why forecast? There are many circumstances and reasons, but forecasting is inevitable in developing plans to satisfy future demand. Most firms cannot wait until orders are actually received before they start to plan what to produce. Customers usually demand delivery in reasonable time, and manufacturers must anticipate future demand for products or services and plan to provide the capacity and resources to meet that demand. Firms that make standard products need to have saleable goods immediately available or at least to have materials and subassemblies available to shorten the delivery time. Firms that make to order cannot begin making a product before a customer places an order but must have the resources of labor and equipment available to meet demand.

Many factors influence the demand for a firm's products and services. Although it is not possible to identify all of them, or their effect on demand, it is helpful to consider some major factors:

- General business and economic conditions.
- Competitive factors.

- Market trends such as changing demand.
- The firm's own plans for advertising, promotion, pricing, and product changes.

DEMAND MANAGEMENT

The prime purpose of an organization is to serve the customer. Marketing focuses on meeting customer needs, but operations, through materials management, must provide the resources. The coordination of plans by these two parties is demand management.

Demand management is the function of recognizing and managing all demands for products. It occurs in the short, medium, and long term. In the long term, demand projections are needed for strategic business planning of such things as facilities. In the medium term, the purpose of demand management is to project aggregate demand for production planning. In the short run, demand management is needed for items and is associated with master production scheduling. We are most concerned with the latter.

If material and capacity resources are to be planned effectively, all sources of demand must be identified. These include domestic and foreign customers, other plants in the same corporation, branch warehouses, service parts and requirements, promotions, distribution inventory, and consigned inventory in customers' locations.

Demand management includes four major activities:

- Forecasting.
- Order processing.
- Making delivery promises. The concept of available-to-promise was discussed in Chapter 3.
- Interfacing between manufacturing planning and control and the marketplace. Figure 8.1 shows this relationship graphically.

Order processing. When a customer's order is received, the product may be delivered from finished goods inventory or it may be made or assembled to order. If goods are sold from inventory, a sales order is produced authorizing the goods to be

Figure 8.1 Demand management and the manufacturing planning and control system.

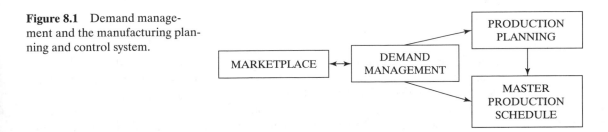

shipped from inventory. If the product is made or assembled to order, the sales department must write up a sales order specifying the product. This may be relatively simple if the product is assembled from standard components but can be a lengthy, complex process if the product requires extensive engineering. A copy of the sales order stating the terms and conditions of acceptance of the order is sent to the customer. Another copy, sent to the master planner, is authorization to go ahead and plan for manufacture. The master planner must know what to produce, how much, and when to deliver. The sales order must be written in language that makes this information clear.

DEMAND FORECASTING

Forecasts depend upon what is to be done. They must be made for the strategic business plan, the production plan, and the master production schedule. As discussed in Chapter 2, the purpose, planning horizons, and level of detail vary for each.

The **strategic business plan** is concerned with overall markets and the direction of the economy over the next two to ten years or more. Its purpose is to provide time to plan for those things that take long to change. For production, the strategic business plan should provide sufficient time for resource planning: plant expansion, capital equipment purchase, and anything requiring a long lead time to purchase. The level of detail is not high, and usually forecasts are in sales units, sales dollars, or capacity. Forecasts and planning will probably be reviewed quarterly or yearly.

Production planning is concerned with manufacturing activity for the next one to three years. For manufacturing, it means forecasting those items needed for production planning, such as budgets, labor planning, long lead time, procurement items, and overall inventory levels. Forecasts are made for groups or families of products rather than specific end items. Forecasts and plans will probably be reviewed monthly.

Master production scheduling is concerned with production activity from the present to a few months ahead. Forecasts are made for individual items, as found on a master production schedule, individual item inventory levels, raw materials and component parts, labor planning, and so forth. Forecasts and plans will probably be reviewed weekly.

CHARACTERISTICS OF DEMAND

In this chapter, the term "demand" is used rather than "sales." The difference is that sales implies what is actually sold whereas demand shows the need for the item. Sometimes demand cannot be satisfied, and sales will be less than demand.

Before discussing forecasting principles and techniques, it is best to look at some characteristics of demand that influence the forecast and the particular techniques used.

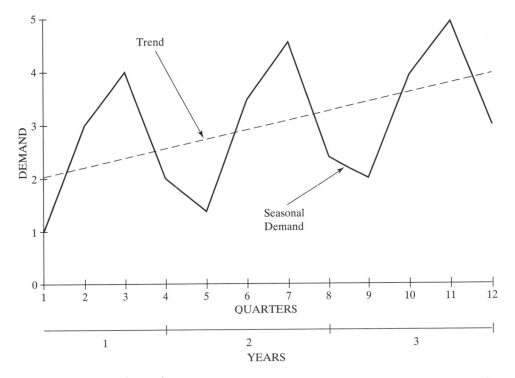

Figure 8.2 Demand over time.

Demand Patterns

If historical data for demand is plotted against a time scale, it will show any shapes or consistent patterns that exist. A pattern is the general shape of a time series. Although some individual data points will not fall exactly on the pattern, they tend to cluster around it.

Figure 8.2 shows a hypothetical historical demand pattern. The pattern shows that actual demand varies from period to period. There are four reasons for this: trend, seasonality, random variation, and cycle.

Trend.   Figure 8.2 shows that demand is increasing in a steady pattern of demand from year to year. This graph illustrates a linear trend, but there are different shapes, such as geometric or exponential. The trend can be level, having no change from period to period, or it can rise or fall.

Seasonality.   The demand pattern in Figure 8.2 shows each year's demand fluctuating depending on the time of year. This fluctuation may be the result of the weather, holiday seasons, or particular events that take place on a seasonal basis. Seasonality is usually thought of as occurring on a yearly basis, but it can also occur on a weekly or even daily basis. A restaurant's demand varies with the hour of the day, and supermarket sales vary with the day of the week.

Figure 8.3 Stable and dynamic demand.

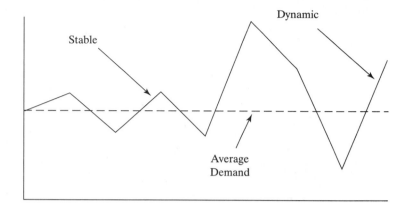

Random variation. Many factors affect demand during specific periods and occur on a random basis. The variation may be small, with actual demand falling close to the pattern, or it may be large, with the points widely scattered. The pattern of variation can usually be measured, and this will be discussed in the section on tracking the forecast.

Cycle. Over a span of several years and even decades, wavelike increases and decreases in the economy influence demand. However, forecasting of cycles is a job for economists and is beyond the scope of this text.

Stable versus Dynamic

The shapes of the demand patterns for some products or services change over time whereas others do not. Those that retain the same general shape are called **stable** and those that do not are called **dynamic.** Dynamic changes can affect the trend, seasonality, or randomness of the actual demand. The more stable the demand, the easier it is to forecast. Figure 8.3 shows a graphical representation of stable and dynamic demand. Notice the average demand is the same for both stable and dynamic patterns. It is usually the average demand that is forecast.

Dependent versus Independent Demand

Chapter 4 discussed dependent and independent demand. It was said that demand for a product or service is independent when it is not related to the demand for any other product or service. Dependent demand for a product or service occurs where the demand for the item is derived from that of a second item. Requirements for dependent demand items need not be forecast but are calculated from that of the independent demand item.

Only independent demand items need be forecast. These are usually end items or finished goods but should also include service parts and items supplied to other plants in the same company (inter-company transfers).

PRINCIPLES OF FORECASTING

Forecasts have four major characteristics or principles. An understanding of these will allow us to make more effective use of forecasts. They are simple and, to some extent, common sense.

1. Forecasts are usually wrong. Forecasts attempt to look into the unknown future and, except by sheer luck, will be wrong to some degree. Errors are inevitable and must be expected.

2. Every forecast should include an estimate of error. Since forecasts are expected to be wrong, the real question is, "By how much?" Every forecast should include an estimate of error often expressed as a percentage (plus and minus) of the forecast or as a range between maximum and minimum values. Estimates of this error can be made statistically by studying the variability of demand about the average demand.

3. Forecasts are more accurate for families or groups. The behavior of individual items in a group is random even when the group has very stable characteristics. For example, the marks for individual students in a class are more difficult to forecast accurately than the class average. High marks average out with low marks. This means that forecasts are more accurate for large groups of items than for individual items in a group.

    For production planning, families or groups are based on the similarity of process and equipment used. For example, a firm forecasting the demand for knit socks as a product group might forecast men's socks as one group and women's as another since the markets are different. However, production of men's and women's ankle socks will be done on the same machines and knee socks on another. For production planning, the forecast should be for (a) men's and women's ankle socks and (b) men's and women's knee socks.

4. Forecasts are more accurate for nearer time periods. The near future holds less uncertainty than the far future. Most people are more confident in forecasting what they will be doing over the next week than a year from now. As someone once said, tomorrow is expected to be pretty much like today.

    In the same way, demand for the near term is easier for a company to forecast than for a time in the distant future. This is extremely important for long-lead-time items and especially so if their demand is dynamic. Anything that can be done to reduce lead time will improve forecast accuracy.

COLLECTION AND PREPARATION OF DATA

Forecasts are usually based on historical data manipulated in some way using either judgment or a statistical technique. Thus, the forecast is only as good as the data on which it is based. To get good data, three principles of data collection are important.

1. *Record data in the same terms as needed for the forecast.*   This is a problem in determining the purpose of the forecast and what is to be forecast. There are three dimensions to this:

   a. If the purpose is to forecast demand on production, data based on demand, not shipments, are needed. Shipments show when goods were shipped and not necessarily when the customer wanted them. Thus shipments do not necessarily give a true indication of demand.

   b. The forecast period, in weeks, months, or quarters, should be the same as the schedule period. If schedules are weekly, the forecast should be for the same time interval.

   c. The items forecast should be the same as those controlled by manufacturing. For example, if there are a variety of options that can be supplied with a particular product, the demand for the product and for each option should be forecast.

   Suppose a firm makes a bicycle that comes in three frame sizes, three possible wheel sizes, a 3-, 5-, or 10-speed gear changer, and with or without deluxe trim. In all, there are 54 ($3 \times 3 \times 3 \times 2$) individual end items sold. If each were forecast, there would be 54 forecasts to make. A better approach is to forecast (a) total demand and (b) the percentage of the total that requires each frame size, wheel size, and so on. That way there need be only 12 forecasts (three frames, three wheels, five gears, and the bike itself).

   In this example, the lead time to make the components would be relatively long in comparison to the lead time to assemble a bike. Manufacturing can make the components according to component forecast and can then assemble bikes according to customer orders. This would be ideal for situations where final assembly schedules are used.

2. *Record the circumstances relating to the data.*   Demand is influenced by particular events, and these should be recorded along with the demand data. For instance, artificial bumps in demand can be caused by sales promotions, price changes, changes in the weather, or a strike at a competitor's factory. It is vital that these factors be related to the demand history so they may be included or removed for future conditions.

3. *Record the demand separately for different customer groups.*   Many firms distribute their goods through different channels of distribution, each having its own demand characteristics. For example, a firm may sell to a number of wholesalers that order relatively small quantities regularly and also sell to a major retailer that buys a large lot twice a year. Forecasts of average demand would be meaningless, and each set of demands should be forecast separately.

FORECASTING TECHNIQUES

There are many forecasting methods, but they can usually be classified into three categories: qualitative, extrinsic, and intrinsic.

Qualitative Techniques

Qualitative techniques are projections based on judgment, intuition, and informed opinions. By their nature, they are subjective. Such techniques are used to forecast general business trends and the potential demand for large families of products over an extended period of time. As such, they are used mainly by senior management. Production and inventory forecasting is usually concerned with the demand for particular end items, and qualitative techniques are seldom appropriate.

When attempting to forecast the demand for a new product, there is no history on which to base a forecast. In these cases, the techniques of market research and historical analogy might be used. Market research is a systematic, formal, and conscious procedure for testing to determine customer opinion or intention. Historical analogy is based on a comparative analysis of the introduction and growth of similar products in the hope that the new product behaves in a similar fashion. Another method is to test market a product.

There are several other methods of qualitative forecasting. One, called the Delphi method, uses a panel of experts to give their opinion on what is likely to happen.

Extrinsic Techniques

Extrinsic forecasting techniques are projections based on external (extrinsic) indicators which relate to the demand for a company's products. Examples of such data would be housing starts, birth rates, and disposable income. The theory is that the demand for a product group is directly proportional, or correlates, to activity in another field. Examples of correlation are:

- Sales of bricks are proportional to housing starts.
- Sales of automobile tires are proportional to gasoline consumption.

Housing starts and gasoline consumption are called **economic indicators.** They describe economic conditions prevailing during a given time period. Some commonly used economic indicators are construction contract awards, automobile production, farm income, steel production, and gross national income. Data of this kind are compiled and published by various government departments, financial papers and magazines, trade associations, and banks.

The problem is to find an indicator that correlates with demand and one that preferably leads demand, that is, one that occurs before the demand does. For example, the number of construction contracts awarded in one period may determine the building material sold in the next period. When it is not possible to find a leading indicator, it may be possible to use a nonleading indicator for which the government or an organization forecasts. In a sense, it is basing a forecast on a forecast.

Extrinsic forecasting is most useful in forecasting the total demand for a firm's products or the demand for families of products. As such, it is used most often in business and production planning rather than the forecasting of individual end items.

Intrinsic Techniques

Intrinsic forecasting techniques use historical data to forecast. These data are usually recorded in the company and are readily available. Intrinsic forecasting techniques are based on the assumption that what happened in the past will happen in the future. This assumption has been likened to driving a car by looking out the rear-view mirror. While there is some obvious truth to this, it is also true that lacking any other "crystal ball," the best guide to the future is what has happened in the past.

Since intrinsic techniques are so important, the next section will discuss some of the more important techniques. They are often used as input to master production scheduling where end-item forecasts are needed for the planning horizon of the plan.

SOME IMPORTANT INTRINSIC TECHNIQUES

Assume that the monthly demand for a particular item over the past year is as shown in Figure 8.4.

Suppose it is the end of December, and we want to forecast demand for January of the coming year. Several rules can be used:

- *Demand this month will be the same as last month.* January demand would be forecast at 84, the same as December. This may appear too simple, but if there is little change in demand month to month, it probably will be quite usable.

- *Demand this month will be the same as demand the same month last year.* Forecast demand would be 92, the same as January last year. This rule is adequate if demand is seasonal and there is little up or down trend.

Rules such as these, based on a single month or past period, are of limited use when there is much random fluctuation in demand. Usually methods that average out history are better because they dampen out some effects of random variation.

Figure 8.4 A 12-month demand history.

January	92	July	84
February	83	August	81
March	66	September	75
April	74	October	63
May	75	November	91
June	84	December	84

As an example, the average of last year's demand can be used as an estimate for January demand. Such a simple average would not be responsive to trends or changes in level of demand. A better method would be to use a moving average.

Average demand. This raises the question of what to forecast. As discussed earlier, demand can fluctuate because of random variation. It is best to forecast the average demand rather than second guess what the effect of random fluctuation will be. The second principle of forecasting discussed earlier said that a forecast should include an estimate of error. As we will see later, this range can be estimated. Thus, a forecast of average demand should be made, and the estimate of error applied to it.

Moving Averages

One simple way to forecast is to take the average demand for, say, the last three or six periods and use that figure as the forecast for the next period. At the end of the next period, the first-period demand is dropped and the latest period demand added to determine a new average to be used as a forecast. This forecast would always be based on the average of the actual demand over the specified period.

For example, suppose it was decided to use a three-month moving average on the data shown in Figure 8.4. Our forecast for January, based on the demand in October, November, and December, would be:

$$\frac{63 + 91 + 84}{3} = 79$$

Now suppose that January demand turned out to be 90 instead of 79. The forecast for February would be calculated as:

$$\frac{91 + 84 + 90}{3} = 88$$

▼ ──

EXAMPLE PROBLEM

Demand over the past three months has been 120, 135, and 114 units. Using a three-month moving average, calculate the forecast for the fourth month.

Answer

$$\text{Forecast for month 4} = \frac{120 + 135 + 114}{3} = \frac{369}{3} = 123$$

Actual demand for the fourth month turned out to be 129. Calculate the forecast for the fifth month.

$$\text{Forecast for month 5} = \frac{135 + 114 + 129}{3} = 126$$

──

In the previous discussion, the forecast for January was 79, and the forecast for February was 88. The forecast has risen, reflecting the higher January value and the dropping of the low October value. If a longer period, such as six months, is used, the forecast does not react as quickly. The fewer months included in the moving average, the more weight is given to the latest information, and the faster the forecast reacts to trends. However, the forecast will always lag behind a trend. For example, consider the following demand history for the past five periods:

Period	Demand
1	1000
2	2000
3	3000
4	4000
5	5000

There is a rising trend to demand. If a five-period moving average is used, the forecast for period 6 is (1000 + 2000 + 3000 + 4000 + 5000) ÷ 5 = 3000. It does not look very accurate since the forecast is lagging actual demand by a large amount. However, if a three-month moving average is used, the forecast is (3000 + 4000 + 5000) ÷ 3 = 4000. Not perfect, but somewhat better. The point is that a moving average always lags a trend, and the more periods included in the average, the greater the lag will be.

On the other hand, if there is no trend but actual demand fluctuates considerably due to random variation, a moving average based on a few periods reacts to the fluctuation rather than forecasts the average. Consider the following demand history:

Period	Demand
1	2000
2	5000
3	3000
4	1000
5	4000

The demand has no trend and is random. If a five-month moving average is used, the forecast for the next month is 3000. This reflects all the values. If a two-month average is taken, the forecasts for the third, fourth, fifth, and sixth months are:

$$
\begin{aligned}
\text{Forecast for third month} &= (2000 + 5000) \div 2 = 3500 \\
\text{Forecast for fourth month} &= (5000 + 3000) \div 2 = 4000 \\
\text{Forecast for fifth month} &= (3000 + 1000) \div 2 = 2000 \\
\text{Forecast for sixth month} &= (1000 + 4000) \div 2 = 2500
\end{aligned}
$$

With a two-month moving average the forecast reacts very quickly to the latest demand and thus is not stable.

Moving averages are best used for forecasting products with stable demand where there is little trend or seasonality. Moving averages are also useful to filter out random fluctuations. This has some common sense since periods of high demand are often followed by periods of low demand.

One drawback to using moving averages is the need to retain several periods of history for each item to be forecast. This will require a great deal of computer storage or clerical effort. Also, the calculations are cumbersome. A common forecasting technique, called **exponential smoothing,** gives the same results as a moving average but without the need to retain as much data and with easier calculations.

Exponential Smoothing

It is not necessary to keep months of history to get a moving average because the previously calculated forecast has already allowed for this history. Therefore, the forecast can be based on the old calculated forecast and the new data.

Using the data in Figure 8.4, suppose an average of the demand of the last six months (80 units) is used to forecast January demand. If at the end of January, actual demand is 90 units, we must drop July's demand and pick up January's demand to determine the new forecast. However, if an average of the old forecast (80) and the actual demand for January (90) is taken, the new forecast for February is 85 units. This formula puts as much weight on the most recent month as on the old forecast (all previous months). If this does not seem suitable, less weight could be put on the latest actual demand and more weight on the old forecast. Perhaps putting only 10% of the weight on the latest month's demand and 90% of the weight on the old forecast would be better. In that case,

$$\text{February forecast} = 0.1\,(90) + 0.9(80) = 81$$

Notice that this forecast did not rise as much as our previous calculation in which the old forecast and the latest actual demand were given the same weight. One advantage to exponential smoothing is that the new data can be given any weight wanted.

The weight given to latest actual demand is called a **smoothing constant** and is represented by the Greek letter alpha (α). It is always expressed as a decimal from 0 to 1.0.

In general, the formula for calculating the new forecast is

New forecast = (α)(latest demand) + (1 - α)(previous forecast)

EXAMPLE PROBLEM

The old forecast for May was 220, and the actual demand for May was 190. If alpha (α) is 0.15, calculate the forecast for June. If June demand turns out to be 218, calculate the forecast for July.

Answer

$$\text{June forecast} = (0.15)(190) + (1 - 0.15)(220) = 215.5$$
$$\text{July forecast} = (0.15)(218) + (0.85)(215.5) = 215.9$$

Exponential smoothing provides a routine method for regularly updating item forecasts. It works quite well when dealing with stable items. Generally, it has been found satisfactory for short-range forecasting. It is not satisfactory where the demand is low or intermittent.

Exponential smoothing will detect trends, although the forecast will lag actual demand if a definite trend exists. Figure 8.5 shows a graph of the exponentially smoothed forecast lagging the actual demand where a positive trend exists. Notice the forecast with the larger α follows actual demand more closely.

If a trend exists, it is possible to use a slightly more complex formula called **double exponential smoothing.** This technique uses the same principles but notes whether each successive value of the forecast is moving up or down on a trend line. Double exponential smoothing is beyond the scope of this text.

A problem exists in selecting the "best" alpha factor. If a low factor such as 0.1 is used, the old forecast will be heavily weighted, and changing trends will not be picked up as quickly as might be desired. If a larger factor such as 0.4 is used, the forecast will react sharply to changes in demand and will be erratic if there is a sizable random fluctuation. A good way to get the best alpha factor is to use computer simulation. Using past actual demand, forecasts are made with different alpha factors to see which one best suits the historical demand pattern for particular products.

Figure 8.5   Exponential forecast where trend exists.

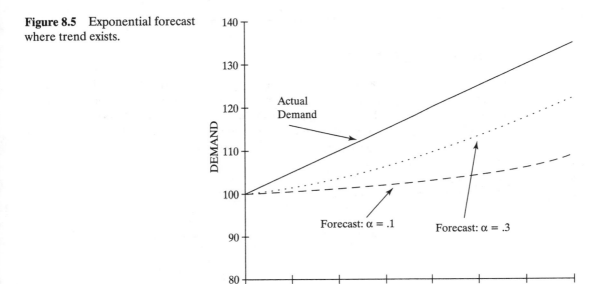

SEASONALITY

Many products have a seasonal or periodic demand pattern: skis, lawnmowers, bathing suits, and Christmas tree lights are examples. Less obvious are products whose demand varies by the time of day, week, or month. Examples of these might be electric power usage during the day or grocery shopping during the week. Power usage peaks between 4 and 7 p.m., and supermarkets are most busy toward the end of the week or before certain holidays.

Seasonal Index

A useful indication of the degree of seasonal variation for a product is the **seasonal index.** This index is an estimate of how much the demand during the season will be above or below the average demand for the product. For example, swimsuit demand might average 100 per month, but in July the average is 175 and in September, 35. The index for July demand would be 1.75 and for September, 0.35.

The formula for the seasonal index is:

$$\text{Seasonal index} = \frac{\text{period average demand}}{\text{average demand for all periods}}$$

The period can be daily, weekly, monthly, or quarterly depending on the basis for the seasonality of demand.

The average demand for all periods is a value that averages out seasonality. This is called the **deseasonalized demand.** The previous equation can be rewritten as:

$$\text{Seasonal index} = \frac{\text{period average demand}}{\text{deseasonalized demand}}$$

EXAMPLE PROBLEM

A product that is seasonal based on quarterly demand and the demand for the past three years is shown in Figure 8.6. There is no trend, but there is definite seasonality. Average quarterly demand is 100 units. Figure 8.6 also shows a graph of actual seasonal demand and average quarterly demand. The average demand shown is the historical average demand for all periods. Remember we forecast average demand, not seasonal demand.

Answer

The seasonal indices can now be calculated as follows:

$$\text{Seasonal index} = \frac{128}{100} = 1.28 \text{ (quarter 1)}$$

$$= \frac{102}{100} = 1.02 \text{ (quarter 2)}$$

$$= \frac{75}{100} = 0.75 \text{ (quarter 3)}$$

$$= \frac{95}{100} = 0.95 \text{ (quarter 4)}$$

$$\text{Total of seasonal indices} = 4.00$$

Note that the total of all the seasonal indices equals the number of periods. This is a good way to check whether the calculations are correct.

Figure 8.6 Seasonal sales history.

Year	Quarter				
	1	2	3	4	Total
1	122	108	81	90	401
2	130	100	73	96	399
3	132	98	71	99	400
Average	128	102	75	95	400

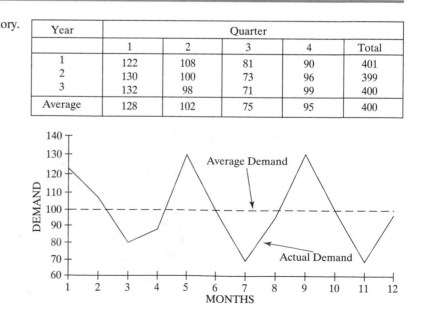

Figure 8.7 Seasonal demand.

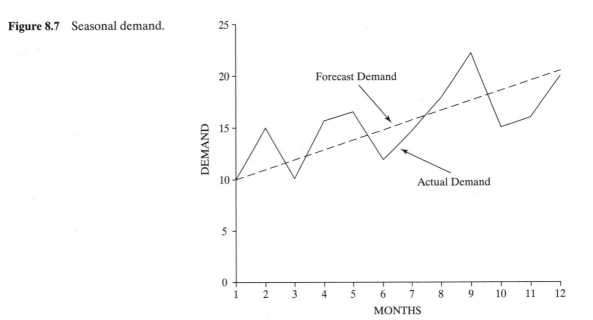

Seasonal Forecasts

The equation for developing seasonal indices is also used to forecast seasonal demand. If a company forecasts average demand for all periods, the seasonal indices can be used to calculate the seasonal forecasts. Changing the equation around we get:

$$\text{Seasonal demand} = (\text{seasonal index})(\text{deseasonalized demand})$$

▼ EXAMPLE PROBLEM

The company in the previous problem forecasts an annual demand next year of 420 units. Calculate the forecast for quarterly sales.

Answer

$$\text{Forecast average quarterly demand} = \frac{420}{4} = 105 \text{ units}$$

$$\begin{aligned}
\text{Expected quarter demand} &= (\text{seasonal index})(\text{forecast quarterly demand}) \\
\text{Expected first-quarter demand} &= 1.28 \times 105 = 134.4 \text{ units} \\
\text{Expected second-quarter demand} &= 1.02 \times 105 = 107.1 \text{ units} \\
\text{Expected third-quarter demand} &= 0.75 \times 105 = 78.75 \text{ units} \\
\text{Expected fourth-quarter demand} &= 0.95 \times 105 = \underline{99.75} \text{ units} \\
\text{Total forecast demand} &= 420 \text{ units}
\end{aligned}$$

Deseasonalized Demand

Forecasts do not consider random variation. They are made for average demand, and seasonal demand is calculated from the average using seasonal indices. Figure 8.7 shows both actual demand and forecast average demand. The forecast average demand is also the deseasonalized demand. Historical data are of actual seasonal demand, and they must be deseasonalized before they can be used to develop a forecast of average demand.

Also, if comparisons are made between sales in different periods, they are meaningless unless deseasonalized data are used. For example, a company selling tennis rackets finds demand is usually largest in the summer. However, some people play indoor tennis, so there is demand in the winter months as well. If demand in January was 5200 units and in June was 24,000 units, how could January demand be compared to June demand to see which was the better demand month? If there is seasonality, comparison of actual demand would be meaningless. Deseasonalized data are needed to make a comparison.

The equation to calculate deseasonalized demand is derived from the previous seasonal equation and is as follows:

$$\text{Deseasonalized demand} = \frac{\text{actual seasonal demand}}{\text{seasonal index}}$$

▼

EXAMPLE PROBLEM

A company selling tennis rackets has a January demand of 5200 units and a July demand of 24,000 units. If the seasonal indices for January were 0.5 and for June were 2.5, calculate the deseasonalized January and July demand. How do the two months compare?

Answer

$$\text{Deseasonalized January demand} = 5200 \div 0.5 = 10,400 \text{ units}$$
$$\text{Deseasonalized June demand} = 24,000 \div 2.5 = 9600 \text{ units}$$

June and January demand can now be compared. On a deseasonalized basis, January demand is greater than June demand.

Deseasonalized data must be used for forecasting. Forecasts are made for average demand, and the forecast for seasonal demand is *calculated* from the average demand using the appropriate season index.

The rules for forecasting with seasonality are:

- Only use deseasonalized data to forecast.
- Forecast deseasonalized demand, not seasonal demand.
- Calculate the seasonal forecast by applying the seasonal index to the base forecast.

▼ ━━

EXAMPLE PROBLEM

A company uses exponential smoothing to forecast demand for its products. For April, the deseasonalized forecast was 1000, and the actual seasonal demand was 1250 units. The seasonal index for April is 1.2 and for May is 0.7. If α is 0.1, calculate:

   a. The deseasonalized actual demand for April.

   b. The deseasonalized May forecast.

   c. The seasonal forecast for May.

Answer

a. Deseasonalized actual demand for April $= \dfrac{1250}{1.2} = 1042$

b. Deseasonalized May forecast $= \alpha(\text{latest actual}) + (1 - \alpha)(\text{previous forecast})$
$= 0.1(1042) + 0.9(1000) = 1004$

c. Seasonalized May forecast $= (\text{seasonal index})(\text{deseasonalized forecast})$
$= 0.7(1004) = 703$

━━

TRACKING THE FORECAST

As noted in the discussion on the principles of forecasting, forecasts are usually wrong. There are several reasons for this, some of which are related to human involvement and others to the behavior of the economy. If there were a method of determining how good a forecast is, forecasting methods could be improved and better estimates could be made accounting for the error. There is no point in continuing with a plan based on poor forecast data. We need to track the forecast. Tracking the forecast is the process of comparing actual demand with the forecast.

Forecast Error

Forecast error is the difference between actual demand and forecast demand. Error can occur in two ways: bias and random variation.

Bias. Cumulative actual demand may not be the same as forecast. Consider the data in Figure 8.8. Actual demand varies from forecast, and over the six-month period, cumulative demand is 120 units greater than expected.

Bias exists when the cumulative actual demand varies from the cumulative forecast. This means the forecast average demand has been wrong. In the example in Figure 8.8, the forecast average demand was 100, but the actual average demand was $720 \div 6 = 120$ units. Figure 8.9 shows a graph of cumulative forecast and actual demand.

Month	Forecast		Actual	
	Monthly	Cumulative	Monthly	Cumulative
1	100	100	110	110
2	100	200	125	235
3	100	300	120	355
4	100	400	125	480
5	100	500	130	610
6	100	600	110	720
Total	600	600	720	720

Figure 8.8   Forecast and actual sales with bias.

Figure 8.9   Forecast and actual demand with bias.

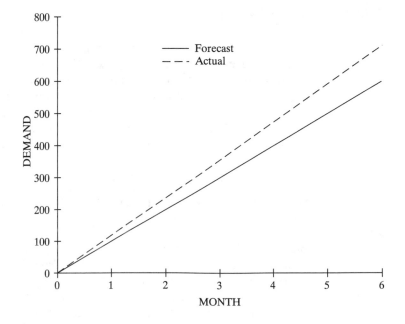

Bias is a systematic error in which the actual demand is consistently above or below the forecast demand. When bias exists, the forecast should be changed to improve its accuracy.

The purpose of tracking the forecast is to be able to react to forecast error by planning around it or by reducing it. When an unacceptably large error or bias is observed, it should be investigated to determine its cause.

Often there are exceptional one-time reasons for error. Examples are machine breakdown, customer shutdown, large one-time orders, and sales promotions. These reasons relate to the discussion on collection and preparation of data and the need to record the circumstances relating to the data. On these occasions, the demand history must be adjusted to consider the exceptional circumstances.

Errors can also occur because of timing. For example, an early or late winter will affect the timing of demand for snow shovels although the cumulative demand will be the same.

Tracking cumulative demand will confirm timing errors or exceptional one-time events. The following example illustrates this. Note that in April the cumulative demand is back in a normal range.

Month	Forecast	Actual	Cumulative Forecast	Cumulative Actual
January	100	95	100	95
February	100	110	200	205
March*	100	155	300	360
April	100	45	400	405
May	100	90	500	495

*Customer foresaw a possible strike and stockpiled.

Random variation. In a given period, actual demand will vary about the average demand. The variability will depend upon the demand pattern of the product. Some products will have a stable demand, and the variation will not be large. Others will be unstable and will have a large variation.

Consider the data in Figure 8.10, showing forecast and actual demand. Notice there is much random variation, but the average error is zero. This shows that the average forecast was correct and there was no bias. The data are plotted in Figure 8.11.

Mean Absolute Deviation

Forecast error must be measured before it can be used to revise the forecast or to help in planning. There are several ways to measure error, but one commonly used is **mean absolute deviation (MAD).**

Consider the data on variability in Figure 8.10. Although the total error (variation) is zero, there is still considerable variation each month. Total error would be useless to measure the variation. One way to measure the variability is to calculate the total error ignoring the plus and minus signs and take the average. This is called mean absolute deviation:

Figure 8.10   Forecast and actual sales without bias.

Month	Forecast	Actual	Variation (error)
1	100	105	5
2	100	94	−6
3	100	98	−2
4	100	104	4
5	100	103	3
6	100	96	−4
Total	600	600	0

Figure 8.11   Forecast and actual sales without bias.

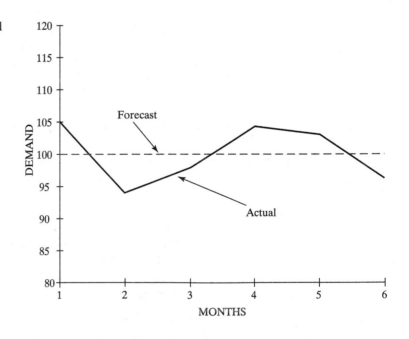

- *mean* implies an average,
- *absolute* means without reference to plus and minus
- *deviation* refers to the error:

$$\text{MAD} = \frac{\text{sum of absolute deviations}}{\text{number of observations}}$$

▼

EXAMPLE PROBLEM

Given the data shown in Figure 8.10, calculate the mean absolute deviation.

Answer

$$\text{Sum of absolute deviations} = 5 + 6 + 2 + 4 + 3 + 4 = 24$$

$$\text{MAD} = \frac{24}{6} = 4$$

Normal distribution. The mean absolute deviation measures the difference (error) between actual demand and forecast. Usually, actual demand is close to the forecast but sometimes is not. A graph of the number of times (frequency) actual demand is of a particular value produces a bell-shaped curve. This distribution is called a **normal distribution** and is shown in Figure 8.12. Chapter 11 gives a more detailed discussion of normal distributions and their characteristics.

There are two important characteristics to normal curves: the central tendency, or average, and the dispersion, or spread, of the distribution. In Figure 8.12, the central tendency is the forecast. The dispersion, the fatness or thinness of the normal curve, is measured by the standard deviation. The greater the dispersion, the larger the standard deviation. The mean absolute deviation is an approximation of the standard deviation and is used because it is easy to calculate and apply.

From statistics we know that the error will be within:

±1 MAD of the average about 60% of the time

±2 MAD of the average about 90% of the time

±3 MAD of the average about 98% of the time.

Uses of mean absolute deviation. Mean absolute deviation has several uses. Some of the most important follow.

Tracking signal. Bias exists when cumulative actual demand varies from forecast. The problem is in guessing whether the variance is due to random variation or bias. If the variation is due to random variation, the error will correct itself, and nothing should be

Figure 8.12 Normal distribution curve.

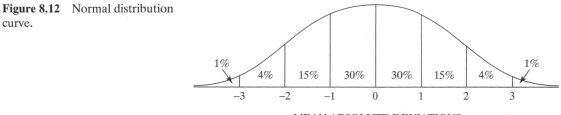

MEAN ABSOLUTE DEVIATIONS

done to adjust the forecast. However, if the error is due to bias, the forecast should be corrected. Using the mean absolute deviation, we can make some judgment about the reasonableness of the error. Under normal circumstances, the actual period demand will be within ±3 MAD of the average 98% of the time. If actual period demand varies from the forecast by more than 3 MAD, we can be about 98% sure that the forecast is in error.

A tracking signal can be used to monitor the quality of the forecast. There are several procedures used, but one of the simpler is based on a comparison of the cumulative sum of the forecast errors to the mean absolute deviation. Following is the equation:

$$\text{Tracking signal} \ = \ \frac{\text{algebraic sum of forecast errors}}{\text{MAD}}$$

EXAMPLE PROBLEM

The forecast is 100 units a week. The actual demand for the past six weeks has been 105, 110, 103, 105, 107, and 115. If MAD is 5, calculate the sum of the forecast error and the tracking signal.

Answer

Sum of forecast error = 5 + 10 + 3 + 5 + 7 + 15 = 45

Tracking signal = 45 ÷ 5 = 9

EXAMPLE PROBLEM

A company uses a trigger of ±4 to decide whether a forecast should be reviewed. Given the following history, determine in which period the forecast should be reviewed. MAD for the item is 2.

Period	Forecast	Actual	Deviation	Cumulative Deviation	Tracking Signal
				5	2.5
1	100	96			
2	100	98			
3	100	104			
4	100	110			

Answer

Period	Forecast	Actual	Deviation	Cumulative Deviation	Tracking Signal
				5	2.5
1	100	96	−4	1	0.5
2	100	98	−2	−1	−0.5
3	100	104	4	3	1.5
4	100	110	10	13	6.5

The forecast should be reviewed in period 4.

Contingency planning. Suppose a forecast is made that demand for door slammers will be 100 units and that capacity for making them is 110 units. Mean absolute deviation of actual demand about the forecast historically has been calculated at 10 units. This means there is a 60% chance that actual demand will be between 90 and 110 units and a 40% chance that they will not. With this information, manufacturing management might be able to devise a contingency plan to cope with the possible extra demand.

Safety stock. The data can be used as a basis for setting safety stock. This will be discussed in detail in Chapter 11.

SUMMARY

Forecasting is an inexact science that is, nonetheless, an invaluable tool if the following is kept in mind:

- Forecasts should be tracked.
- There should be a measure of reasonableness of error.
- When actual demand exceeds the reasonableness of error, an investigation should be made to discover the cause of the error.
- If there is no apparent cause of error, the method of forecasting should be reviewed to see if there is a better way to forecast.

QUESTIONS

1. What is demand management? What functions does it include?
2. Why must we forecast?
3. What factors influence the demand for a firm's products?
4. Describe the purpose of forecasting for strategic business planning, production planning, and master production scheduling.
5. The text describes three characteristics of demand. Name and describe each.
6. Describe trend, seasonality, random variation, and cycle as applied to forecasting.
7. The text discusses four principles of forecasting. Name and describe each.
8. Name and describe the three principles of data collection.
9. Describe the characteristics and differences between qualitative, extrinsic, and intrinsic forecasting techniques.
10. Describe and give the advantages and disadvantages of (a) moving averages and (b) exponential smoothing.
11. What is a seasonal index? How is it calculated?
12. What is meant by the term *deseasonalized demand?*
13. What is meant by the term *tracking the forecast?* In which two ways can forecasts go wrong?
14. What is bias error in forecasting? What are some of the causes?
15. What is random variation?
16. What is the mean absolute deviation (MAD)? Why is it useful in forecasting?
17. What action should be taken when unacceptable error is found in tracking a forecast?

PROBLEMS

8.1 Over the past three months, the demand for a product has been 250, 274, and 235. Calculate the three-month moving average forecast for month 4. If the actual demand in month 4 is 229, calculate the forecast for month 5.

Answer.   253, 246

8.2 Given the following data, calculate the three-month moving average forecasts for months 4, 5, 6, and 7.

Month	Actual Demand	Forecast
1	67	
2	75	
3	43	
4	50	
5	77	
6	65	
7		

8.3 Monthly demand over the past ten months is given in what follows.

    a. Graph the demand.

    b. What is your best guess for the demand for month 11?

    c. Using a three-month moving average calculate the forecasts for months 4, 5, 6, 7, 8, 9, 10, and 11.

Month	Actual Demand	Forecast
1	102	
2	91	
3	95	
4	105	
5	94	
6	100	
7	109	
8	92	
9	101	
10	98	
11		

8.4 If the forecast for February was 122 and actual demand was 130, what would be the forecast for March if the smoothing constant (α) is 0.15? Use exponential smoothing for your calculation.

    Answer. Forecast = 123

8.5 If the old forecast is 100 and the latest actual demand is 85, what is the exponentially smoothed forecast for the next period? Alpha is 0.2.

8.6 Using exponential smoothing, calculate the forecasts for months 2, 3, 4, 5, and 6. The smoothing constant is 0.2, and the old forecast for month 1 is 245.

Month	Actual Demand	Forecast Demand
1	250	
2	230	
3	225	
4	245	
5	250	
6		

8.7 Using exponential smoothing, calculate the forecasts for the same months as in problem 8.3c. The old average for month 3 was 96 and $\alpha = 0.2$. What is the difference between the two forecasts for month 11?

Month	Actual Demand	Forecast
1	102	
2	91	
3	95	
4	105	
5	94	
6	100	
7	109	
8	92	
9	101	
10	98	
11		

8.8 Weekly demand for an item averaged 100 units over the past year. Actual demand for the next eight weeks is shown in what follows:

a. Plot the data on graph paper.

b. Letting $\alpha = 0.2$, calculate the smoothed forecast for each week.

c. Comment on how well the forecast is tracking actual demand. Is it lagging or leading actual demand?

Week	Actual Demand	Forecast
1	103	100
2	112	
3	113	
4	120	
5	126	
6	128	
7	138	
8	141	
9		

8.9 If the average demand for the first quarter was 122.5 and the average demand for all quarters was 175, what is the seasonal index for the first quarter?

Answer. Seasonal index = 0.70

8.10 Using the data in problem 8.9, if the forecast for next year is 800, calculate the forecast for first quarterly demand next year.

Answer. Forecast for first quarter = 140

8.11 The average demand for January has been 90, and the average annual demand has been 1800. Calculate the seasonal index for January. If the company forecasts annual demand next year at 2000 units, what is the forecast for January next year?

8.12 Given the following average demand for each month, calculate the seasonal indices for each month.

Month	Average Demand	Seasonal Index
January	20	
February	40	
March	75	
April	100	
May	115	
June	235	
July	245	
August	125	
September	100	
October	80	
November	50	
December	15	
Total		

8.13 Using the data in problem 8.12 and the seasonal indices you have calculated, calculate expected monthly demand if the annual forecast is 2000 units.

Month	Seasonal Index	Forecast
January		
February		
March		
April		
May		
June		
July		
August		
September		
October		
November		
December		

8.14 If the actual demand for April was 1200 units and the seasonal index was 2.5, what would be the deseasonalized April demand?

Answer. Deseasonalized demand = 480 units

8.15 Calculate the deseasonalized demands for the following:

Quarter	Actual Demand	Seasonal Index	Deseasonalized Demand
1	130	0.62	
2	170	1.04	
3	375	1.82	
4	90	0.52	
Total			

8.16 The old deseasonalized forecast is 100 units, $\alpha = 0.2$, and the actual demand for the last month was 130 units. If the seasonal index for the last month is 1.2 and the next month is 0.9, calculate:

a. The deseasonalized actual demand for the last month.

b. The deseasonalized forecast for next month using exponential smoothing.

c. The forecast of actual demand for the next month.

Answer. **a.** Deseasonalized last month's demand = 108

b. Deseasonalized forecast for next month = 101.6

c. Forecast of seasonal demand = 91.4

8.17 The Fast Track Ski Shoppe sells ski goggles during the four months of the ski season. Average demand follows:

a. Calculate the deseasonalized sales and the seasonal index for each of the four months.

b. If next year's demand is forecast at 1200 pairs of goggles, what will be the forecast sales for each month?

Month	Average Past Demand	Seasonal Index	Forecast Demand Next Year
December	300		
January	200		
February	150		
March	150		
Total			

8.18 Given the following forecast and actual demand, calculate the mean absolute deviation.

Period	Forecast	Actual Demand	Absolute Deviation
1	100	85	
2	100	105	
3	100	120	
4	100	100	
5	100	90	
Total			

Answer. MAD = 10

8.19 For the following data, calculate the mean absolute deviation.

Period	Forecast	Actual Demand	Absolute Deviation
1	100	110	
2	105	90	
3	110	90	
4	115	130	
5	120	140	
6	125	115	
Total	675	675	

8.20 A company uses a tracking signal trigger of ±4 to decide whether a forecast should be reviewed. Given the following history, determine in which period the forecast should be reviewed. MAD for the item is 15. Is there any previous indication that the forecast should be reviewed?

Period	Forecast	Actual	Deviation	Cumulative Deviation	Tracking Signal
1	100	110			
2	105	90			
3	110	85			
4	115	110			
5	120	105			
6	125	95			

9

Inventory Fundamentals

INTRODUCTION

Inventories are materials and supplies that a business or institution carries either for sale or to provide inputs or supplies to the production process. All businesses and institutions require inventories. Often they are a substantial part of total assets.

Financially, inventories are very important to manufacturing companies. On the balance sheet, they usually represent from 20% to 60% of total assets. As inventories are used, their value is converted into cash, which improves cash flow and return on investment. There is a cost for carrying inventories, which increases operating costs and decreases profits. Good inventory management is essential.

Inventory management is responsible for planning and controlling inventory from the raw material stage to the customer. Since inventory either results from production or supports it, the two cannot be managed separately and, therefore, must be coordinated. Inventory must be considered at each of the planning levels and is thus part of production planning, master production scheduling, and material requirements planning. Production planning is concerned with overall inventory, master planning with end items, and material requirements planning with component parts and raw material.

AGGREGATE INVENTORY MANAGEMENT

Aggregate inventory management deals with managing inventories according to their classification (raw material, work-in-process, and finished goods) and the function they perform rather than at the individual item level. It is financially oriented and is concerned with the costs and benefits of carrying the different classifications of inventories. As such, aggregate inventory management involves:

- Flow and kinds of inventory needed.
- Supply and demand patterns.
- Functions that inventories perform.
- Objectives of inventory management.
- Costs associated with inventories.

ITEM INVENTORY MANAGEMENT

Inventory is not only managed at the aggregate level but also at the item level. Management must establish decision rules about inventory items so the staff responsible for inventory control can do their job effectively. These rules include the following:

- Which individual inventory items are most important.
- How individual items are to be controlled.
- How much to order at one time.
- When to place an order.

This chapter will study aggregate inventory management and some factors influencing inventory management decisions, which include:

- Types of inventory based on the flow of material.
- Supply and demand patterns.
- Functions performed by inventory.
- Objectives of inventory management.
- Inventory costs.

Finally, this chapter will conclude with a study of the first two decisions, deciding the importance of individual end items and how they are controlled. Subsequent chapters will discuss the question of how much stock to order at one time and when to place orders.

INVENTORY AND THE FLOW OF MATERIAL

There are many ways to classify inventories. One often-used classification is related to the flow of materials into, through, and out of a manufacturing organization, as shown in Figure 9.1.

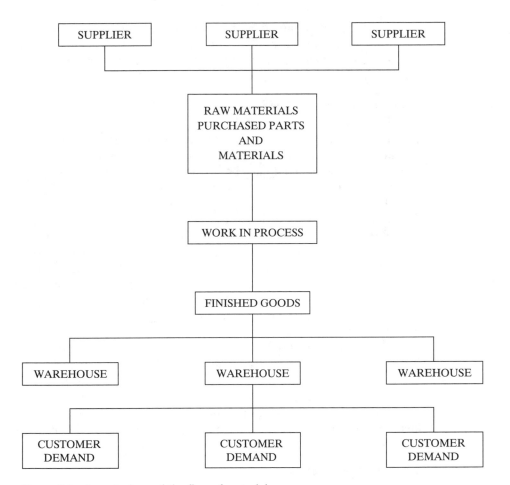

Figure 9.1 Inventories and the flow of materials.

- *Raw materials.*   These are purchased items received which have not entered the production process. They include purchased materials, component parts, and subassemblies.

- *Work-in-process (WIP).*   Raw materials that have entered the manufacturing process and are being worked on or waiting to be worked on.

- *Finished goods.*   The finished products of the production process that are ready to be sold as completed items. They may be held at a factory or central warehouse or at various points in the distribution system.

- *Distribution inventories.*   Finished goods located in the distribution system.

- *Maintenance, repair, and operational supplies (MROs).*   Items used in production that do not become part of the product. These include hand tools, spare parts, lubricants, and cleaning supplies.

Classification of an item into a particular inventory depends on the production environment. For instance, sheet steel or tires are finished goods to the supplier but are raw materials and component parts to the car manufacturer.

SUPPLY AND DEMAND PATTERNS

If supply met demand exactly, there would be little need for inventory. Goods could be made at the same rate as demand, and no inventory would build up. For this situation to exist, demand must be predictable, stable, and relatively constant over a long time period.

If this is so, manufacturing can produce goods on a line-flow basis, matching production to demand. Using this system, raw materials are fed to production as required, work flow from one workstation to another is balanced so little work-in-process inventory is required, and goods are delivered to the customer at the rate the customer needs them. Flow manufacturing systems were discussed in Chapter 1. Because the variety of products they can make is so limited, demand has to be large enough to justify economically setting up the system. These systems are characteristic of just-in-time manufacturing and will be discussed in Chapter 15.

Demand for most products is neither sufficient nor constant enough to warrant setting up a line-flow system, and these products are usually made in lots or batches. Workstations are organized by function, for example, all machine tools in one area, all welding in another, and assembly in another. Work moves in lots from one workstation to another as required by the routing. By the nature of the system, inventory will build up in raw materials, work-in-process, and finished goods.

FUNCTIONS OF INVENTORIES

In batch manufacturing, the basic purpose of inventories is to decouple supply and demand. Inventory serves as a buffer between:

- Supply and demand.
- Customer demand and finished goods.
- Finished goods and component availability.
- Requirements for an operation and the output from the preceding operation.
- Parts and materials to begin production and the suppliers of materials.

Based on this, inventories can be classified according to the function they perform.

Anticipation Inventory

These inventories are built up in anticipation of future demand. For example, they are created ahead of a peak selling season, a promotion program, vacation shutdown, or possibly the threat of a strike. They are built up to help level production and to reduce the costs of changing production rates.

Fluctuation Inventory (Safety Stock)

Inventory is held to cover random unpredictable fluctuations in supply and demand or lead time. If demand or lead time is greater than forecast, a stockout will occur. Safety stock is carried to protect against this possibility. Its purpose is to prevent disruptions in manufacturing or deliveries to customers. Safety stock is also called buffer stock or reserve stock.

Lot-Size Inventory

Items purchased or manufactured in quantities greater than needed immediately create lot-size inventories. This is to take advantage of quantity discounts, to reduce shipping, clerical, and setup costs, and in cases where it is impossible to make or purchase items at the same rate they will be used or sold. Lot-size inventory is sometimes called **cycle stock.** It is the portion of inventory that depletes gradually as customers' orders come in and is replenished cyclically when suppliers' orders are received.

Transportation Inventory

These inventories exist because of the time needed to move goods from one location to another such as from a plant to a distribution center or a customer. They are sometimes called **pipeline** or **movement inventories.** The average amount of inventory in transit is:

$$I = \frac{tA}{365}$$

where I is the average annual inventory in transit, t is transit time in days, and A is annual demand. Notice that the transit inventory does not depend upon the shipment size but on the transit time and the annual demand. The only way to reduce the inventory in transit, and its cost, is to reduce the transit time.

▼

EXAMPLE PROBLEM

Delivery of goods from a supplier is in transit for ten days. If the annual demand is 5200 units, what is the average annual inventory in transit?

Answer

$$I = \frac{10 \times 5200}{365} = 142.5 \text{ units}$$

The problem can be solved in the same way using dollars instead of units.

Hedge Inventory

Some products such as minerals and commodities, for example, grains or animal products, are traded on a worldwide market. The price for these products fluctuates according to world supply and demand. If buyers expect prices to rise, they can purchase hedge inventory when prices are low. Hedging is complex and beyond the scope of this text.

Maintenance, Repair, and Operating Supplies (MRO)

MROs are items used to support general operations and maintenance but which do not become directly part of a product. They include maintenance supplies, spare parts, and consumables such as cleaning compounds, lubricants, pencils, and erasers.

OBJECTIVES OF INVENTORY MANAGEMENT

A firm wishing to maximize profit will have at least the following objectives:

- Maximum customer service.
- Low-cost plant operation.
- Minimum inventory investment.

Customer Service

In broad terms, **customer service** is the ability of a company to satisfy the needs of customers. In inventory management, the term is used to describe the availability of items when needed and is a measure of inventory management effectiveness. The customer can be a purchaser, a distributor, another plant in the organization, or the workstation where the next operation is to be performed.

There are many different ways to measure customer service, each with its strengths and weaknesses, but there is no one best measurement. Some measures are percentage of orders shipped on schedule, percentage of line items shipped on schedule, and order-days out of stock.

Inventories help to maximize customer service by protecting against uncertainty. If we could forecast exactly what customers want and when, we could plan to meet demand with no uncertainty. However, demand and the lead time to get an item are often uncertain, possibly resulting in stockouts and customer dissatisfaction. For these reasons, it may be necessary to carry extra inventory to protect against uncertainty. This inventory is called **safety stock** and will be discussed in Chapter 11.

Operating Efficiency

Inventories help make a manufacturing operation more productive in four ways:

1. Inventories allow operations with different rates of production to operate separately and more economically. If two or more operations in a sequence have

different rates of output and are to be operated efficiently, inventories must build up between them.

2. Chapter 2 discussed production planning for seasonal products in which demand is nonuniform throughout the year. One strategy discussed was to level production and build anticipation inventory for sale in the peak periods. This would result in the following:

- Lower overtime costs.
- Lower hiring and firing costs.
- Lower training costs.
- Lower subcontracting costs.
- Lower capacity required.

By leveling production, manufacturing can continually produce an amount equal to the average demand. The advantage of this strategy is that the costs of changing production levels are avoided. Figure 9.2 shows this strategy.

3. Inventories allow manufacturing to run longer production runs, which result in the following:

- *Lower setup costs per item.*   The cost to make a lot or batch depends upon the setup costs and the run costs. The setup costs are fixed, but the run costs vary with the number produced. If larger lots are run, the setup costs are absorbed over a larger number, and the average (unit) cost is lower.

- *An increase in production capacity due to production resources being used a greater portion of the time for processing as opposed to setup.*   Time on a work center is taken up by setup and by run time. Output occurs only when an item is being worked on and not when setup is taking place. If larger quantities are produced at one time, there are fewer setups required to produce a given annual output and thus more time is available for producing goods. This is most important with bottleneck resources. Time lost on setup on these resources is lost throughput (total production) and lost capacity.

Figure 9.2   Operation leveling.

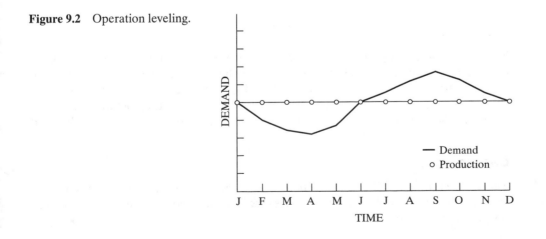

4. Inventories allow manufacturing to purchase in larger quantities, which results in lower ordering costs per unit and quantity discounts.

But all of this is at a price. The problem is to balance inventory investment with the following:

1. *Customer service.*   The lower the inventory, the higher the likelihood of a stockout and the lower the level of customer service. The higher the inventory level, the higher customer service will be.

2. *Costs associated with changing production levels.*   Excess equipment capacity, overtime, hiring, training, and layoff costs will all be higher if production fluctuates with demand.

3. *Cost of placing orders.*   Lower inventories can be achieved by ordering smaller quantities more often, but this practice results in higher annual ordering costs.

4. *Transportation costs.*   Goods moved in small quantities cost more to move per unit than those moved in large quantities. However, moving large lots implies higher inventory.

If inventory is carried, there has to be a benefit that exceeds the costs of carrying that inventory. Someone once said that the only good reason for carrying inventory beyond current needs is if it costs less to carry it than not. This being so, we should turn our attention to the costs associated with inventory.

INVENTORY COSTS

The following costs are used for inventory management decisions:

- Item cost.
- Carrying costs.
- Ordering costs.
- Stockout costs.
- Capacity-related costs.

Item Cost

The price paid for a purchased item consists of the cost of the item and any other direct costs associated in getting the item into the plant. These could include such things as transportation, custom duties, and insurance. The inclusive cost is often called the **landed price**. For an item manufactured in-house, the cost includes direct material, direct labor, and factory overhead. These costs can usually be obtained from either purchasing or accounting.

Carrying Costs

These costs include all expenses incurred by the firm because of the volume of inventory carried. As inventory increases, so do these costs. They can be broken down into three categories:

1. *Capital costs.*   Money invested in inventory is not available for other uses and as such represents a lost opportunity cost. The minimum cost would be the interest lost by not investing the money at the prevailing interest rate, and it may be much higher depending on investment opportunities for the firm.

2. *Storage costs.*   Storing inventory requires space, workers and equipment. As inventory increases, so do these costs.

3. *Risk costs.*   The risks in carrying inventory are:

    a. Obsolescence; loss of product value resulting from a model or style change or technological development.

    b. Damage; inventory damaged while being held or moved.

    c. Pilferage; goods lost, strayed, or stolen.

    d. Deterioration; inventory that rots or dissipates in storage or whose shelf life is limited.

What does it cost to carry inventory? Actual figures vary from industry to industry and company to company. Capital costs may vary depending upon interest rates, the credit rating of the firm, and the opportunities the firm may have for investment. Storage costs vary with location and type of storage needed. Risk costs can be very low or can be close to 100% of the value of the item for perishable goods. The carrying cost is usually defined as a percentage of the dollar value of inventory per unit of time (usually one year). Textbooks tend to use a figure of 20–30% in manufacturing industries. This is realistic in many cases but not with all products. For example, the possibility of obsolescence with fad or fashion items is high, and the cost of carrying such items is greater.

EXAMPLE PROBLEM

A company carries an average annual inventory of $2,000,000. If they estimate the cost of capital is 10%, storage costs are 7%, and risk costs are 6%, what does it cost per year to carry this inventory?

Answer

Total cost of carrying inventory $= 10\% + 7\% + 6\% = 23\%$

Annual cost of carrying inventory $= 0.23 \times \$2,000,000 = \$460,000$

Ordering Costs

Ordering costs are those associated with placing an order either with the factory or a supplier. The cost of placing an order does not depend upon the quantity ordered.

Whether a lot of 10 or 100 is ordered, the costs associated with placing the order are essentially the same. However, the annual cost of ordering depends upon the number of orders placed in a year.

Ordering costs in a factory include the following:

- *Production control costs.*   The annual cost and effort expended in production control depend on the number of orders placed, not on the quantity ordered. The fewer orders per year, the less cost. The costs incurred are those of issuing and closing orders, scheduling, loading, dispatching, and expediting.

- *Setup and teardown costs.*   Every time an order is issued, work centers have to set up to run the order and tear down the setup at the end of the run. These costs do not depend upon the quantity ordered but on the number of orders placed per year.

- *Lost capacity cost.*   Every time an order is placed at a work center, the time taken to set up is lost as productive output time. This represents a loss of capacity and is directly related to the number of orders placed. It is particularly important and costly with bottleneck work centers.

- *Purchase order cost.*   Every time a purchase order is placed, costs are incurred to place the order. These costs include order preparation, follow-up, expediting, receiving, authorizing payment, and the accounting cost of receiving and paying the invoice. The annual cost of ordering depends upon the number of orders placed.

The annual cost of ordering depends upon the number of orders placed in a year. This can be reduced by ordering more at one time, resulting in the placing of fewer orders. However, this drives up the inventory level and the annual cost of carrying inventory.

EXAMPLE PROBLEM

Given the following annual costs, calculate the average cost of placing one order.

Production control salaries = $60,000

Supplies and operating expenses for production control department = $15,000

Cost of setting up work centers for an order = $120

Orders placed each year = 2000

Answer

$$\text{Average cost} = \frac{\text{fixed costs}}{\text{number of orders}} + \text{variable cost}$$

$$= \frac{\$60,000 + \$15,000}{2000} + \$120 = \$157.50$$

Stockout Costs

If demand during the lead time exceeds forecast, we can expect a stockout. A stockout can potentially be expensive because of back-order costs, lost sales, and possibly lost customers. Stockouts can be reduced by carrying extra inventory to protect against those times when the demand during lead time is greater than forecast.

Capacity-Associated Costs

When output levels must be changed, there may be costs for overtime, hiring, training, extra shifts, and layoffs. These costs can be avoided by leveling production, that is, by producing items in slack periods for sale in peak periods. However, this builds inventory in the slack periods.

EXAMPLE PROBLEM

A company makes and sells a seasonal product. Based on a sales forecast of 2000, 3000, 6000, and 5000 per quarter, calculate a level production plan, quarterly ending inventory, and average quarterly inventory.

If inventory carrying costs are $3 per unit per quarter, what is the annual cost of carrying inventory? Opening and ending inventories are zero.

Answer

	Quarter 1	Quarter 2	Quarter 3	Quarter 4	Total
Forecast Demand	2000	3000	6000	5000	16,000
Production	4000	4000	4000	4000	16,000
Ending Inventory 0	2000	3000	1000	0	
Average Inventory	1000	2500	2000	500	
Inventory Cost (dollars)	3000	7500	6000	1500	18,000

FINANCIAL STATEMENTS AND INVENTORY

The two major financial statements are the balance sheet and the income statement. The balance sheet shows assets, liabilities, and owners' equity. The income statement shows the revenues made and the expenses incurred in achieving that revenue.

Balance Sheet

An **asset** is something that has value and is expected to benefit the future operation of the business. An asset may be tangible such as cash, inventory, machinery, and buildings, or may be intangible such as accounts receivable or a patent.

Liabilities are obligations or amounts owed by a company. Accounts payable, wages payable, and long-term debt are examples of liabilities.

Owners' equity is the difference between assets and liabilities. After all the liabilities are paid, it represents what is left for the owners of the business. Owners' equity is created either by the owners investing money in the business or through the operation of the business when it earns a profit. It is decreased when owners take money out of the business or when the business loses money.

The accounting equation. The relationship between assets, liabilities, and owners' equity is expressed by the balance sheet equation:

$$\text{Assets} = \text{liabilities} + \text{owners' equity}$$

This is a basic accounting equation. Given two of the values the third can always be found.

EXAMPLE PROBLEM

a. If the owners' equity is $1,000 and liabilities are $800, what are the assets?

b. If the assets are $1,000 and liabilities are $600, what is the owners' equity?

Answer

a. Assets = Liabilities + owners' equity
Assets = $800 + $1,000 = $1,800

b. Owners' equity = assets − liabilities
= $1,000 − $600 = $400

Balance Sheet. The balance sheet is usually shown with the assets on the left side and the liabilities and owners' equity on the right side as follows.

Assets		Liabilities	
Cash	$100,000	Notes Payable	$5,000
Accounts receivable	$300,000	Accounts payable	$20,000
Inventory	$500,000	Long-term debt	$500,000
Fixed assets	$1,000,000	Total liabilities	$525,000
		Owners' equity	
		Capital	$1,000,000
		Retained earnings	$375,000
Total assets	$1,900,000	Total liabilities and owners' equity	$1,900,000

Capital is the amount of money the owners have invested in the company.

Retained earnings are increased by the revenues a company makes and decreased by the expenses incurred. The summary of revenues and expenses is shown on the income statement.

Income Statement

Income (Profit). The primary purpose of a business is to increase the owners' equity by making a profit. For this reason owners' equity is broken down into a series of accounts, called revenue accounts, which show what increased owners' equity, and expense accounts, which show what decreased owners' equity.

$$\text{Income} = \text{revenue} - \text{expenses}$$

Revenue comes from the sale of goods or services. Payment is sometimes immediate in the form of cash, but often is made as a promise to pay at a later date, called an account receivable.

Expenses are the costs incurred in the process of making revenue. They are usually categorized into the cost of goods sold and general and administrative expenses.

Cost of goods sold are costs that are incurred to make the product. They include direct labor, direct material, and factory overhead. Factory overhead is all other factory costs except direct labor and direct material.

General and administrative expenses include all other costs in running a business. Examples of these are advertising, insurance, property taxes, and wages and benefits other than factory.

The following is an example of an income statement.

Revenue		$1,000,000
Cost of goods sold		
Direct labor	$200,000	
Direct material	400,000	
Factory overhead	200,000	$800,000
Gross margin (profit)		$200,000
General and administrative expenses		$100,000
Net income (profit)		$100,000

EXAMPLE PROBLEM

Given the following data, calculate the gross margin and the net income.

Revenue	=	$1,500,000
Direct labor	=	$ 300,000
Direct material	=	$ 500,000
Factory overhead	=	$ 400,000
General and administrative expenses	=	$ 150,000

How much would profits increase if, through better materials management, material costs are reduced by $50,000?

Revenue		$1,500,000
Cost of goods sold		
Direct labor	$300,000	
Direct material	500,000	
Overhead	400,000	$1,200,000
Gross margin (gross profit)		$ 300,000
General and administrative expenses		$ 150,000
Net income (profit)		$ 150,000

If material costs are reduced by $50,000, income increases by $50,000. Materials management can have a direct impact on the bottom line—net income.

Cash Flow Analysis

When inventory is purchased as raw material, it is recorded as an asset. When it enters production, it is recorded as work in process inventory (WIP) and, as it is processed, its value increases by the amount of direct labor applied to it and the overhead attributed to its processing. The material is said to absorb overhead. When the goods are ready for sale, they do not become revenue until they are sold. However, the expenses incurred in producing the goods must be paid for. This raises another financial issue: Businesses must have the cash to pay their bills. Cash is generated by sales and the flow of cash into a business must be sufficient to pay bills as they become due. Businesses develop financial statements showing the cash flows into and out of the business. Any shortfall of cash must be provided for, perhaps by borrowing or in some other way. This type of analysis is called cash flow analysis.

Financial Inventory Performance Measures

From a financial point of view, inventory is an asset and represents money that is tied up and cannot be used for other purposes. As we saw earlier in this chapter, inventory has a carrying cost—the costs of capital, storage, and risk. Finance wants as little inventory as possible and needs some measure of the level of inventory. Total inventory investment is one measure, but in itself does not relate to sales. Two measures that do relate to sales are the inventory turns ratio and days of supply.

Inventory turns. Ideally, a manufacturer carries no inventory. This is impractical, since inventory is needed to support manufacturing and often to supply customers. How much inventory is enough? There is no one answer. A convenient measure of how effectively inventories are being used is the inventory turns ratio:

$$\text{Inventory turns} \ = \ \frac{\text{annual cost of goods sold}}{\text{average inventory in dollars}}$$

The calculation of average inventory can be complicated and is a subject for cost accounting. In this text it will be taken as a given.

For example, if the annual cost of goods sold is $1 million and the average inventory is $500,000, then

$$\text{Inventory turns} \ = \ \frac{\$1,000,000}{\$500,000} = 2$$

What does this mean? At the very least, it means that with $500,000 of inventory a company is able to generate $1 million in sales. If, through better materials management, the firm is able to increase its turns ratio to 10, the same sales are generated with only $100,000 of average inventory. If the annual cost of carrying inventory is 25% of the inventory value, the reduction of $400,000 in inventory results in a cost reduction (and profit increase) of $80,000.

▼ ━━

EXAMPLE PROBLEM

a. What will be the inventory turns ratio if the annual cost of goods sold is $24 million a year and the average inventory is $6 million?

Answer

$$\text{Inventory turns} = \frac{\text{annual cost of goods sold}}{\text{average inventory in dollars}}$$

$$= \frac{24,000,000}{6,000,000} = 4$$

b. What would be the reduction in inventory if inventory turns were increased to 12 times per year?

Answer

$$\text{Average inventory} = \frac{\text{annual cost of goods sold}}{\text{inventory turns}}$$

$$= \frac{24,000,000}{12}$$

$$= \$2,000,000$$

Reduction in inventory $= 6,000,000 - 2,000,000 = \$4,000,000$

c. If the cost of carrying inventory is 25% of the average inventory, what will the savings be?

Answer

$$\text{Reduction in inventory} = \$4,000,000$$
$$\text{Savings} \qquad = \$4,000,000 \times 0.25 = \$1,000,000$$

Days of supply.   This is a measure of the equivalent number of days of inventory on hand, based on usage. The equation to calculate the days of supply is

$$\text{Days of supply} = \frac{\text{inventory on hand}}{\text{average daily usage}}$$

EXAMPLE PROBLEM

A company has 9000 units on hand and the annual usage is 48,000 units. There are 240 working days in the year. What is the days of supply?

Answer

$$\text{Average daily usage} = \frac{48,000}{240} = 200 \text{ units}$$

$$\text{Days of supply} = \frac{\text{inventory on hand}}{\text{average daily usage}} = \frac{9000}{200} = 45 \text{ days}$$

ABC INVENTORY CONTROL

Control of inventory is exercised by controlling individual items called **stock-keeping units (SKUs).** In controlling inventory, four questions must be answered:

1. What is the importance of the inventory item?
2. How are they to be controlled?
3. How much should be ordered at one time?
4. When should an order be placed?

The ABC inventory classification system answers the first two questions by determining the importance of items and thus allowing different levels of control based on the relative importance of items.

Most companies carry a large number of items in stock. To have better control at a reasonable cost, it is helpful to classify the items according to their importance. Usually this is based on annual dollar usage, but other criteria may be used.

The ABC principle is based on the observation that a small number of items often dominate the results achieved in any situation. This observation was first made by an Italian economist, Vilfredo Pareto, and is called **Pareto's law.** As applied to inventories, it is usually found that the relationship between the percentage of items and the percentage of annual dollar usage follows a pattern in which:

A   About 20% of the items account for about 80% of the dollar usage.

B   About 30% of the items account for about 15% of the dollar usage.

C   About 50% of the items account for about 5% of the dollar usage.

The percentages are approximate and should not be taken as absolute. This type of distribution can be used to help control inventory.

Steps in Making an ABC Analysis

1. Establish the item characteristics that influence the results of inventory management. This is usually annual dollar usage but may be other criteria, such as scarcity of material.

2. Classify items into groups based on the established criteria.

3. Apply a degree of control in proportion to the importance of the group.

The factors affecting the importance of an item include annual dollar usage, unit cost, and scarcity of material. For simplicity, only annual dollar usage is used in this text. The procedure for classifying by annual dollar usage is as follows:

1. Determine the annual usage for each item.

2. Multiply the annual usage of each item by its cost to get its total annual dollar usage.

3. List the items according to their annual dollar usage.

4. Calculate the cumulative annual dollar usage and the cumulative percentage of items.

5. Examine the annual usage distribution and group the items into A, B, and C groups based on percentage of annual usage.

EXAMPLE PROBLEM

A company manufactures a line of ten items. Their usage and unit cost are shown in the following table along with the annual dollar usage. The latter is obtained by multiplying the unit usage by the unit cost.

    a. Calculate the annual dollar usage for each item.

    b. List the items according to their annual dollar usage.

    c. Calculate the cumulative annual dollar usage and the cumulative percent of items.

    d. Group items into an A, B, C classification.

Answer

a. Calculate the annual dollar usage for each item.

Part Number	Unit Usage	Unit Cost $	Annual $ Usage
1	1100	2	2200
2	600	40	24,000
3	100	4	400
4	1300	1	1300
5	100	60	6000
6	10	25	250
7	100	2	200
8	1500	2	3000
9	200	2	400
10	500	1	500
Total	5510		$38,250

b., c., and **d.**

Part Number	Annual $ Usage	Cumulative $ Usage	Cumulative % $ Usage	Cumulative % of Items	Class
2	24,000	24,000	62.75	10	A
5	6000	30,000	78.43	20	A
8	3000	33,000	86.27	30	B
1	2200	35,200	92.03	40	B
4	1300	36,500	95.42	50	B
10	500	37,000	96.73	60	C
9	400	37,400	97.78	70	C
3	400	37,800	98.82	80	C
6	250	38,050	99.48	90	C
7	200	38,250	100.00	100	C

The percentage of value and the percentage of items is often shown as a graph such as in Figure 9.3.

Control Based on ABC Classification

Using the ABC approach, there are two general rules to follow:

1. *Have plenty of low-value items.* C items represent about 50% of the items but account for only about 5% percent of the total inventory value. Carrying extra stock of C items adds little to the total value of the inventory. C items are really only important if there is a shortage of one of them—when they become extremely important—so a supply should always be on hand. For example, order a year's supply at a time and carry plenty of safety stock. That way there is only once a year when a stockout is even possible.

2. *Use the money and control effort saved to reduce the inventory of high-value items.* A items represent about 20% of the items and account for about 80% of the value. They are extremely important and deserve the tightest control and the most frequent review.

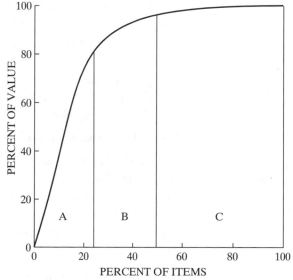

Figure 9.3 ABC curve: percentage of value versus percentage of items.

Different controls used with different classifications might be the following:

- *A Items: high priority.* Tight control including complete accurate records, regular and frequent review by management, frequent review of demand forecasts, and close follow-up and expediting to reduce lead time.

- *B Items: medium priority.* Normal controls with good records, regular attention, and normal processing.

- *C Items: lowest priority.* Simplest possible controls—make sure there are plenty. Simple or no records; perhaps use a two-bin system or periodic review system. Order large quantities and carry safety stock.

SUMMARY

There are benefits as well as costs to having inventory. The problem is to balance the cost of carrying inventory with the following:

- *Customer service.* The lower the inventory level, the higher the likelihood of a stockout and the potential cost of back orders, lost sales, and lost customers. The higher the inventory level, the higher the level of customer service.

- *Operating efficiency.* Inventories decouple one operation from another and allow manufacturing to operate more efficiently. They allow leveling production and avoid the costs of changing production levels. Carrying inventory allows longer production runs and reduces the number of setups. Finally, inventories let manufacturing purchase in larger quantities. The ABC inventory classification system prioritizes individual items so that inventory and costs can be better controlled.

- *Cost of placing orders.*   Inventory can be reduced by ordering less each time an order is placed. However, this increases the annual cost of ordering.

- *Transportation and handling costs.*   The more often goods have to be moved and the smaller the quantities moved, the greater the transportation and material handling costs.

Inventory management is influenced by several factors:

- The classification of the inventory, whether raw material, work-in-process, or finished goods.

- The functions that inventory serves: anticipation, fluctuation, lot size, or transportation.

- Supply and demand patterns.

- The costs associated with carrying (or not carrying) inventory.

Besides managing inventory at the aggregate level, it must also be managed at the item level. Management needs to establish decision rules about inventory items so inventory control personnel can do their job effectively.

QUESTIONS

1. What are inventories? Why are they important to manufacturing companies?
2. What are the responsibilities of inventory management?
3. What is aggregate inventory management? With what is it concerned?
4. What are decision rules? Why are they necessary?
5. According to the flow of material, what are the four classifications of inventories?
6. Why is less inventory needed in a line-flow manufacturing system than in lot or batch manufacturing?
7. What is the basic purpose of inventories? In what four areas do they provide a buffer?
8. Describe the function and purpose of the following kinds of inventories:
    a. Anticipation.
    b. Fluctuation.
    c. Lot size.
    d. Transportation.
9. Describe how inventories influence each of the following:
    a. Customer service.
    b. Plant operations.
10. What are the five costs associated with inventories?
11. Name and describe the categories of inventory-carrying costs.

12. Name and describe the categories of ordering costs found in a factory.

13. What are stockout costs and capacity-associated costs? What is their relationship to inventories?

14. What are the balance sheet equation and the income statement equation?

15. What is the purpose of cash flow analysis?

16. What do inventory turns and days of supply measure?

17. What is the basic premise of ABC analysis? What are the three steps in making an ABC inventory analysis?

18. What are the five steps in the procedure for classifying inventory by annual dollar usage?

PROBLEMS

9.1 If the transit time is ten days and the annual demand for an item is 10,000 units, what is the average annual inventory in transit?

Answer. 274 units

9.2 A company is using a carrier to deliver goods to a major customer. The annual demand is $2,000,000, and the average transit time is 10 days. Another carrier promises to deliver in 8 days. What is the reduction in transit inventory?

9.3 Given the following percentage costs of carrying inventory, calculate the annual carrying cost if the average inventory is $1 million. Capital costs are 10%, storage costs are 5%, and risk costs are 8%.

Answer. $230,000

9.4 A florist carries an average inventory of $10,000 in cut flowers. The flowers require special storage and are highly perishable. The florist estimates capital cost at 10%, storage cost at 25%, and risk costs at 50%. What is the annual carrying cost?

9.5 Annual purchasing salaries are $75,000, operating expenses for the purchasing department are $25,000, and inspecting and receiving costs are $30 per order. If the purchasing department places 10,000 orders a year, what is the average cost of ordering? What is the annual cost of ordering?

Answer.

Average ordering cost = $40

Annual cost = $400,000

9.6 An importer operates a small warehouse that has the following annual costs. Wages for purchasing are $45,000, purchasing expenses are $30,000, customs and brokerage costs are $25 per order, the cost of financing the inventory is 8%, storage costs are 6%, and the risk costs are 10%. The average inventory is $250,000, and 5000 orders are placed in a year. What are the annual ordering and carrying costs?

9.7 A company manufactures and sells a seasonal product. Based on the sales forecast that follows, calculate a level production plan, quarterly ending inventories, and average quarterly inventories. Assume that the average quarterly inventory is the average of the starting and ending inventory for the quarter. If inventory carrying costs are $2 per unit per quarter, what is the annual cost of carrying this anticipation inventory? Opening and ending inventories are zero.

	Quarter 1	Quarter 2	Quarter 3	Quarter 4	Totals $
Sales	1000	2000	3000	2000	
Production					
Ending Inventory					
Average Inventory					
Inventory Cost					

Answer. Annual inventory cost = $4000

9.8 Given the following data, calculate a level production plan, quarterly ending inventory, and average quarterly inventory. If inventory carrying costs are $6 per unit per quarter, what is the annual carrying cost? Opening and ending inventory are zero.

	Quarter 1	Quarter 2	Quarter 3	Quarter 4	Totals $
Forecast Demand	5000	6000	9000	8000	
Production					
Ending Inventory					
Average Inventory					
Inventory Cost					

If the company always carries 100 units of safety stock, what is the annual cost of carrying it?

9.9 Given the following data, calculate a level production plan, quarterly ending inventory, and average quarterly inventory. If inventory carrying costs are $3 per unit per quarter, what is the annual carrying cost? Opening and ending inventory are zero.

	Quarter 1	Quarter 2	Quarter 3	Quarter 4	Totals $
Forecast Demand	3000	4000	7000	6000	
Production					
Ending Inventory					
Average Inventory					
Inventory Cost					

9.10 If the assets are $2,000,000 and liabilities are $1,500,000, what is the owners' equity?

    Answer. $500,000

9.11 If the liabilities are $3,000,000 and the owners' equity is $1,200,000, what are the assets worth?

9.12 Given the following data, calculate the gross margin and the net income.

Revenue	=	$2,500,000
Direct labor	=	$500,000
Direct material	=	$800,000
Factory overhead	=	$700,000
General and administrative expense	=	$300,000

    Answer. Gross margin = $500,000

              Net income   = $200,000

9.13 In question 9.12, how much would profit increase if the materials costs are reduced by $200,000?

9.14 If the annual cost of goods sold is $10 million and the average inventory is $2.5 million:

    a. What is the inventory turns ratio?

    b. What would be the reduction in average inventory if, through better materials management, inventory turns were increased to 10 times per year?

    c. If the cost of carrying inventory is 20% of the average inventory, what is the annual savings?

    Answer. a. 4

              b. $1,500,000

              c. $300,000

9.15 If the annual cost of goods sold is $30 million and the average inventory is $10 million:

a. What is the inventory turns ratio?

b. What would be the reduction in average inventory, if, through better materials management, inventory turns were increased to 10 times per year?

c. If the cost of carrying inventory is 20% of the average inventory, what is the annual savings?

9.16 A company has 500 units on hand and the annual usage is 6000 units. There are 240 working days in the year. What is the days of supply?

Answer. 20 days

9.17 Over the past year, a company has sold the following ten items. The following table shows the annual sales in units and the cost of each item.

a. Calculate the annual dollar usage of each item.

b. List the items according to their total annual dollar usage.

c. Calculate the cumulative annual dollar usage and the cumulative percentage of items.

d. Group the items into A, B, and C groups based on percentage of annual dollar usage.

Part Number	Annual Unit Usage	Unit Cost $	Annual $ Usage
1	21,000	1	
2	5000	40	
3	1600	3	
4	12,000	1	
5	1000	100	
6	50	50	
7	800	2	
8	10,000	3	
9	4000	1	
10	5000	1	

Answer. A items   2, 5
               B items   8, 1, 4
               C items   10, 3, 9, 6, 7

9.18 Analyze the following data to produce an ABC classification based on annual dollar usage.

Part Number	Annual Unit Usage	Unit Cost $	Annual $ Usage
1	200	10	
2	15,000	4	
3	60,000	6	
4	15,000	15	
5	1400	10	
6	100	50	
7	25,000	2	
8	700	3	
9	25,000	1	
10	7500	1	

10

Order Quantities

INTRODUCTION

The objectives of inventory management are to provide the required level of customer service and to reduce the sum of all costs involved. To achieve these objectives, two basic questions must be answered:

1. How much should be ordered at one time?
2. When should an order be placed?

Management must establish decision rules to answer these questions so inventory management personnel know when to order and how much. Lacking any better knowledge, decision rules are often made based on what seems reasonable. Unfortunately, such rules do not always produce the best results.

This chapter will examine methods of answering the first question, and the next chapter will deal with the second question. First, we must decide what we are ordering and controlling.

Stock-Keeping Unit (SKU)

Control is exercised through individual items in a particular inventory. These are called a **stock-keeping unit (SKU).** Two white shirts in the same inventory but of different sizes or styles would be two different SKUs. The same shirt in two different inventories would be two different SKUs.

Lot-size Decision Rules

The eighth edition of the APICS dictionary defines a lot, or batch, as a quantity produced together and sharing the same production costs and specifications. Following are some common decision rules for determining what lot size to order at one time.

Lot-for-lot.   This rule says to order exactly what is needed—no more—no less. The order quantity changes whenever requirements change. This technique requires time-phased information such as provided by a material requirements plan or a master production schedule. Since items are ordered only when needed, this system creates no unused lot-size inventory. Because of this, it is the best method for planning "A" items and is also used in a just-in-time environment.

Fixed-order quantity.   These types of decision rules specify the number of units to be ordered each time an order is placed for an individual item or SKU. The quantity is usually arbitrary, such as 200 units at a time. The advantage to this type of rule is that it is easily understood. The disadvantage is that it does not minimize the costs involved.

A variation on the fixed-order quantity system is the **min-max system.** In this system, an order is placed when the quantity available falls below the order point (discussed in the next chapter). The quantity ordered is the difference between the actual quantity available at the time of order and the maximum. For example, if the order point is 100 units, the maximum is 300 units, and the quantity actually available when the order is placed is 75, the order quantity is 225 units. If the quantity actually available is 80 units, an order for 220 units is placed.

One commonly used method of calculating the quantity to order is the economic-order quantity, which is discussed in the next section.

Order "n" periods supply.   Rather than ordering a fixed quantity, inventory management can order enough to satisfy future demand for a given period of time. The question is how many periods should be covered? The answer is given later in this chapter in the discussion on the period-order quantity system.

Costs

As shown in the last chapter, the cost of ordering and the cost of carrying inventory both depend on the quantity ordered. Ideally, the ordering decision rules used will minimize the sum of these two costs. The best known system is the economic-order quantity.

ECONOMIC-ORDER QUANTITY (EOQ)

Assumptions

The assumptions on which the EOQ is based are as follows:

1. Demand is relatively constant and is known.
2. The item is produced or purchased in lots or batches and not continuously.

3. Order preparation costs and inventory-carrying costs are constant and known.

4. Replacement occurs all at once.

These assumptions are usually valid for finished goods whose demand is independent and fairly uniform. However, there are many situations where the assumptions are not valid and the EOQ concept is of no use. For instance, there is no reason to calculate the EOQ for made-to-order items in which the customer specifies the order quantity, the shelf life of the product is short, or the length of the run is limited by tool life or raw material batch size. In material requirements planning, the lot-for-lot decision rule is often used, but there are also several rules used that are variations of the economic-order quantity.

Development of the EOQ Formula

Under the assumptions given, the quantity of an item in inventory decreases at a uniform rate. Suppose for a particular item, the order quantity is 200 units, and the usage rate is 100 units a week. Figure 10.1 shows how inventory would behave.

The vertical lines represent stock arriving all at once as the stock on hand reaches zero. The quantity of units in inventory then increases instantaneously by Q, the quantity ordered. This is an accurate representation of the arrival of purchased parts or manufactured parts where all parts are received at once.

From the preceding,

$$\text{Average lot size inventory} \ = \ \frac{\text{order quantity}}{2} \ = \ \frac{200}{2} \ = \ 100 \text{ units}$$

$$\text{Number of orders per year} \ = \ \frac{\text{annual demand}}{\text{order quantity}} \ = \ \frac{100 \ \times \ 52}{200}$$

$$= \ 26 \text{ times per year}$$

Figure 10.1   Inventory on hand over time.

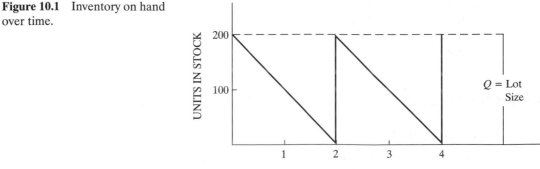

EXAMPLE PROBLEM

The annual demand for an SKU is 10,075 units, and it is ordered in quantities of 650 units. Calculate the average inventory and the number of orders placed per year.

Answer

$$\text{Average inventory} = \frac{\text{order quantity}}{2} = \frac{650}{2} = 325 \text{ units}$$

$$\text{Number of orders per year} = \frac{\text{annual demand}}{\text{order quantity}} = \frac{10{,}075}{650} = 15.5$$

Notice in the example problem the number of orders per year is rounded neither up nor down. It is an average figure, and the actual number of orders per year will vary from year to year but will average to the calculated figure. In the example, 16 orders will be placed in one year and 15 in the second.

Relevant costs.   The relevant costs are as follows:

- Annual cost of placing orders.
- Annual cost of carrying inventory.

As the order quantity increases, the average inventory and the annual cost of carrying inventory increase, but the number of orders per year and the ordering cost decrease. It is a bit like a seesaw where one cost can be reduced only at the expense of increasing the other. The trick is to find the particular order quantity in which the total cost of carrying inventory and the cost of ordering will be a minimum.
    Let:

A = annual usage in units
S = ordering cost in dollars per order
i = annual carrying cost rate as a decimal of a percentage
c = unit cost in dollars
Q = order quantity in units

    Then:

Annual ordering cost $=$ number of orders $\times$ costs per order

$$= \frac{A}{Q} \times S$$

Annual carrying cost $=$ average inventory $\times$ cost of carrying one unit for one year

$=$ average inventory $\times$ unit cost $\times$ carrying cost

$$= \frac{Q}{2} \times c \times i$$

Total annual costs $=$ annual ordering costs $+$ annual carrying costs

$$= \frac{A}{Q} \times S + \frac{Q}{2} \times c \times i$$

EXAMPLE PROBLEM

The annual demand is 10,000 units, the ordering cost $30 per order, the carrying cost 20%, and the unit cost $15. The order quantity is 600 units. Calculate:

    a. Annual ordering cost

    b. Annual carrying cost

    c. Total annual cost

Answer

$A = $ 10,000 units

$S = $ $30

$i = $ 0.20

$c = $ $15

$Q = $ 600 units

    a. annual ordering cost $= \dfrac{A}{Q} \times S = \dfrac{10,000}{600} \times \$30 = \$500$

    b. annual carrying cost $= \dfrac{Q}{2} \times c \times i = \dfrac{600}{2} \times \$15 \times 0.2 = \underline{\$900}$

    c. total annual cost $= \$1400$

Ideally, the total cost will be a minimum. For any situation in which the annual demand (A), the cost of ordering (S), and the cost of carrying inventory (i) are given, the total cost will depend upon the order quantity (Q).

Trial-and-Error Solution

Consider the following example:

A hardware supply distributor carries boxes of 3-inch bolts in stock. The annual usage is 1000 boxes, and demand is relatively constant throughout the year. Ordering costs are $20 per order, and the cost of carrying inventory is estimated to be 20%. The cost per unit is $5.

Let:

A = 1000 units

S = $20 per order

c = $5 per unit

i = 20% = 0.20

Then:

$$\text{Annual ordering cost} = \frac{A}{Q} \times S = \frac{1000}{Q} \times 20$$

$$\text{Annual carrying cost} = \frac{Q}{2} \times c \times i = \frac{Q}{2} \times 5 \times 0.20$$

$$\text{Total annual cost} = \text{annual ordering cost} + \text{annual carrying cost}$$

Figure 10.2 is a tabulation of the costs for different order quantities. The results from the table in Figure 10.2 are represented on the graph of Figure 10.3.

Figures 10.2 and 10.3 show the following important facts:

1. There is an order quantity in which the sum of the ordering costs and carrying costs is a minimum.
2. This EOQ occurs when the cost of ordering equals the cost of carrying.
3. The total cost varies little for a wide range of lot sizes about EOQ.

The last point is important for two reasons. First, it is usually difficult to determine accurately the cost of carrying inventory and the cost of ordering. Since the total cost is relatively flat around the EOQ, it is not critical to have exact values.

Figure 10.2 Costs for different lot sizes.

Order Quantity (Q)	Ordering Costs (AS/Q)	Carrying Costs ($Qci/2$)	Total Cost
50	$400	$25	$425
100	200	50	250
150	133	75	208
200	100	100	200
250	80	125	205
300	67	150	217
350	57	175	232
400	50	200	250

Figure 10.3   Cost versus lot size.

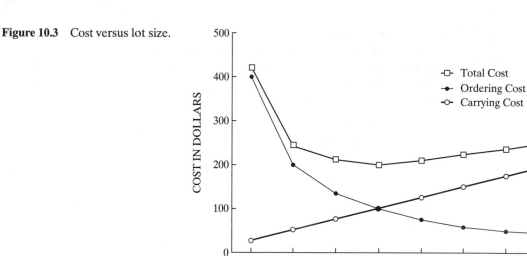

Good approximations are sufficient. Second, parts are often ordered in convenient packages such as pallet loads, cases, or dozens, and it is adequate to pick the closest package quantity to the EOQ.

Economic-Order Quantity Formula

The previous section showed that the EOQ occurred at an order quantity in which the ordering costs equal the carrying costs. If these two costs are equal, the following formula can be derived:

$$\text{Carrying costs} = \text{ordering costs}$$
$$\frac{Qic}{2} = \frac{AS}{Q}$$

Solving for Q gives

$$Q^2 = \frac{2AS}{ic}$$

$$Q = \sqrt{\frac{2AS}{ic}}$$

This value for the order quantity is the economic-order quantity. Using the formula to calculate the EOQ in the preceding example yields:

$$\text{EOQ} = \sqrt{\frac{2AS}{ic}} = \sqrt{\frac{2 \times 20 \times 1000}{0.20 \times 5}} = 200 \text{ units}$$

How to Reduce Lot Size

Looking at the EOQ formula, there are four variables. The EOQ will increase as the annual demand (A) and the cost of ordering (S) increase, and it will decrease as the cost of carrying inventory (i) and the unit cost (c) increase.

The annual demand (A) is a condition of the marketplace and is beyond the control of manufacturing. The cost of carrying inventory (i) is determined by the product itself and the cost of money to the company. As such, it is beyond the control of manufacturing.

The unit cost (c) is either the purchase cost of the SKU or the cost of manufacturing the item. Ideally, both costs should be as low as possible. In any event, as the unit cost decreases, the EOQ increases.

The cost of ordering (S) is either the cost of placing a purchase order or the cost of placing a manufacturing order. The cost of placing a manufacturing order is made up from production control costs and setup costs. Anything that can be done to reduce these costs reduces the EOQ.

Just-in-time manufacturing emphasizes reduction of setup time. There are several reasons why this is desirable, and the reduction of order quantities is one. Chapter 15 discusses just-in-time manufacturing further.

VARIATIONS OF THE EOQ MODEL

There are several modifications that can be made to the basic EOQ model to fit particular circumstances. Two that are often used are the monetary unit lot-size model and the noninstantaneous receipt model.

Monetary Unit Lot Size

The EOQ can be calculated in monetary units rather than physical units. The same EOQ formula given in the preceding can be used, but the annual usage changes from units to dollars.

A_D = annual usage in dollars

S = ordering costs in dollars

i = carrying cost rate as a decimal of a percent

Because the annual usage is expressed in dollars, the unit cost is not needed in the modified EOQ equation.

The EOQ in dollars is:

$$\text{EOQ} = \sqrt{\frac{2A_D S}{i}}$$

EXAMPLE PROBLEM

An item has an annual demand of $5000, preparation costs of $20 per order, and a carrying cost of $20. What is the EOQ in dollars?

A_D = $5000

S = $20

i = 20% = 0.20

$$\text{EOQ} = \sqrt{\frac{2 A_D S}{i}} = \sqrt{\frac{2 \times 5000 \times 20}{0.2}} = \$1000$$

QUANTITY DISCOUNTS

When material is purchased, suppliers often give a discount on orders over a certain size. This can be done because larger orders reduce the supplier's costs; to get larger orders, they are willing to offer volume discounts. The buyer must decide whether to accept the discount, and in doing so, must consider the relevant costs:

- Purchase cost.
- Ordering costs.
- Carrying costs.

EXAMPLE PROBLEM

An item has an annual demand of 25,000 units, a unit cost of $10, an order preparation cost of $10, and a carrying cost of 20%. It is ordered on the basis of an EOQ, but the supplier has offered a discount of 2% on orders of $10,000 or more. Should the offer be accepted?

Answer

A_D = 25,000 × $10 = $250,000

S = $10

i = 20%

$$EOQ = \sqrt{\frac{2 \times 250{,}000 \times 10}{0.2}} = \$5000$$

Discounted order quantity = $10{,}000 \times 0.98 = \$9{,}800$

	No discount	Discount lot size
Unit Price	$10	$9.80
Lot Size	$5000	$9800
Average Lot Size Inventory ($Qc \div 2$)	$2500	$4900
Number of Orders per Year	50	25
Purchase Cost	$250,000	$245,000
Inventory-Carrying Cost (20%)	500	980
Order Preparation Cost ($10 each)	500	250
Total Cost	$251,000	$246,230

From the preceding example problem, it can be said that taking the discount results in the following:

- There is a saving in purchase cost.
- Ordering costs are reduced because fewer orders are placed since larger quantities are being ordered.
- Inventory-carrying costs rise because of the larger order quantity.

The buyer must weigh the first two against the last and decide what to do. What counts is the total cost. Depending on the figures, it may or may not be best to take the discount.

USE OF EOQ WHEN COSTS ARE NOT KNOWN

The EOQ formula depends upon the cost of ordering and the cost of carrying inventory. In practice, these costs are not necessarily known or easy to determine. However, the formula can still be used to advantage when applied to a family of items.

For a family of items, the ordering costs and the carrying costs are generally the same for each item. For instance, if we were ordering hardware items—nuts, bolts, screws, nails, and so on—the carrying costs would be virtually the same (storage, capital, and risk costs) and the cost of placing an order with the supplier would be the same for each item. In cases such as this, the cost of placing an order (S) is the same for all items in the family as is cost of carrying inventory (i).

Now

$$Q = \sqrt{\frac{2A_D S}{i}}$$

where A (annual demand) is in *dollars*.

Since S is the same for all the items and i is the same for all items, the ratio $2S \div i$ must be the same for all items in the family. For convenience, let:

$$K = \sqrt{\frac{2S}{i}}$$

Then $\qquad\qquad Q = K\sqrt{A_D}$

Also $\qquad\qquad Q = \dfrac{\text{annual demand}}{\text{orders per year}} = \dfrac{A_D}{N}$

Therefore $\quad K\sqrt{A_D} = \dfrac{A_D}{N}$

$$K = \frac{\sqrt{A_D}}{N}$$

Let's see how this can be applied.

EXAMPLE PROBLEM

Suppose there were a family of items for which the decision rule was to order each item four times a year. Since the cost of ordering (S) and the cost of carrying inventory (i) are not known, ordering four times a year is not based on an EOQ. Can we come up with a better decision rule even if the EOQ cannot be calculated?

Item	Annual Usage	Orders per Year	Present Lot Size	$\sqrt{A_D}$	$K = \dfrac{\sqrt{A_D}}{N}$
A	$10,000	4	$2500	$100	25
B	400	4	100	20	5
C	144	4	36	12	3
		12	$2636	$132	33
Average inventory =			$1318		

The sum of all the lots is $2636. Since the average inventory is equal to half the order quantity, the average inventory is $2636 ÷ 2 = $1318.

Since this is a family of items where the preparation costs are the same and the carrying costs are the same, the values for $K = (2S \div i)^{1/2}$ should be the same for all items. The preceding calculations show that they are not. The correct value for K is not known, but a better value would be the average of all the values:

$$K = \frac{\Sigma \sqrt{A_D}}{N}$$

$$= \frac{132}{12}$$

$$= 11$$

This value of K can be used to recalculate the order quantities for each item.

Item	Annual Usage	Present Orders per Year	Present Lot Size	$\sqrt{A_D}$	New Lot Size = $K\sqrt{A_D}$	New Orders per Year $N = A_D/Q$
A	$10,000	4	$2500	$100	$1100	9.09
B	400	4	100	20	220	1.82
C	144	4	36	12	132	1.09
	$10,544	12	$2636	$132	$1452	12.00
Average inventory			$1318		$726	

Item A: New lot size $= K\sqrt{A_D} = 11 \times 100 = \$1100$

New orders per year $= \dfrac{A_D}{Q} = \dfrac{10{,}000}{1100} = 9.09$

The average inventory has been reduced from $1318 to $726 while the number of orders per year (12) remains the same. Thus, the total costs associated with inventory have been reduced.

PERIOD ORDER QUANTITY (POQ)

The economic-order quantity attempts to minimize the total cost of ordering and carrying inventory and is based on the assumption that demand is uniform. Often demand is not uniform, particularly in material requirements planning, and using the EOQ does not produce a minimum cost.

The period-order quantity lot-size rule is based on the same theory as the economic-order quantity. It uses the EOQ formula to calculate an economic **time between orders.** This is calculated by dividing the EOQ by the demand rate. This produces a time interval for which orders are placed. Instead of ordering the same quantity (EOQ), orders are placed to satisfy requirements for the calculated time interval. The number of orders placed in a year is the same as for an economic-order quantity, but the amount ordered each time varies. Thus, the ordering cost is the same but, because the order quantities are determined by actual demand, the carrying cost is reduced.

$$\text{Period order quantity} = \frac{\text{EOQ}}{\text{average weekly usage}}$$

EXAMPLE PROBLEM

The EOQ for an item is 2800 units, and the annual usage is 52,000 units. What is the period-order quantity?

Answer

$$\text{Average weekly usage} = 52,000 \div 52 = 1000 \text{ per week}$$

$$\text{Period order quantity} = \frac{\text{EOQ}}{\text{average weekly usage}}$$

$$= \frac{2800}{1000} = 2.8 \text{ weeks} \longrightarrow 3 \text{ weeks}$$

When an order is placed it will cover the requirements for the next three weeks.

Notice the calculation is approximate. Precision is not important.

EXAMPLE PROBLEM

Given the following MRP record and an EOQ of 250 units, calculate the planned order receipts using the economic-order quantity. Next, calculate the period-order quantities and the planned order receipts. In both cases, calculate the ending inventory and the total inventory carried over the ten weeks.

Week	1	2	3	4	5	6	7	8	9	10	Total
Net Requirements	100	50	150		75	200	55	80	150	30	890
Planned Order Receipt											

Answer

EOQ = 250 units

Week	1	2	3	4	5	6	7	8	9	10	Total
Net Requirements	100	50	150		75	200	55	80	150	30	890
Planned Order Receipt	250		250			250			250		
Ending Inventory	150	100	200	200	125	175	120	40	140	110	1360

Period order quantity:
Weekly average demand = 890 ÷ 10 = 89 units
POQ = 250 ÷ 89 = 2.81 ➤ 3 weeks

Week	1	2	3	4	5	6	7	8	9	10	Total
Net Requirements	100	50	150		75	200	55	80	150	30	890
Planned Order Receipt	300				330			260	180		
Ending Inventory	200	150	0	0	255	55	0	180	30	0	870

Notice in the example problem the total inventory is reduced from 1360 units to 870 units over the ten-week period.

SUMMARY

The economic-order quantity is based on the assumption that demand is relatively uniform. This is appropriate for some inventories, and the EOQ formula can be used with reasonable results. One problem in using the EOQ formula is in determining the cost of ordering and the cost of carrying inventory. Since the total cost curve is flat at the bottom, good guesses very often will produce an order quantity that is economical. We have also seen that the EOQ concept can be used effectively with groups of items when the costs of carrying and ordering are not known.

The two costs influenced by the order quantity are the cost of ordering and the cost of carrying inventory. All methods of calculating order quantities attempt to minimize the sum of these two costs. The period-order quantity does this. It has the advantage over the EOQ in that it is better for lumpy demand because it looks forward to see what is actually needed.

Practical Considerations When Using the EOQ

Lumpy demand.   The EOQ assumes that demand is uniform and replenishment occurs all at once. When this is not true, the EOQ will not produce the best results. It is better to use the period-order quantity.

Anticipation inventory.   Demand is not uniform, and stock must be built ahead. It is better to plan a buildup of inventory based on capacity and future demand.

Minimum order.   Some suppliers require a minimum order. This minimum may be based on the total order rather than on individual items. Often these are C items where the rule is to order plenty, not an EOQ.

Transportation inventory.   As will be discussed in Chapter 13, carriers give rates based on the amount shipped. A full load costs less per ton to ship than a part load. This is similar to the price break given by suppliers for large quantities. The same type of analysis can be used.

Multiples.   Sometimes, order size is constrained by package size. For example, a supplier may ship only in skid-load lots. In these cases, the unit used should be the minimum package size.

QUESTIONS

1. What are the two basic questions in inventory management discussed in the text?
2. What are decision rules? What is their purpose?
3. What is an SKU?

4. What is the lot-for-lot decision rule? What is its advantage? Where would it be used?

5. What are the four assumptions on which economic-order quantities are based? For what kind of items are these assumptions valid? When are they not?

6. Under the assumptions on which EOQs are based, what are the formulas for average lot size and the number of orders per year?

7. What are the relevant costs associated with the two formulas? As the order quantities increase, what happens to each cost? What is the objective in establishing a fixed-order quantity?

8. Define each of the following in your own words and as a formula:

   a. Annual ordering cost.

   b. Annual carrying cost.

   c. Total annual cost.

9. What is the economic-order quantity (EOQ) formula? Define each term and give the units used. How do the units change when monetary units are used?

10. What is the EOQ formula for the noninstantaneous receipt case? What do the two additional terms mean?

11. What are the relevant costs to be considered when deciding whether to take a quantity discount? On what basis should the decision be made?

12. What is the period-order quantity? How is it established? When can it be used?

13. How do each of the following influence inventory decisions?

    a. Lumpy demand.

    b. Minimum orders.

    c. Transportation costs.

    d. Multiples.

PROBLEMS

10.1 An SKU costing $10 is ordered in quantities of 500 units, annual demand is 5200 units, carrying costs are 20%, and the cost of placing an order is $40. Calculate the following:

   a. Average inventory.

   b. Number of orders placed per year.

   c. Annual inventory-carrying cost.

   d. Annual ordering cost.

   e. Annual total cost.

   Answer. a. 250 units

   b. 10.4 orders per year

   c. Inventory-carrying cost = $500

   d. Annual ordering cost = $416

   e. Annual total cost = $916

10.2 If the order quantity is increased to 1000 units, recalculate problems 10.1a to 10.1e and compare the results.

10.3 A company decides to establish an EOQ for an item. The annual demand is 100,000 units, each costing $8, ordering costs are $32 per order, and inventory-carrying costs are 20%. Calculate the following:

a. The EOQ in units.

b. Number of orders per year.

c. Cost of ordering, cost of carrying inventory, and total cost.

> *Answer.* **a.** 2000 units
>
> **b.** Number of orders per year = 50
>
> **c.** Annual cost of ordering = $1600
>
> Annual cost of carrying = $1600
>
> Annual total cost = $3200

10.4 A company wishes to establish an EOQ for an item for which the annual demand is $800,000, the ordering cost is $32, and the cost of carrying inventory is 20%. Calculate the following:

a. The EOQ in dollars.

b. Number of orders per year.

c. Cost of ordering, cost of carrying inventory, and total cost.

d. Compare your answers to those in problem 10.2.

> *Answer.* **a.** $16,000
>
> **b.** Number of orders per year = 50
>
> **c.** Annual cost of ordering = $1600
>
> Annual cost of carrying = $1600
>
> Annual total cost = $3200
>
> **d.** Results are the same.

10.5 An SKU has an annual demand of 5200 units, each costing $10, ordering costs are $200 per order, and the cost of carrying inventory is 20%. Calculate the EOQ in units and then convert to dollars.

10.6 A company is presently ordering on the basis of an EOQ. The demand is 10,000 units a year, unit cost is $10, ordering cost is $30, and the cost of carrying inventory is 20%. The supplier offers a discount of 3% on orders of 1000 units or more. What will be the saving (loss) of accepting the discount?

> *Answer.* Savings = $2825.45

10.7 Refer to problem 10.3. The supplier offers a 3% discount on orders of 5000 units. Calculate the purchase cost, the cost of ordering, the cost of carrying, and the total cost if orders of 5000 are placed. Compare the results and calculate the savings if the discount is taken.

10.8 Calculate the new lot size for the following if $K = 5$.

Item	Annual Demand	$\sqrt{A_D}$	New Lot Size
1	1600		
2	400		
3	144		

Answer. Item 1   200
Item 2   100
Item 3   60

10.9 Calculate K for the following data:

Item	Annual Demand	Orders per Year	$\sqrt{A_D}$
11	$10,000	5	
2	3600	5	
3	1600	5	
Totals			

Answer. $K = 13.33$

10.10 A company manufactures three sizes of lightning rods. Ordering costs and carrying costs are not known, but it is known that they are the same for each size. Each size is produced six times per year. If the demand for each size is as follows, calculate order quantities to minimize inventories and maintain the same total number of runs. Calculate the old and new average inventories. Is there any change in the number of orders per year?

Item	Annual Usage	Present Orders per Year	Present Lot Size	$\sqrt{A_D}$	New Lot Size $= K\sqrt{A_D}$	New Orders per Year $N = A_D/Q$
1	$22,500	6				
2	$5625	6				
3	$1600	6				
Totals						
Average Inventory						

Answer.   Average inventory with present lot sizes = $2477.08

Average inventory with new lot sizes = $1950.40

10.11 A company manufactures five sizes of screwdrivers. Ordering costs and carrying costs are not known, but it is known that they are the same for each size. At present, each size is produced four times per year. If the demand for each size is as follows, calculate order quantities to minimize inventories and maintain the same total number of runs. Calculate the old and new average inventories. Is there any change in the number of orders per year?

Item	Annual Usage	Present Orders per Year	Present Lot Size	$\sqrt{A_D}$	New Lot Size $= K\sqrt{A_D}$	New Orders per Year $N = A_D/Q$
1	$10,000	4				
2	$6400	4				
3	$3600	4				
4	$1600	4				
5	$400	4				
Totals						
Average Inventory						

10.12 The EOQ for an item is 950 units, and the annual usage is 13,000 units. What is the periodic order quantity?

Answer. 4 weeks

10.13 Given the following net requirements, calculate the planned order receipts based on the period-order quantity. The EOQ is 300 units, and the annual demand is 4200 units.

Week	1	2	3	4	5	6	7	8	
Net Requirements	100	75	90	90	85	70	80	40	630
Planned Order Receipts									

Answer. POQ is 4 weeks

Planned order period 1 = 355 units

Planned order period 5 = 275 units

10.14 Given the following MRP record and an EOQ of 200 units, calculate the planned order receipts using the economic-order quantity. Next, calculate the period-order quantities and the planned order receipts. In both cases, calculate the ending inventory and the total inventory carried over the ten weeks.

Week	1	2	3	4	5	6	7	8	9	10	Total
Net Requirements	75	100	60	0	100	80	70	60	0	60	
Planned Order Receipt											
Ending Inventory											

Answer. EOQ total ending inventory = 1150 units

POQ total ending inventory = 560 units

11

Independent Demand Ordering Systems

INTRODUCTION

The concept of an economic-order quantity, covered in the last chapter, addresses the question of how much to order at one time. Another important question is when to place a replacement order. If stock is not reordered soon enough, there will be a stockout and a potential loss in customer service. However, stock ordered earlier than needed will create extra inventory. The problem then is how to balance the costs of carrying extra inventory against the costs of a stockout.

No matter what the items are, some rules for reordering are needed and can be as simple as order when needed, order every month, or order when stock falls to a predetermined level. We all use rules of some sort in our own lives, and they vary depending on the significance of the item. A homemaker uses some intuitive rules to make up the weekly shopping list. Order enough meat for a week, order salt when the box is empty, order vanilla extract if it will be needed over the next week, and so on.

In industry there are many inventories that involve a large investment and where stockout costs are high. Controlling these inventories requires effective reorder systems. Three basic systems are used to determine when to order:

- Order point system.
- Periodic review system.
- Material requirements planning.

The first two are for independent demand items; the last is for dependent demand items.

ORDER POINT SYSTEM

When the quantity of an item on hand in inventory falls to a predetermined level, called an **order point,** an order is placed. The quantity ordered is usually precalculated and based on economic-order-quantity concepts.

Using this system, an order must be placed when there is enough stock on hand to satisfy demand from the time the order is placed until the new stock arrives (called the **lead time).** Suppose that for a particular item the average demand is 100 units a week and the lead time is four weeks. If an order is placed when there are 400 units on hand, on the average there will be enough stock on hand to last until the new stock arrives. However, demand during any one lead-time period probably varies from the average—sometimes more and sometimes less than the 400. Statistically, half the time the demand is greater than average, and there is a stockout; half the time the demand is less than average, and there is extra stock. If it is necessary to provide some protection against a stockout, safety stock can be added. The item is ordered when the quantity on hand falls to a level equal to the demand during the lead time plus the safety stock:

$$OP = DDLT + SS$$

where

OP = order point

DDLT = demand during the lead time

SS = safety stock

It is important to note that it is the *demand during the lead time that is important.* The only time a stockout is possible is during the lead time. If demand during the lead time is greater than expected, there will be a stockout unless sufficient safety stock is carried.

EXAMPLE PROBLEM

Demand is 200 units a week, the lead time is three weeks, and safety stock is 300 units. Calculate the order point.

Answer

OP = DDLT + SS

$\quad = 200 \times 3 + 300$

$\quad = 900$ units

Figure 11.1   Quantity on hand versus time: independent demand item.

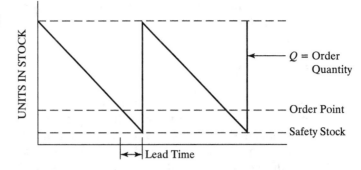

Figure 11.1 shows the relationship between safety stock, lead time, order quantity, and order point. With the order point system:

- Order quantities are usually fixed.
- The order point is determined by the average demand during the lead time. If the average demand or the lead time changes and there is no corresponding change in the order point, effectively there has been a change in safety stock.
- The intervals between replenishment are not constant but vary depending on the actual demand during the reorder cycle.

- Average inventory $= \dfrac{\text{order quantity}}{2} + \text{safety stock} = \dfrac{Q}{2} + \text{SS}$

▼

EXAMPLE PROBLEM

Order quantity is 1000 units and safety stock is 300 units. What is the average annual inventory?

Answer

$$\text{Average inventory} = \frac{Q}{2} + \text{SS}$$

$$= \frac{1000}{2} + 300$$

$$= 800 \text{ units}$$

Determining the order point depends on the demand during the lead time and the safety stock required.

Methods of estimating the demand during the lead time were discussed in Chapter 8. We now look at the factors to consider when determining safety stock.

DETERMINING SAFETY STOCK

Safety stock is intended to protect against uncertainty in supply and demand. Uncertainty may occur in two ways: quantity uncertainty and timing uncertainty. Quantity uncertainty occurs when the amount of supply or demand varies; for example, if the demand is greater or less than expected in a given period. Timing uncertainty occurs when the time of receipt of supply or demand differs from that expected. A customer or a supplier may change a delivery date, for instance.

There are two ways to protect against uncertainty: carry extra stock, called safety stock, or order early, called safety lead time. **Safety stock** is a calculated extra amount of stock carried and is generally used to protect against quantity uncertainty. **Safety lead time** is used to protect against timing uncertainty by planning order releases and order receipts earlier than required. Safety stock and safety lead time both result in extra inventory, but the methods of calculation are different.

Safety stock is the most common way of buffering against uncertainty and is the one described in this text. The safety stock required depends on the following:

- Variability of demand during the lead time.
- Frequency of reorder.
- Service level desired.
- Length of the lead time. The longer the lead time, the more safety stock has to be carried to provide a specified service level. This is one reason it is important to reduce lead times as much as possible.

Variation in Demand During Lead Time

Chapter 8 discussed forecast error and said that actual demand varies from forecast for two reasons: bias error in forecasting the average demand, and random variations in demand about the average. It is the latter with which we are concerned in determining safety stock.

Suppose two items, A and B, have a ten-week sales history, as shown in Figure 11.2. Average demand over the lead time of one week is 1000 per week for both items. However, the weekly demand for A has a range from 700 to 1400 units a week and that for B is from 200 to 1600 units per week. The demand for B is more erratic than that for A. If the order point is 1200 units for both items, there will be one stockout for A and four for B. If the same service level is to be provided (the same chance of stockout for all items), some method of estimating the randomness of item demand is needed.

Variation in Demand About the Average

Suppose over the past 100 weeks a history of weekly demand for a particular item shows an average demand of 1000 units. As expected, most of the demands are

Figure 11.2 Actual demand for two items.

Week	Item A	Item B
1	1200	400
2	1000	600
3	800	1600
4	900	1300
5	1400	200
6	1100	1100
7	1100	1500
8	700	800
9	1000	1400
10	800	1100
Total	10,000	10,000
Average	1000	1000

around 1000; a smaller number would be farther away from the average and still fewer would be farthest away. If the weekly demands are classified into groups or ranges about the average, a picture of the distribution of demand about the average appears. Suppose the demand is distributed as follows:

Weekly Demand	Number of Weeks
725–774	2
775–824	3
825–874	7
875–924	12
925–974	17
975–1024	20
1025–1074	17
1075–1124	12
1125–1174	7
1175–1224	3
1225–1274	2

These data are plotted to give the results shown in Figure 11.3. This is a **histogram.**

Normal distribution. Everything in life varies—even identical twins in some respects. The pattern of demand distribution about the average will differ for different products and markets. Some method is needed to describe the distribution—its shape, center, and spread.

The shape of the histogram in Figure 11.3 indicates that although there is variation in the distribution, it follows a definite pattern, as shown by the smooth curve. Such a natural pattern shows predictability. As long as the demand conditions remain the same, we can expect the pattern to remain very much the same. If the demand is erratic, so is the demand pattern, making it difficult to predict with any accuracy. Fortunately, most demand patterns are stable and predictable.

The most common predictable pattern is similar to the one outlined by the histogram in Figure 11.3 and is called a **normal curve,** or **bell curve,** because its shape resembles a bell. The shape of a perfectly normal distribution is shown in Figure 11.4.

The normal distribution has most of the values clustered near a central point with progressively fewer results occurring away from the center. It is symmetrical about this central point in that it spreads out evenly on both sides.

Figure 11.3 Histogram of actual demand.

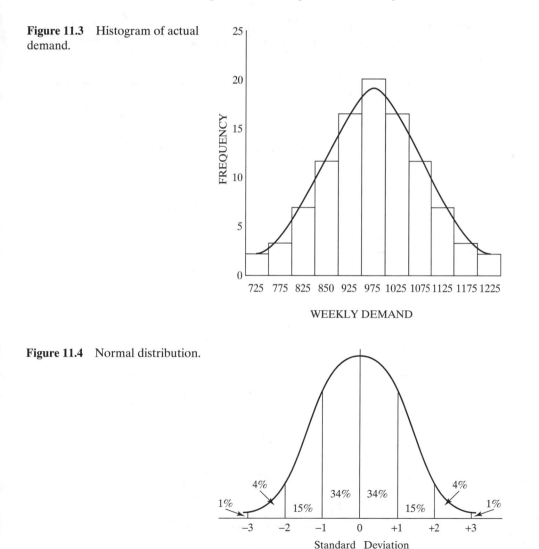

Figure 11.4 Normal distribution.

The normal curve is described by two characteristics. One relates to its central tendency, or average, and the other to the variation, or dispersion, of the actual values about the average.

Average or mean.   This value is at the high point of the curve. It is the central tendency of the distribution. The symbol for the mean is $\bar{x}$ (pronounced "x bar"). It is calculated by adding the data and dividing by the total number of data. In mathematical terms, it can be written as:

$$\bar{x} = \frac{\Sigma x}{n}$$

Where x stands for the individual data (in this case, the individual demands), Σ (capital Greek letter sigma) is the summation sign, and n is the number of data (demands).

▼

EXAMPLE PROBLEM

Given the following actual demands for a ten-week period, Calculate the average $\bar{x}$ of the distribution.

Period	Actual Demand
1	1200
2	1000
3	800
4	900
5	1400
6	1100
7	1100
8	700
9	1000
10	800
Total	10,000

Answer

$$\bar{x} = \frac{\Sigma x}{n} = \frac{10,000}{10} = 1000 \text{ units}$$

Dispersion.   The variation, or dispersion, of actual demands about the average refers to how closely the individual values cluster around the mean or average. It can be measured in several ways:

- As a range of the maximum minus the minimum value.
- As the mean absolute deviation (MAD), which is a measure of the average forecast error. (Calculation of MAD was discussed in Chapter 8.)
- As a standard deviation.

Standard Deviation (Sigma)

This is a statistical value that measures how closely the individual values cluster about the average. It is represented by the Greek letter sigma (σ). The standard deviation is calculated as follows:

1. Calculate the deviation for each period by subtracting the actual demand from the forecast demand.
2. Square each deviation.
3. Add the squares of the deviations.
4. Divide the value in step 3 by the number of periods to determine the average of the squared deviations.
5. Calculate the square root of the value calculated in step 4. This is the standard deviation.

It is important to note that the deviations in demand are for the same time intervals as the lead time. If the lead time is one week, then the variation in demand over a one-week period is needed to determine the safety stock.

EXAMPLE PROBLEM

Given the data from the previous example problem, calculate the standard deviation (sigma).

Answer

Period	Forecast demand	Actual demand	Deviation	Deviation squared
1	1000	1200	200	40000
2	1000	1000	0	0
3	1000	800	−200	40000
4	1000	900	−100	10000
5	1000	1400	400	140000
6	1000	1100	100	10000
7	1000	1100	100	10000
8	1000	700	−300	90000
9	1000	1000	0	0
10	1000	800	−200	40000
Total	10000	10000	0	380000

Average of the squares of the deviation = 380,000 ÷ 10 = 38,000

Sigma = $\sqrt{38,000}$ = 194.9 = 195 units

From statistics, we can determine that:

The actual demand will be within ± 1 sigma of the forecast average approximately 68% of the time.

The actual demand will be within ± 2 sigma of the forecast average approximately 98% of the time.

The actual demand will be within ± 3 sigma of the forecast average approximately 99.88% of the time.

Determining the Safety Stock and Order Point

Now that we have calculated the standard deviation, we must decide how much safety stock is needed.

One property of the normal curve is that it is symmetrical about the average. This means that half the time the actual demand is less than the average and half the time it is greater. Safety stocks are needed to cover only those periods in which the demand during the lead time is greater than the average. Thus, a service level of 50% can be attained with no safety stock. If a higher service level is needed, safety stock must be provided to protect against those times when the actual demand is greater than the average.

As stated earlier, we know from statistics that the error is within ± 1 sigma of the forecast about 68% of the time (34% of the time less and 34% of the time greater than the forecast).

Suppose the standard deviation of demand during the lead time is 100 units and this amount is carried as safety stock. This much safety stock provides protection against stockout for the 34% of the time that actual demand is greater than expected. In total, there is enough safety stock to provide protection for the 84% of the time (50% + 34% = 84%) that a stockout is possible.

The service level is a statement of the percentage of time there is no stockout. But what exactly is meant by supplying the customer 84% of the time? It means being able to supply when a stockout is possible, and a stockout is possible only at the time an order is to be placed. If we order 100 times a year, there are 100 chances of a stockout. With safety stock equivalent to one mean absolute deviation, on the average we would expect no stockouts about 84 of the 100 times.

▼ ───

EXAMPLE PROBLEM

Using the figures in the last example problem in which the sigma was calculated as 195 units,

    a. Calculate the safety stock and the order point for an 84% service level.

    b. If a safety stock equal to two standard deviations is carried, calculate the safety stock and the order point.

Answer

a. Safety stock = 1 sigma

$$= 1 \times 195$$

$$= 195 \text{ units}$$

Order point = DDLT + SS

$$= 1000 + 195 = 1195 \text{ units}$$

where DDLT and SS are as defined previously. With this order point and level of safety stock, on the average there are no stockouts 84% of the time when a stockout is possible.

b. SS $= 2 \times 195$

$= 390$ units

OP $=$ DDLT $+$ SS

$= 1000 + 390$

$= 1390$ units

Safety factor.   The service level is directly related to the number of standard deviations provided as safety stock and is usually called the **safety factor.**

Figure 11.5 shows safety factors for various service levels. Note that the service level is the percentage of order cycles without a stockout.

Figure 11.5   Table of safety factors.

Service Level (%)	Safety Factor
50	0.00
75	0.67
80	0.84
85	1.04
90	1.28
94	1.56
95	1.65
96	1.75
97	1.88
98	2.05
99	2.33
99.86	3.00
99.99	4.00

EXAMPLE PROBLEM

If the standard deviation is 200 units, what safety stock should be carried to provide a service level of 90%? If the expected demand during the lead time is 1500 units, what is the order point?

Answer

From Figure 11.5, the safety factor for a service level of 90% is 1.28. Therefore,

Safety stock = sigma × safety factor

$$= 200 \times 1.28$$

$$= 256 \text{ units}$$

Order point = DDLT + SS

$$= 1500 + 256$$

$$= 1756 \text{ units}$$

DETERMINING SERVICE LEVELS

Theoretically, we want to carry enough safety stock on hand so the cost of carrying the extra inventory plus the cost of stockouts is a minimum. Stockouts cost money for the following reasons:

- Back-order costs.
- Lost sales.
- Lost customers.

The cost of a stockout varies depending on the item, the market served, the customer, and competition. In some markets, customer service is a major competitive tool, and a stockout can be very expensive. In others, it may not be a major consideration. Stockout costs are difficult to establish. Usually the decision about what the service level should be is a senior management decision and is part of the company's corporate and marketing strategy. As such, it is beyond the scope of this text.

The only time it is possible for a stockout to occur is when stock is running low, and this happens every time an order is to be placed. Therefore, the chances of a stockout are directly proportional to the frequency of reorder. The more often stock is reordered, the more often there is a chance of a stockout. Figure 11.6 shows the

Figure 11.6 Exposures to stockout.

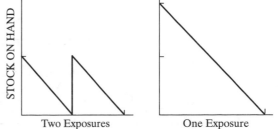

Two Exposures   One Exposure

effect of the order quantity on the number of exposures per year. Note also that when the order quantity is increased, exposure to stockout decreases. The safety stock needed decreases, but because of the larger order quantity, the average inventory increases.

It is the responsibility of management to determine the number of stockouts per year that are tolerable. Then the service level, safety stock, and order point can be calculated.

▼

EXAMPLE PROBLEM

Suppose management stated that it could tolerate only one stockout per year for a specific item.

For this particular item, the annual demand is 52,000 units, it is ordered in quantities of 2600, and the standard deviation of demand during the lead time is 100 units. The lead time is one week. Calculate:

   a. Number of orders per year

   b. Service level

   c. Safety stock

   d. Order point

Answer

   a. Number of orders per year $= \dfrac{\text{annual demand}}{\text{order quantity}}$

   $$= \dfrac{52,000}{2600} = 20 \text{ times per year}$$

   b. Since one stockout per year is tolerable, there must be no stockouts 19 (20 − 1) times per year.

   $$\text{Service level} = \dfrac{20 - 1}{20} = 95\%$$

   c. From Figure 11.5

   $$\text{Safety factor} = 1.65$$
   $$\text{Safety stock} = \text{safety factor} \times \text{sigma}$$
   $$= 1.65 \times 100 = 165 \text{ units}$$

   d. Demand during lead time $= (1 \text{ week}) \dfrac{(52,000)}{52} = 1000 \text{ units}$

   $$\text{Order point} = \text{demand during lead time} + \text{SS}$$
   $$= 1000 + 165 = 1165 \text{ units}$$

DIFFERENT FORECAST AND LEAD-TIME INTERVALS

Usually, there are many items in an inventory, each with different lead times. Records of actual demand and forecasts are normally made on a weekly or monthly basis for all items regardless of what the individual lead times are. It is almost impossible to measure the variation in demand about the average for each of the lead times. Some method of adjusting standard deviation for the different time intervals is needed.

If the lead time is zero, the standard deviation of demand is zero. As the lead time increases, the standard deviation increases. However, it will not increase in direct proportion to the increase in time. For example, if the standard deviation is 100 for a lead time of one week, then for a lead time of four weeks it will not be 400, since it is very unlikely that the deviation would be high for four weeks in a row. As the time interval increases, there is a smoothing effect, and the longer the time interval, the more smoothing takes place.

The following adjustment can be made to the standard deviation or the safety stock to compensate for differences between lead time interval (LTI) and forecast interval (FI). While not exact, the formula gives a good approximation:

$$\text{sigma for LTI} = (\text{sigma for FI}) \sqrt{\frac{\text{LTI}}{\text{FI}}}$$

EXAMPLE PROBLEM

The forecast interval is four weeks, the lead time interval is two weeks, and sigma for the forecast interval is 150 units. Calculate the standard deviation for the lead time interval.

Answer

$$\text{sigma for LTI} = 150 \sqrt{\tfrac{2}{4}} = 150 \times 0.707 = 106 \text{ units}$$

The preceding relationship is also useful where there is a change in the LTI. Now it is probably more convenient to work directly with the safety stock rather than the mean absolute deviation. The relationship is as follows:

$$\text{New safety factor} = \text{old safety factor} \sqrt{\frac{\text{new interval}}{\text{old interval}}}$$

▼

EXAMPLE PROBLEM

The safety stock for an item is 150 units, and the lead time is two weeks. If the lead time changes to three weeks, calculate the new safety stock.

Answer

$$SS \text{ (new)} = 150 \sqrt{\tfrac{3}{2}}$$
$$= 150 \times 1.22 \quad = \quad 183 \text{ units}$$

DETERMINING WHEN THE ORDER POINT IS REACHED

There must be some method to show when the quantity of an item on hand has reached the order point. In practice, there are many systems, but they all are inclined to be variations or extensions of two basic systems: the two-bin system and the perpetual inventory system.

Two-Bin System

A quantity of an item equal to the order point quantity is set aside (frequently in a separate or second bin) and not touched until all the main stock is used up. When this stock needs to be used, the production control or purchasing department is notified and a replenishment order is placed.

There are variations on this system, such as the red-tag system, where a tag is placed in the stock at a point equal to the order point. Book stores frequently use this system. A tag or card is placed in a book that is in a stack in a position equivalent to the order point. When a customer takes that book to the checkout, the store is effectively notified that it is time to reorder that title.

The two-bin system is a simple way of keeping control of C items. Because they are of low value, it is best to spend the minimum amount of time and money controlling them. However, they do need to be managed, and someone should be assigned to ensure that when the reserve stock is reached an order must be placed. When it is out of stock, a C item becomes an A item.

Perpetual Inventory Record System

A perpetual inventory record is a continual account of inventory transactions as they occur. At any instant, it holds an up-to-date record of transactions. At a minimum, it contains the balance on hand, but it may also contain the quantity on order but not received, the quantity allocated but not issued, and the available balance. The accuracy of the record depends upon the speed with which transactions are recorded

426254 SCREW				ORDER QUANTITY 500		ORDER POINT 100
DATE	ORDERED	RECEIVED	ISSUED	ON HAND	ALLOCATED	AVAILABLE
01				500		500
02				500	400	100
03	500			500		
04			400	100	0	100
05		500		600	0	600

Figure 11.7 Perpetual inventory record.

and the accuracy of the input. Because manual systems rely on the input of humans, they are more likely to have slow response and inaccuracies. Computer-based systems have a higher transaction speed and reduce the possibility of human error.

An inventory record contains variable and permanent information. Figure 11.7 shows an example of a perpetual inventory record.

"Permanent" information is shown at the top of Figure 11.7. Although not absolutely permanent, this information does not change frequently. Any alteration is usually the result of an engineering change, manufacturing process change, or inventory management change. It includes data such as the following:

- Part number, name, and description
- Storage location
- Order point
- Order quantity
- Lead time
- Safety stock
- Suppliers

Variable information is information that changes with each transaction and includes the following:

- Quantities ordered: dates, order numbers, and quantities
- Quantities received: dates, order numbers, and quantities
- Quantities issued: dates, order numbers, and quantities
- Balance on hand
- Allocated: dates, order numbers, and quantities
- Available balance

The information depends on the needs of the company and the particular circumstances.

Figure 11.8   Periodic review system: units in stock versus time.

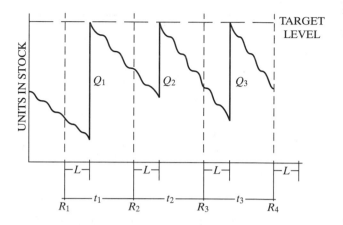

PERIODIC REVIEW SYSTEM

In the order point system, an order is placed when the quantity on hand falls to a predetermined level called the **order point.** The quantity ordered is usually predetermined on some basis such as the economic-order quantity. The interval between orders varies depending on the demand during any particular cycle.

Using the periodic review system, the quantity on hand of a particular item is determined at specified, fixed-time intervals, and an order is placed. Figure 11.8 illustrates this system.

Figure 11.8 shows that the review intervals (t_1, t_2, and t_3) are equal and that Q_1, Q_2, and Q_3 are not necessarily the same. Thus the review period is fixed, and the order quantity is allowed to vary. The quantity on hand plus the quantity ordered must be sufficient to last until the *next shipment* is received. That is, the quantity on hand plus the quantity ordered must equal the sum of the demand during the lead time plus the demand during the review period plus the safety stock.

Target Level or Maximum-Level Inventory

The quantity equal to the demand during the lead time plus the demand during the review period plus safety stock is called the **target level** or **maximum-level inventory:**

$$T = D(R + L) + \text{SS}$$

where

       T = target (maximum) inventory level

       D = demand per unit of time

       L = lead-time duration

       R = review period duration

       SS = safety stock

The order quantity is equal to the maximum-inventory level minus the quantity on hand at the review period:

$$Q = T - I$$

where

Q = order quantity

I = inventory on hand

The periodic review system is useful for the following:

- Where there are many small issues from inventory, and posting transactions to inventory records are very expensive. Supermarkets and retailers are in this category.
- Where ordering costs are small. This occurs when many different items are ordered from one source. A regional distribution center may order most or all of its stock from a central warehouse.
- Where many items are ordered together to make up a production run or fill a truckload. A good example of this is a regional distribution center that orders a truckload once a week from a central warehouse.

EXAMPLE PROBLEM

A hardware company stocks nuts and bolts and orders them from a local supplier once every two weeks (ten working days). Lead time is two days. The company has determined that the average demand for ½-inch bolts is 150 per week (five working days), and it wants to keep a safety stock of three days' supply on hand. An order is to be placed this week, and stock on hand is 130 bolts.

    a. What is the target level?

    b. How many ½-inch bolts should be ordered this time?

Answer

Let D = demand per unit of time = $150 \div 5 = 30$ per working day

          L = lead time duration = 2 days

          R = review period duration = 10 days

          SS = safety stock = 3 days' supply = 90 units

          I = inventory on hand = 130 units

Then

    Target level $T = D(R + L) + \text{SS}$

               = 30(10 + 2) + 90

               = 450 units

    Order quantity $Q = T - I$

               = 450 − 130 = 320 units

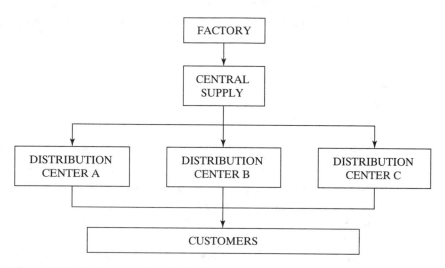

Figure 11.9 Schematic of a distribution system.

DISTRIBUTION INVENTORY

Distribution inventory includes all the finished goods held anywhere in the distribution system. The purpose of holding inventory in distribution centers is to improve customer service by locating stock near the customer and to reduce transportation costs by allowing the manufacturer to ship full loads rather than partial loads over long distances. This will be studied in Chapter 13.

The objectives of distribution inventory management are to provide the required level of customer service, to minimize the costs of transportation and handling, and to be able to interact with the factory to minimize scheduling problems.

Distribution systems vary considerably, but in general they have a central supply facility that is supported by a factory, a number of distribution centers, and, finally, customers. Figure 11.9 is a schematic of such a system. The customers may be the final consumer or some intermediary in the distribution chain.

Unless a firm delivers directly from factory to customer, demand on the factory is created by central supply. In turn, demand on central supply is created by the distribution centers. This can have severe repercussions on the pattern of demand on central supply and the factory. Although the demand from customers may be relatively uniform, the demand on central supply is not, because it is dependent on when the distribution centers place replenishment orders. In turn, the demand on the factory depends on when central supply places orders. Figure 11.10 shows the process schematically.

The distribution system is the factory's customer, and the way the distribution system interfaces with the factory has a significant effect on the efficiency of factory operations.

Figure 11.10   Distribution inventory.

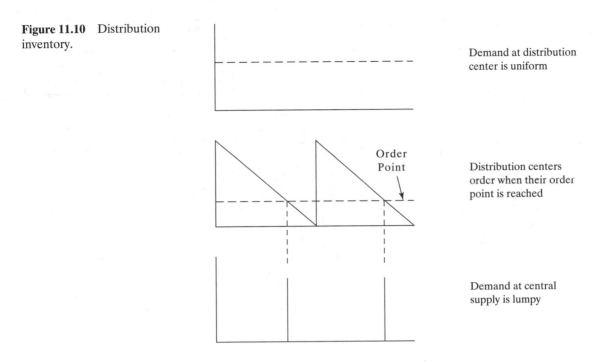

Demand at distribution center is uniform

Order Point

Distribution centers order when their order point is reached

Demand at central supply is lumpy

Distribution inventory management systems can be classified into decentralized, centralized, and distribution requirements planning.

Decentralized System

In a decentralized system, each distribution center first determines what it needs and when, and then places orders on central supply. Each center orders on its own without regard for the needs of other centers, available inventory at central supply, or the production schedule of the factory.

The advantage of the decentralized system is that each center can operate on its own and thus reduce communication and coordination expense. The disadvantage is the lack of coordination and the effect this may have on inventories, customer service, and factory schedules. Because of these deficiencies, many distribution systems have moved toward more central control.

A number of ordering systems can be used, including the order point and periodic review systems. The decentralized system is sometimes called the pull system because orders are placed on central supply and "pulled" through the system.

Centralized System

In a centralized system, all forecasting and order decisions are made centrally. Stock is "pushed" out into the system from central supply. Distribution centers have no say about what they receive.

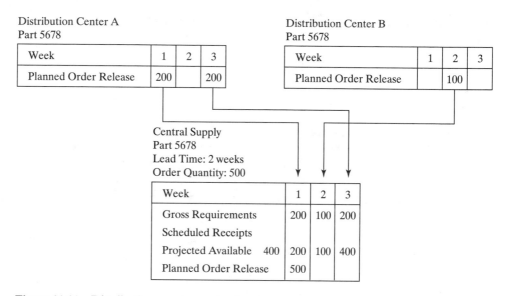

Figure 11.11   Distribution requirements planning.

Different ordering systems can be used, but generally an attempt is made to replace the stock that has been sold and to provide for special situations such as seasonality or sales promotions. These systems attempt to balance the available inventory with the needs of each distribution center.

The advantage of these systems is the coordination between factory, central supply, and distribution center needs. The disadvantage is the inability to react to local demand, thus lowering the level of service.

Distribution Requirements Planning

Distribution requirements planning is a system that forecasts when the various demands will be made by the system on central supply. This gives central supply and the factory the opportunity to plan for the goods that will actually be needed and when. It is able both to respond to customer demand and coordinate planning and control.

The system translates the logic of material requirements planning to the distribution system. Planned order releases from the various distribution centers become the input to the material plan of central supply. The planned order releases from central supply become the forecast of demand for the factory master production schedule. Figure 11.11 shows the system schematically. The records shown are all for the same part number.

EXAMPLE PROBLEM

A company making lawnmowers has a central supply attached to their factory and two distribution centers. Distribution center A forecasts demand at 25, 30, 55, 50, and 30 units over the next five weeks and has 100 lawnmowers in transit that are due in week 2. The transit time is two weeks, the order quantity is 100 units, and there are 50 units on hand. Distribution center B forecasts demand at 95, 85, 100, 70, and 50 over the next five weeks. Transit time is one week, the order quantity is 200 units, and there are 100 units on hand. Calculate the gross requirements, projected available, and planned order releases for the two distribution centers, and the gross requirements, projected available, and planned order releases for the central warehouse.

Answer

Distribution Center A
Transit Time: 2 weeks
Order Quantity: 100 units

Week	1	2	3	4	5
Forecast	25	30	55	50	30
In Transit		100			
Projected Available 50	25	95	40	90	60
Planned Order Release		100			

Distribution Center B
Transit Time: 1 week
Order Quantity: 200 units

Week	1	2	3	4	5
Forecast	95	85	100	70	50
In Transit					
Projected Available 100	5	120	20	150	100
Planned Order Release	200		200		

Central Supply
Lead Time: 2 weeks
Order Quantity: 500 units

Week	1	2	3	4	5
Forecast	200	100	200		
Scheduled Receipts					
Projected Available 400	200	100	400		
Planned Order Release	500				

QUESTIONS

1. What are independent demand items? What two basic ordering systems are used for these items? What are dependent demand items? What system should be used to order these items?

2. **a.** Using the order point system, when must an order be placed?

   b. Why is safety stock carried?

   c. What is the formula for the order point?

   d. On what two things does the order point depend?

   e. Why is the demand during the lead time important?

3. What are four characteristics of the order point system?

4. What are the five factors that can influence the amount of safety stock that should be carried? How does the length of the lead time affect the safety stock carried?

5. What is a normal distribution? What two characteristics define it? Why is it important in determining safety stock?

6. What is the standard deviation of demand during the lead time? If the standard deviation for the lead-time interval is 100 units, what percentage of the time would the actual demand be equal to ±100 units? To ±200 units? To ±300 units?

7. What is service level?

8. What are the three categories of stockout costs? What do these costs depend upon in any company?

9. Why does the service level depend upon the number of orders per year?

10. If the lead time increases from one week to four weeks, will the standard deviation of demand during the lead time increase four times? If not, why not?

11. Describe the two-bin system.

12. What kinds of information are shown on a perpetual inventory record?

13. What are the differences between the order point system and the periodic review system regarding when orders are placed and the quantity ordered at any one time?

14. Define the target level used in the periodic review system.

15. What are the objectives of distribution inventory management?

16. If a factory does not supply the customer directly, from where does demand on the factory come? Is it independent or dependent demand?

17. Describe and compare the pull and push systems of inventory management.

18. Describe distribution requirements planning.

PROBLEMS

11.1 For a particular SKU, the lead time is four weeks, the average demand is 150 units per week, and safety stock is 100 units. What is the average inventory if 1500 units are ordered at one time? What is the order point?

Answer. Average inventory = 850 units

Order point = 700 units

11.2 For a particular SKU, the lead time is six weeks, the average demand is 100 units a week, and safety stock is 200 units. What is the average inventory if ten weeks' supply is ordered at one time? What is the order point?

11.3   Given the following data, calculate the average $\bar{x}$ of the distribution and the standard deviation (sigma).

Period	Actual Demand	Deviation	Deviation Squared
1	500		
2	550		
3	475		
4	450		
5	600		
6	575		
7	375		
8	475		
9	525		
10	475		
Total			

Answer.   Average demand $\bar{x} = 500$ units
Sigma $= 62$ units

11.4 Given the following data, calculate the average demand and the standard deviation of demand about the average.

Period	Actual Demand	Deviation	Deviation Squared
1	1700		
2	2100		
3	1900		
4	2200		
5	2000		
6	1800		
7	2100		
8	2300		
9	2100		
10	1800		
Total			

11.5 If the sigma is 150 units, and the demand during the lead time is 200 units, calculate the safety stock and order point for:

a. A 50% service level

b. An 80% service level

Use the table in Figure 11.5 to help you calculate your answer.

Answer. **a.** Safety stock = zero

Order point = 200 units

b. Safety stock = 126 units

Order point = 326 units

11.6 The standard deviation of demand during the lead time is 100 units.

    a. Calculate the safety stock required for the following service levels: 75%, 80%, 85%, 90%, 95%, 99.99%.

    b. Calculate the *change* in safety stock required to increase the service levels from 75% to 80%, 80% to 85%, 85% to 90%, 90% to 95%, and 95% to 99.99%. What conclusion do you reach?

11.7 For an SKU, the standard deviation of demand during the lead time is 150 units, the annual demand is 10,000 units, and the order quantity is 750 units. Management says it will tolerate only two stockouts per year. What safety stock should be carried? What is the average inventory? If the lead time is two weeks, what is the order point?

    Answer. Safety stock = 156 units

                Average inventory = 531 units

                Order point = 541 units

11.8 A company stocks an SKU with a weekly demand of 250 units and a lead time of three weeks. If sigma for the lead time is 175 and the order quantity is 800 units, what is the average inventory and the order point? Management will tolerate one stockout per year.

11.9 A company stocks an SKU with a weekly demand of 500 units and a lead time of four weeks. If sigma for the lead time is 100 and the order quantity is 2,500 units, what is the average inventory and the order point? Management will tolerate two stockouts per year.

11.10 If the standard deviation is calculated from weekly demand data at 100 units, what is the equivalent sigma for a three-week lead time?

    Answer. 173 units

11.11 If the safety stock for an item is 150 units and the lead time is three weeks, what should the safety stock become if the lead time is extended to five weeks?

    Answer. 194 units

11.12 If the weekly standard deviation is 150 units, what is it if the lead time is four weeks?

11.13 The safety stock on an SKU is set at 200 units. The supplier says it can reduce the lead time from eight weeks to six. What should be the new safety stock?

    Answer. 173

11.14 The safety stock on an SKU is set at 200 units. The supplier says it has to increase the lead time from six weeks to eight. What should be the new safety stock?

11.15 Management has stated that it will tolerate one stockout per year. The forecast of annual demand for a particular SKU is 100,000 units, and it is ordered in quantities of 10,000 units. The lead time is two weeks. Sales history for the past ten weeks follows. Calculate:

a. Sigma for the history time interval

b. Sigma for the lead time interval

c. The service level

d. The safety stock required for this service level

e. The order point

Week	Actual Demand	Deviation	Deviation Squared
1	2100		
2	1700		
3	2600		
4	1400		
5	1800		
6	2300		
7	2200		
8	1600		
9	2100		
10	2200		
Total			

11.16 If in problem 11.15, management said that it is considering increasing the service level to one stockout every two years, what would the new safety stock be? If the cost of carrying inventory on this item is $10 per unit per year, what is the cost of increasing the inventory from one stockout per year to one every two years?

11.17 The annual demand for an item is 10,000 units, the order quantity is 250, and the service level is 90%. Calculate the probable number of stockouts per year.

   Answer.   Four stockouts per year

11.18 A company that manufactures stoves has one plant and two distribution centers (DCs). Given the following information for the two distribution centers, calculate the gross requirements, projected available, and planned order releases for the two DCs and the gross requirements, projected available, and planned order releases for the central warehouse.

Distribution Center A
Transit Time: 2 weeks
Order Quantity: 100 units

Week	1	2	3	4	5
Forecast	50	50	35	50	110
In Transit		100			
Projected Available 75					
Planned Order Release					

Distribution Center B
Transit Time: 1 week
Order Quantity: 200 units

Week	1	2	3	4	5
Forecast	95	100	115	80	70
In Transit	200				
Projected Available 50					
Planned Order Release					

Central Supply
Lead Time: 2 weeks
Order Quantity: 500 units

Week	1	2	3	4	5
Forecast					
Scheduled Receipts					
Projected Available 400					
Planned Order Release					

Answer.   Planned order release from central supply: 500 in week 2.

11.19 A company that manufactures snow shovels has one plant and two distribution centers (DCs). Given the following information for the two distribution centers, calculate the gross requirements, projected available, and planned order releases for the two DCs and the gross requirements, projected available, and planned order releases for the central warehouse.

Distribution Center A
Transit Time: 2 weeks
Order Quantity: 500 units

Week	1	2	3	4	5
Forecast	300	200	150	175	200
In Transit	500				
Projected Available 200					
Planned Order Release					

Distribution Center B
Transit Time: 2 weeks
Order Quantity: 200 units

Week	1	2	3	4	5
Forecast	50	75	100	125	150
In Transit					
Projected Available 150					
Planned Order Release					

Central Supply
Lead Time: 1 week
Order Quantity: 600 units

Week	1	2	3	4	5
Forecast					
Scheduled Receipts					
Projected Available 400					
Planned Order Release					

11.20 A firm orders a number of items from a regional warehouse every two weeks. Delivery takes one week. Average demand is 100 units per week, and safety stock is held at two weeks' supply.

   a. Calculate the target level.

   b. If 350 units are on hand, how many should be ordered?

   Answer.   Target level = 500 units

   Order quantity = 150 units

11.21 A regional warehouse orders items once a week from a central warehouse. The truck arrives three days after the order is placed. The warehouse operates five days a week. For a particular brand and size of chicken soup, the demand is fairly steady at 20 cases per day. Safety stock is sct at two days' supply.

   a. What is the target level?

   b. If the quantity on hand is 90 cases, how many should be ordered?

12

Physical Inventory and Warehouse Management

INTRODUCTION

Because inventory is stored in warehouses, the physical management of inventory and warehousing are intimately connected. In some cases, inventory may be stored for an extended time. In other situations, inventory is turned over rapidly, and the warehouse functions as a distribution center.

In a factory, "stores" perform the same functions as warehouses and contain raw materials, work-in-process inventory, finished goods, supplies, and possibly repair parts. Since they perform the same functions, stores and warehouses are treated alike in this chapter.

WAREHOUSING MANAGEMENT

As with other elements in a distribution system, the objective of a warehouse is to minimize cost and maximize customer service. To do this, efficient warehouse operations perform the following:

1. Provide timely customer service.
2. Keep track of items so they can be found readily and correctly.
3. Minimize the total physical effort and thus the cost of moving goods into and out of storage.

4. Provide communication links with customers.

The costs of operating a warehouse can be broken down into capital and operating costs. Capital costs are those of space and materials handling equipment. The space needed depends on the peak quantities that must be stored, the methods of storage, and the need for ancillary space for aisles, docks, offices, and so on.

The major operating cost is labor, and the measure of labor productivity is the number of units (for example, pallets) that an operator can move in a day. This depends on the type of material handling equipment used, the location and accessibility of stock, warehouse layout, stock location system, and the order-picking system used.

Warehouse Activities

Operating a warehouse involves several processing activities, and the efficient operation of the warehouse depends upon how well these are performed. These activities are as follows:

1. *Receive goods.* The warehouse accepts goods from outside transportation or an attached factory and accepts responsibility for them. This means the warehouse must:

   a. Check the goods against an order and the bill of lading.

   b. Check the quantities.

   c. Check for damage and fill out damage reports if necessary.

   d. Inspect goods if required.

2. *Identify the goods.* Items are identified with the appropriate stock-keeping unit (SKU) number (part number) and the quantity received recorded.

3. *Dispatch goods to storage.* Goods are sorted and put away.

4. *Hold goods.* Goods are kept in storage and under proper protection until needed.

5. *Pick goods.* Items required from stock must be selected from storage and brought to a marshalling area.

6. *Marshal the shipment.* Goods making up a single order are brought together and checked for omissions or errors. Order records are updated.

7. *Dispatch the shipment.* Orders are packaged, shipping documents prepared, and goods loaded on the right vehicle.

8. *Operate an information system.* A record must be maintained for each item in stock showing the quantity on hand, quantity received, quantity issued, and location in the warehouse. The system can be very simple, depending on a minimum of written information and human memory, or it may be a sophisticated computer-based system.

In various ways, all these activities take place in any warehouse. The complexity depends on the number of SKUs handled, the quantities of each SKU, and the number of orders received and filled. To maximize productivity and minimize cost, warehouse management must work with the following:

1. *Maximum use of space.* Usually the largest capital cost is for space. This means not only floor space but cubic space as well since goods are stored in the space above the floor as well as on it.

2. *Effective use of labor and equipment.* Materials handling equipment represents the second largest capital cost and labor the largest operating cost. There is a tradeoff between the two in that labor costs can be reduced by using more materials handling equipment. Warehouse management will need to:

   - Select the best mix of labor and equipment to maximize the overall productivity of the operation.
   - Provide ready access to all SKUs. The SKUs should be easy to identify and find. This requires a good stock location system and layout.
   - Move goods efficiently. Most of the activity that goes on in a warehouse is materials handling: the movement of goods into and out of stock locations.

   Several factors influence effective use of warehouses. Some are:

   - Cube utilization and accessibility.
   - Stock location.
   - Order picking and assembly.
   - Packaging.

   With the exception of packaging, these are discussed in the following sections.

Cube Utilization and Accessibility

Goods are stored not just on the floor, but in the cubic space of the warehouse. Although the size of a warehouse can be described as so many square feet, warehouse capacity depends on how high goods can be stored.

Space is also required for aisles, receiving and shipping docks, offices, and order picking and assembly. In calculating the space needed for storage, some design figure for maximum inventory is needed. Suppose that a maximum of 90,000 cartons are to be inventoried and 30 cartons fit on a pallet. Space is needed for 3000 pallets. If pallets are stacked three high, 1000 pallet positions are required. A pallet is a platform usually measuring 48″ × 40″ × 4″.

Pallet positions. Suppose a section of a warehouse is as shown in Figure 12.1. Since the storage area is 48″ deep, the 40″ side is placed along the wall. The pallets cannot be placed tight against one another; a 2″ clearance must be allowed between them so they can be moved. This then leaves room for (120′ × 12″) ÷ 42″ = 34.3, or 34, **pallet positions** along each side of the aisle. Since the pallets are stacked three high, there is room for 34 × 3 × 2 = 204 pallets.

Figure 12.1   Cube utilization.

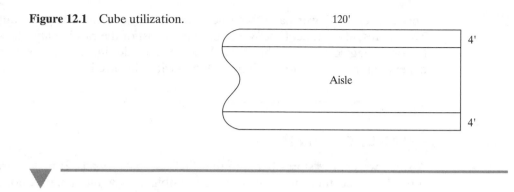

EXAMPLE PROBLEM

A company wants to store an SKU consisting of 13,000 cartons on pallets each containing 30 cartons. How many pallet positions are needed if the pallets are stored three high?

Answer

Number of pallets required = $13,000 \div 30 = 433.33 \rightarrow 434$ pallets

Number of pallet positions = $434 \div 3 = 144.67 \rightarrow 145$ pallet positions

   Notice one pallet position will contain only two pallets.

Accessibility.   Accessibility means being able to get at the goods wanted with a minimum amount of work. For example, if no other goods had to be moved to reach an SKU, the SKU would be 100% accessible. As long as all pallets contain the same SKU, there is no problem with accessibility. The SKU can be reached without moving any other product. When several SKUs are stored in the area, each product should be accessible with a minimum of difficulty.

Cube utilization.   Suppose items are stacked along a wall, as shown in Figure 12.2. There will be excellent accessibility for all items except item 9, but cube utilization is not maximized. There is room for 30 pallets, but only 21 spaces are being used for a cube utilization of 70% ($21 \div 30 \times 100$). Some method must be devised to increase cube utilization and maintain accessibility. One way is to install

Figure 12.2   Cube utilization versus accessibility.

1	1	2	3	4					
1	1	2	3	4					10
1	1	2	3	4	5	6	7	8	9

tiers of racks so lower pallets can be removed without disturbing the upper ones. This represents a tradeoff between the capital cost of the racking and the savings in the operating cost of extra handling. Whether the additional cost is worthwhile will depend on the amount of handling and the savings involved.

▼

EXAMPLE PROBLEM

A small warehouse stores five different SKUs in pallet loads. If pallets are stacked three high and there is to be 100% accessibility, how many pallet positions are needed? What is the cube utilization?

SKU A	4 pallets
SKU B	6 pallets
SKU C	14 pallets
SKU D	8 pallets
SKU E	5 pallets
Total	37 pallets

Answer

SKU	Pallet positions
A: 4 pallets	2
B: 6 pallets	2
C: 14 pallets	5
D: 8 pallets	3
E: 5 pallets	2
Total	14

In 14 pallet positions, there is room to store $14 \times 3 = 42$ pallets.

Number of pallets actually stored = 37

Cube utilization = $37 \div 42 \times 100\% = 88\%$

Stock Location

Stock location, or warehouse layout, is concerned with the location of individual items in the warehouse. There is no single universal stock location system suitable for all occasions, but there are a number of basic systems that can be used. Which system, or mix of systems, is used depends on the type of goods stored, the type of storage facilities needed, the throughput, and the size of orders. Whatever the system, management must maintain enough inventory of safety and working stock to provide the required level of customer service, keep track of items so they can be found easily, and reduce the total effort required to receive goods, store them, and retrieve them for shipment.

The following are some basic systems of locating stock:

- *Group functionally related items together.* Group together items similar in their use (functionally related). For example, put all hardware items in the same area of the warehouse. If functionally related items are ordered together, order picking is easier. Warehouse personnel become familiar with the locations of items.

- *Group fast-moving items together.* If fast-moving items are placed close to the receiving and shipping area, the work of moving them in and out of storage is reduced. Slower moving items can be placed in more remote areas of the warehouse.

- *Group physically similar items together.* Physically similar items often require their own particular storage facilities and handling equipment. Small packaged items may require shelving whereas heavy items, such as tires or drums, require different facilities and handling equipment. Frozen foods need freezer storage space.

- *Locate working stock and reserve stock separately.* Relatively small quantities of working stock—stock from which withdrawals are made—can be located close to the marshalling and shipping area whereas reserve stock used to replenish the working stock can be located more remotely. This allows order picking to occur in a compact area and replenishment of the working stock in bulk by pallet or container load.

There are two basic systems for assigning specific locations to individual stock items: fixed location and floating location. Either system may be used with any of the above location systems.

Fixed location. In this system, an SKU is assigned a permanent location or locations, and no other items are stored there. This system makes it possible to store and retrieve items with a minimum of record keeping. In some small, manual systems, no records are kept at all. It is like always keeping cornflakes on the same shelf in the kitchen cupboard at home. Everything is nice and simple so things are readily found. However, fixed-location systems usually have poor cube utilization. If demand is uniform, presumably the average inventory is half the order quantity, and enough space has to be allocated for a full-order quantity. On the average, only 50% of the cube space is utilized. Fixed-location systems are often used in small warehouses where space is not at a premium, where throughput is small, and where there are a few SKUs.

Floating location. In this system, goods are stored wherever there is appropriate space for them. The same SKU may be stored in several locations at the same time and different locations at different times. The advantage to this system is improved cube utilization. However, it requires accurate and up-to-date information on item location and the availability of empty storage space so items can be put away and retrieved efficiently. Modern warehouses using floating-location systems are usually computer based. The computer assigns free locations to incoming items, remembers what items are on hand and where they are located, and directs the order picker to the right location to find the item. Thus, cube utilization and warehouse efficiency are greatly improved.

Point-of-use storage.   Sometimes, particularly in repetitive manufacturing and in a JIT environment, inventory is stored close to where it will be used. There are several advantages to this technique.

- Materials are readily accessible to users.
- Material handling is reduced or eliminated.
- Central storage costs are reduced.
- Material is accessible at all times.

This method is excellent as long as inventory is kept low and operating personnel can keep control of inventory records. Sometimes C items are issued as "floor stock" where manufacturing is issued a large quantity which is used as needed. Inventory records are adjusted when the stock is issued, not when it is used.

Central storage.   As opposed to point-of-use storage, central storage contains all inventory in one central location. There are several advantages:

- Ease of control.
- Inventory record accuracy is easier to maintain.
- Specialized storage can be used.
- Reduced safety stock, since users do not need to carry their own safety stock.

Order Picking and Assembly

Once an order is received, the items on the order must be retrieved from the warehouse, assembled, and prepared for shipment. All these activities involve labor and the movement of goods. The work should be organized to provide the level of customer service required and at least cost. There are several systems that can be used to organize the work, among which are the following:

1. *Area system.*   The order picker circulates throughout the warehouse selecting the items on the order, much as a shopper would in a supermarket. The items are then taken to the shipping area for shipment. The order is self-marshalling in that when the order picker is finished, the order is complete. This system is generally used in small warehouses where goods are stored in fixed locations.
2. *Zone system.*   The warehouse is broken down into zones, and order pickers work only in their own area. An order is divided up by zone, and each order

picker selects those items in their zone and sends them to the marshalling area where the order is assembled for shipment. Each order is handled separately and leaves the zone before another is handled.

Zones are usually established by grouping related parts together. Parts may be related because of the type of storage needed for them (for example, freezer storage) or because they are often ordered together.

A variation of the zone system is to have the order move to the next zone rather than to the marshalling area. By the time it exits the last zone, it is assembled for shipment.

3. *Multi-order system.* This system is the same as the zone system except that, rather than handling individual orders, a number of orders are gathered together and all the items divided by zone. The pickers then circulate through their area, collecting all the items required for that group of orders. The items are then sent to the marshalling area where they are sorted to individual orders for shipment.

The area system is simple to manage and control, but as the warehouse throughput and size increase, it becomes unwieldy. The zone systems break down the order-filling process into a series of smaller areas that can be better managed individually. The multi-order system is probably most suited to the situation in which there are many items or many small orders with few items.

Working stock and reserve stock. In addition to the above systems, reserve stock and working stock may be separated. This is appropriate when the pick unit for a customer's order may be a box or a case that is stored on pallets. A pallet can be moved into the working area by a lift truck and cartons or boxes picked from it. The working stock is located close to the shipping area so the work in picking is reduced. A separate work force is used to replenish the working stock from the reserve stock.

PHYSICAL CONTROL AND SECURITY

Because inventory consists of tangible things, items have a nasty habit of becoming lost, strayed, or stolen, or of disappearing in the night. It is not that people are dishonest, rather that they are forgetful. What is needed is a system that makes it difficult for people to make mistakes or be dishonest. There are several elements that help.

- *A good part numbering system.* Part numbering was discussed in Chapter 4 on material requirements planning.
- *A simple, well-documented transaction system.* When goods are received, issued, or moved in any way, a transaction occurs. There are four steps in any transaction: identify the item, verify the quantity, record the transaction, and physically execute the transaction.

1. *Identify the item.* Many errors occur because of incorrect identification. When receiving an item, the purchase order, part number, and quantity must be properly identified. When goods are stored, the location must be accurately specified. When issued, the quantity, location, and part number must be recorded.

2. *Verify quantity.* Quantity is verified by a physical count of the item by weighing or by measuring. Sometimes standard-sized containers are useful in counting.

3. *Record the transaction.* Before any transaction is physically carried out, *all* information about the transaction must be recorded.

4. *Physically execute the transaction.* Move the goods in, about, or out of the storage area.

Limited access. Inventory must be kept in a safe, secure place with limited general access. It should be locked except during normal working hours. This is less to prevent theft than to ensure people do not take things without completing the transaction steps. If people can wander into the stores area at any time and take something, the transaction system fails.

A well-trained workforce. Not only should the stores staff be well trained in handling and storing material and in recording transactions, but other personnel who interact with stores must be trained to ensure transactions are recorded properly.

INVENTORY RECORD ACCURACY

The usefulness of inventory record is directly related to its accuracy. Based on the inventory record, a company determines net requirements for an item, releases orders based on material availability, and performs inventory analysis. If the records are not accurate, there will be shortages of material, disrupted schedules, late deliveries, lost sales, low productivity, and excess inventory (of the wrong things).

These three pieces of information must be accurate: part description (part number), quantity, and location. Accurate inventory records enable firms to:

- *Operate an effective materials management system.* If inventory records are inaccurate, gross-to-net calculations will be in error.

- *Maintain satisfactory customer service.* If records show the item is in inventory when it is not, any order promising it will be in error.

- *Operate effectively and efficiently.* Planners can plan, confident that the parts will be available.

- *Analyze inventory.* Any analysis of inventory is only as good as the data it is based on.

Inaccurate inventory records will result in:

- Lost sales.
- Shortages and disrupted schedules.
- Excess inventory (of the wrong things).
- Low productivity.
- Poor delivery performance.
- Excessive expediting, since people will always be reacting to a bad situation rather than planning for the future.

Causes of Inventory Record Errors

Poor inventory record accuracy can be caused by many things, but they all result from poor record-keeping systems and poorly trained personnel. Some examples of causes of inventory record error are:

- Unauthorized withdrawal of material.
- Unsecured stockroom.
- Poorly trained personnel.
- Inaccurate transaction recording. Errors can occur because of inaccurate piece counts, unrecorded transactions, delay in recording transactions, inaccurate material location, and incorrectly identified parts.
- Poor transaction recording systems. Most systems today are computer based and can provide the means to record transactions properly. Errors, when they occur, are usually the fault of human input to the system. The documentation reporting system should be designed to reduce the likelihood of human error.
- Lack of audit capability. Some program of verifying the inventory counts and locations is necessary. The most popular one today is cycle counting, discussed in the next section.

Measuring Inventory Record Accuracy

Inventory accuracy ideally should be 100%. Banks and other financial institutions reach this level. Other companies can move toward this potential.

Figure 12.3 shows ten inventory items, their physical count, and the quantity shown on their record. What is the inventory accuracy? The total of all items is the same, but only two of the ten items are correct. Is the accuracy 100% or 20% or something else?

Tolerance.　　To judge inventory accuracy, a tolerance level for each part must be specified. For some items, this may mean no variance; for others, it may be very difficult or costly to measure and control to 100% accuracy. An example of the latter

Figure 12.3   Inventory record accuracy.

Part Number	Inventory Record	Shelf Count
1	100	105
2	100	100
3	100	98
4	100	97
5	100	102
6	100	103
7	100	99
8	100	100
9	100	97
10	100	99
Total	1000	1000

might be nuts or bolts ordered and used in the thousands. For these reasons, tolerances are set for each item. **Tolerance** is the amount of permissible variation between an inventory record and a physical count.

Tolerances are set on individual items based on value, critical nature of the item, availability, lead time, ability to stop production, safety problems, or the difficulty of getting precise measurement.

Figure 12.4 shows the same data as the previous figure, but includes tolerances. This information tells us exactly what inventory accuracy is.

Part Number	Inventory Record	Shelf Count	Tolerance	Within Tolerance	Outside Tolerance
1	100	105	±5%	X	
2	100	100	±0%	X	
3	100	98	±3%	X	
4	100	97	±2%		X
5	100	102	±2%	X	
6	100	103	±2%		X
7	100	99	±3%	X	
8	100	100	±0%	X	
9	100	97	±5%	X	
10	100	99	±5%	X	
Total	1000	1000			

Figure 12.4 Inventory accuracy with tolerances.

EXAMPLE PROBLEM

Determine which of the following items are within tolerance. Item A has a tolerance of ±5%; item B, ±2%; item C, ±3%; and item D, ±0%.

Part Number	Shelf Count	Inventory Record	Tolerance
A	1500	1550	±5%
B	120	125	±2%
C	225	230	±3%
D	155	155	±0%

Answer

Item A.	With a tolerance of ±5%, variance can be up to ±75 units. Item A is within tolerance.
Item B.	With a tolerance of ±2%, variance can be up to ±2 units. Item B is outside tolerance.
Item C.	With a tolerance of ±3%, variance can be up to ±7 units. Item C is within tolerance.
Item D.	With a tolerance of ±0%, variance can be up to ±0 units. Item D is within tolerance.

Auditing Inventory Records

Errors occur, and they must be detected so inventory accuracy is maintained. There are two basic methods of checking the accuracy of inventory records: periodic (usually annual) counts of all items and cyclic (usually daily) counts of specified items. It is important to audit record accuracy, but it is more important to audit the system to find the causes of record inaccuracy and eliminate them. Cycle counting does this; periodic audits tend not to.

Periodic (annual) inventory.   The primary purpose of an annual physical inventory is to satisfy the financial auditors that the inventory records represent the value of the inventory. To planners, the physical inventory represents an opportunity to correct any inaccuracies in the records. Whereas financial auditors are concerned with the total value of the inventory, planners are concerned with item detail.

The responsibility for taking the physical inventory usually rests with the materials manager who should ensure that a good plan exists and it is followed. George Plossl once said that taking a physical inventory was like painting; the results depend on good preparation.[1] There are three factors in good preparation: housekeeping, identification, and training.

Housekeeping.   Inventory must be sorted and the same parts collected together so they can easily be counted. Sometimes items can be precounted and put into sealed cartons.

Identification.   Parts must be clearly identified and tagged with part numbers. This can, and should, be done before the inventory is taken. Personnel who are familiar with parts identification should be involved and all questions resolved before the physical inventory starts.

Training.   Those who are going to do the inventory must be properly instructed and trained in taking inventory. Physical inventories are usually taken once a year, and the procedure is not always remembered from year to year.

Process.   Taking a physical inventory consists of four steps:

1. Count items and record the count on a ticket left on the item.
2. Verify this count by recounting or by sampling.
3. When the verification is finished, collect the tickets and list the items in each department.
4. Reconcile the inventory records for differences between the physical count and inventory dollars. Financially, this step is the job of accountants, but materials personnel are involved in adjusting item records to reflect what is actually on hand. If major discrepancies exist, they should be checked immediately.

Taking a physical inventory is a time-honored practice in many companies mainly because it has been required for an "accurate" appraisal of inventory value for the annual financial statements. However, taking an annual physical inventory has several problems. Usually the factory has to be shut down, thus losing production; labor and paperwork are expensive; the job is often done hurriedly and poorly since there is much pressure to get it done and the factory running again. In addition, the people doing the inventory are not used to the job and are prone to making errors. As a result, more errors often are introduced into the records than are eliminated.

---

1. George W. Plossl, *Production and Inventory Control, Principles and Techniques,* 2nd ed., Appendix VI: Physical Inventory Techniques. Englewood Cliffs, NJ: Prentice Hall, 1985.

Because of these problems, the idea of cycle counting has developed.

Cycle Counting.   This is a system of counting inventory continually throughout the year. Physical inventory counts are scheduled so that each item is counted on a predetermined schedule. Depending on their importance, some items are counted frequently throughout the year whereas others are not. The idea is to count selected items each day.

The advantages to cycle counting are:

- Timely detection and correction of problems. The purpose of the count is first to find the cause of error and to correct the cause so the error is less likely to happen again.
- Complete or partial reduction of lost production.
- Use of personnel trained and dedicated to cycle counting. This provides experienced inventory takers who will not make the errors "once-a-year" personnel do. Cycle counters are also trained to identify problems and to correct them.

Count frequency.   The basic idea is to count some items each day so all items are counted a predetermined number of times each year. The number of times an item is counted in a year is called its **count frequency.** For an item, the count frequency should increase as the value of the item and number of transactions (chance of error) increase. Several methods can be used to determine the frequency. Three common ones are the ABC method, zone method, and location audit method.

- *ABC method.*   This is a popular method. Inventories are classified according to the ABC system (refer to Chapter 10). Some rule is established for count frequency. For example, A items might be counted weekly or monthly; B items, bimonthly or quarterly; and C items, biannually or once a year. On this basis, a count schedule can be established. Figure 12.5 shows an example of a cycle count scheduled using the ABC system.

Figure 12.5   Scheduling cycle counts.

Classification	Number of Items	Count Frequency	Number of Counts
A	1000	12	12,000
B	1500	4	6000
C	2500	1	2500
Total Counts			20,500
Workdays per Year			250
Counts per Day			82

EXAMPLE PROBLEM

A company has classified its inventory into ABC items. They have decided that A items are to be counted once a month; B items, four times a year; and C items, twice a year. There are 2000 A items, 3000 B items, and 5000 C items in inventory. Develop a schedule of the counts for each class of item.

Answer

Classification	Number of Items	Count Frequency	Number of Counts
A	2000	12	24,000
B	3000	4	12,000
C	5000	2	10,000
Total Counts			46,000
Workdays per Year			250
Counts per Day			184

- *Zone method.* Items are grouped by zones to make counting more efficient. The system is used when a fixed-location system is used, or when work-in-process or transit inventory is being counted.

- *Location audit system.* In a floating-location system, goods can be stored anywhere, and the system records where they are. Because of human error, these locations may not be 100% correct. If material is mislocated, normal cycle counting may not find it. In using location audits, a predetermined number of stock locations are checked each period. The item numbers of the material in each bin are checked against inventory records to verify stock point locations.

A cycle counting program may include all these methods. The zone method is ideal for fast-moving items. If a floating-location system is used, a combination of ABC and location audit is appropriate.

When to count. Cycle counts can be scheduled at regular intervals or on special occasions. Some selection criteria are:

- *When an order is placed.* Items are counted just before an order is placed. This has the advantage of detecting errors before the order is placed and reducing the amount of work by counting at a time when stock is low.

- *When an order is received.* Inventory is at its lowest level.

- *When the inventory record reaches zero.* Again, this method has the advantage of reducing work.
- *When a specified number of transactions have occurred.* Errors occur when transactions occur. Fast-moving items have more transactions and are more prone to error.
- *When an error occurs.* A special count is appropriate when an obvious error is detected. This may be a negative balance on the stock record or when no items can be found although the record shows some in stock.

QUESTIONS

1. What are four objectives of warehouse operation?
2. Name and describe each of the eight warehouse activities.
3. What are cube utilization and accessibility?
4. Why is stock location important in a warehouse? Name and describe four basic systems of stock location.
5. Describe fixed and floating systems for assigning locations to SKUs.
6. Name and describe three order-picking systems.
7. What are three prime objectives of materials handling? Describe the characteristics of conveyors, industrial trucks, and cranes.
8. What are the four steps in any transaction?
9. What are some of the results of poor inventory accuracy?
10. Six causes of poor inventory accuracy are discussed in the text. Name and describe each.
11. How should inventory accuracy be measured? What is tolerance? Why is it necessary?
12. What is the basis for setting tolerance?
13. What are the two major purposes of auditing inventory accuracy?
14. In taking a physical inventory, what are the three factors in preparation? Why is good preparation essential?
15. What are the four steps in taking a physical inventory?
16. Describe cycle counting. On what basis can the count frequency be determined?
17. Why is cycle counting a better way to audit inventory records than an annual physical inventory?
18. When are some good times to count inventory?

PROBLEMS

12.1 A company wants to store an SKU consisting of 5000 cartons on pallets each containing 25 cartons. They are to be stored three high in the warehouse. How many pallet positions are needed?

Answer.   67 pallet positions

12.2 A company has 7000 cartons to store on pallets. Each pallet takes 30 cartons, and the cartons are stored four high. How many pallet positions are needed?

12.3 A company has an area for storing pallets as shown in the following diagram. How many pallets measuring 48″× 40″ can be stored four high if there is a 2″ space between the pallets?

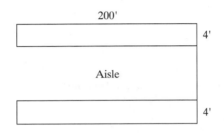

Answer.   456 pallets

12.4 A company has a warehouse with the dimensions shown in the following. How many pallets measuring 48″ × 40″ can be stored three high if there is to be a 2″ space between the pallets?

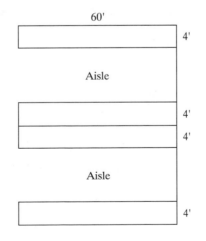

12.5 A company wishes to store the following SKUs so there is 100% accessibility. The items are stored on pallets that can be stacked three high.

a. How many pallet positions are needed?

b. What is the cube utilization?

c. If the company bought racking for storing the pallets, how many pallet positions are needed to give 100% accessibility?

SKU	Number of Pallets	Pallet Positions Required
A	5	
B	6	
C	4	
D	10	
E	14	
Total		

Answer. **a.** Pallet positions needed = 15

b. Cube utilization = 87%

c. 13 pallet positions

12.6 A company wants to store the following ten SKUs so there is 100% accessibility. Items are stored on pallets that are stored four high.

a. How many pallet positions are needed?

b. What is the cube utilization?

c. If the company bought racking for storing the pallets, how many pallet positions are needed to give 100% accessibility?

SKU	Number of Pallets	Pallet Positions Required
A	14	
B	17	
C	40	
D	33	
E	55	
F	22	
G	34	
Total		

12.7 Which of the following items are within tolerance?

Part Number	Shelf Count	Inventory Record	Difference	% Difference	Tolerance	Within Tolerance?
A	650	625			± 3%	
B	1205	1205			± 0%	
C	1350	1500			± 5%	
D	77	80			± 5%	
E	38	40			± 3%	
Total						

Answer. B and D are within tolerance.

12.8 Which of the following items are within tolerance?

Part Number	Shelf Count	Inventory Record	Difference	% Difference	Tolerance	Within Tolerance?
A	78	80			± 3%	
B	120	120			± 0%	
C	1400	1425			± 5%	
D	75	76			± 5%	
E	68	66			± 2%	
Total						

12.9 A company does an ABC analysis of its inventory and calculates that out of 5000 items 22% can be classified as A items, 33% as B items, and the remainder as C items. A decision is made that A items are to be cycle counted once a month, B items every 3 months, and C items twice a year. Calculate the total counts and the counts per day. The company works 5 days per week and 50 weeks per year.

Classification	Number of Items	Count Frequency	Number of Counts
A			
B			
C			
Total Counts			
Workdays per Year			
Counts per Day			

Answer. Counts per day = 97.2

12.10 A company does an ABC analysis of its inventory and calculates that out of 10,000 items 19% can be classified as A items, 30% as B items, and the remainder as C items. A decision is made that A items are to be cycle counted twice a month, B items every 3 months, and C items once a year. Calculate the total counts and the counts per day. There are 250 working days per year.

Classification	Number of Items	Count Frequency	Number of Counts
A			
B			
C			
Total Counts			
Workdays per Year			
Counts per Day			

13

Physical Distribution

INTRODUCTION

Chapter 1 introduced the supply chain concept. It was pointed out that a supply chain is composed of a series of suppliers and customers linked together by a physical distribution system. Usually the supply chain consists of several companies linked in this way. This chapter will discuss the physical distribution aspect of supply chains.

Physical distribution is the movement of materials from the producer to the consumer. It is the responsibility of the distribution department, which is part of an integrated materials management or logistics system. Figure 13.1 shows the relationship of the various functions in this type of system.

In Figure 13.1, the movement of materials is divided into two functions: physical supply and physical distribution. **Physical supply** is the movement and storage of goods from suppliers to manufacturing. Depending on the conditions of sale, the cost may be paid by either the supplier or the customer, but it is ultimately passed on to the customer. **Physical distribution,** on the other hand, is the movement and storage of finished goods from the end of production to the customer. The particular path in which the goods move—through distribution centers, wholesalers, and retailers—is called the **channel of distribution.**

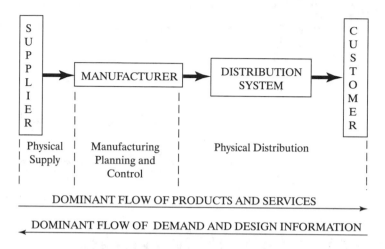

Figure 13.1   Supply chain (logistics system).

Channels of Distribution

A channel of distribution is one or more companies or individuals who participate in the flow of goods and/or services from the producer to the final user or consumer. Sometimes a company delivers directly to its customers, but often it uses other companies or individuals to distribute some or all of its products to the final consumer. These companies or individuals are called intermediaries. Examples of intermediaries are wholesalers, agents, transportation companies, and warehousers.

There are really two related channels involved. The **transaction channel** is concerned with the transfer of ownership. Its function is to negotiate, sell, and contract. The **distribution channel** is concerned with the transfer or delivery of the goods or services. The same intermediary may perform both functions, but not necessarily.

Figure 13.2 shows an example of the separation of distribution and transaction channels. The example might be for a company distributing a major appliance such as a refrigerator or stove. In such a system the retailer usually carries only display models. When the customer orders an appliance, delivery is made from either the regional warehouse or the public warehouse.

In this text we are concerned with the distribution channel.

Although it can be argued that one firm's physical supply is another firm's physical distribution, frequently there are important differences, particularly as they relate to the bulk and physical condition of raw materials and finished goods. The logistics problems that occur in moving and storing iron ore are quite different from those that occur in moving sheet steel. These differences influence the design of a logistics system and are important in deciding the location of distribution centers and factories. This text refers to both physical distribution and physical supply as physical distribution, but the differences for any particular company should be remembered.

Figure 13.2 Separation of distribution and transaction channels.

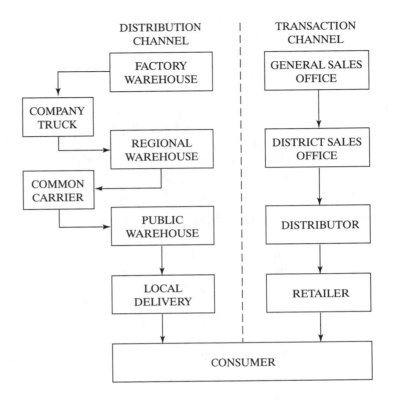

Physical distribution is vital in our lives. Usually, manufacturers, customers, and potential customers are widely dispersed geographically. If manufacturers serve only their local market, they restrict their potential for growth and profit. By extending its market, a firm can gain economies of scale in manufacturing, reduce the cost of purchases by volume discounts, and improve its profitability. However, to extend markets requires a well-run distribution system. Manufacturing adds form value to a product by taking the raw materials and creating something more useful. Bread is made from grain and is far more useful to humans than the grain itself. Distribution adds place value and time value by placing goods in markets where they are available to the consumer at the time the consumer wants them.

The specific way in which materials move depends upon many factors. For example:

- *The channels of distribution that the firm is using.* For example, producer to wholesaler to retailer to consumer.
- *The types of markets served.* Market characteristics such as the geographic dispersion of the market, the number of customers, and the size of orders.

- *The characteristics of the product.*   For example, weight, density, fragility, and perishability.
- *The type of transportation available to move the material.*   For example, trains, ships, planes, and trucks.

All are closely related. For instance, florists selling a perishable product to a local market will sell directly and probably use their own trucks. However, a national canning company selling a nonperishable product to a national market through a distribution channel composed of wholesalers and retailers may use trucks and rail transport.

PHYSICAL DISTRIBUTION SYSTEM

Physical distribution is responsible for delivering to the customer what is wanted on time and at minimum cost. The objective of distribution management is to design and operate a distribution system that attains the required level of customer service and does so at least cost. To reach this objective, all activities involved in the movement and storage of goods must be organized into an integrated system.

Activities in the Physical Distribution System

A system is a set of components or activities that interact with each other. A car engine is a system; if any part malfunctions, the performance of the whole engine suffers. In a distribution system, six interrelated activities affect customer service and the cost of providing it:

1. *Transportation.*   Transportation involves the various methods of moving goods outside the firm's buildings. For most firms, transportation is the single highest cost in distribution, usually accounting for 30% to 60% of distribution costs. Transportation adds place value to the product.

2. *Distribution inventory.*   Distribution inventory includes all finished goods inventory at any point in the distribution system. In cost terms, it is the second most important item in distribution, accounting for about 25% to 30% of the cost of distribution. Inventories create time value by placing the product close to the customer.

3. *Warehouses (distribution centers).*   Warehouses are used to store inventory. The management of warehouses makes decisions on site selection, number of distribution centers in the system, layout, and methods of receiving, storing, and retrieving goods.

4. *Materials handling.*   Materials handling is the movement and storage of goods inside the distribution center. The type of materials handling equipment used

affects the efficiency and cost of operating the distribution center. Materials handling represents a capital cost, and a tradeoff exists between this capital cost and the operating costs of the distribution center.

5. *Protective packaging.* Goods moving in a distribution system must be contained, protected, and identified. In addition, goods are moved and stored in packages and must fit into the dimension of the storage spaces and the transportation vehicles.

6. *Order processing and communication.* Order processing includes all activities needed to fill customer orders. Order processing represents a time element in delivery and is an important part of customer service. Many intermediaries are involved in the movement of goods, and good communication is essential to a successful distribution system.

Total-Cost Concept

The objective of distribution management is to provide the required level of customer service at the *least total system cost.* This does not mean that transportation costs or inventory costs or any one activity cost should be a minimum, but that the total of all costs should be a minimum. What happens to one activity has an effect on other activities, total system cost, and the service level. Management must treat the system as a whole and understand the relationships among the activities.

EXAMPLE PROBLEM

A company normally ships a product by rail. Transport by rail costs $200, and the transit time is 10 days. However, the goods can be moved by air at a cost of $1000 and will take one day to deliver. The cost of inventory in transit for a particular shipment is $100 per day. What are the costs involved in their decision?

Answer

	Rail	Air
Transportation Cost	$ 200	$1000
In-Transit Inventory Carrying Cost	1000	100
Total	$1200	$1100

There are two related principles illustrated here:

1. *Cost tradeoff.*   The cost of transportation increased with the use of air transport, but the cost of carrying inventory decreased. There was a cost tradeoff between the two.

2. *Total cost.*   By considering all of the costs and not just any one cost, the total system cost is reduced. Note also that even though no cost is attributed to it, customer service is improved by reducing the transit time. The total cost should also reflect the effect of the decision on other departments, such as production and marketing.

The preceding example does not mean that using faster transport always results in savings. For example, if the goods being moved are of low value and inventory-carrying cost is only $10 per day, rail will be cheaper. In addition, other costs may have to be considered.

Most of the decisions in distribution, and indeed much of what is done in business and in our own lives, involve tradeoffs and an appreciation of the total costs involved. In this section, the emphasis is on the costs and tradeoffs incurred and on improvement in customer service. Generally, but not always, an increase in customer service requires an increase in cost, which is one of the major tradeoffs.

INTERFACES

By taking the goods produced by manufacturing and delivering them to the customer, physical distribution provides a bridge between marketing and production. As such, there are several important interfaces among physical distribution and production and marketing.

Marketing

Although physical distribution interacts with all departments in a business, its closest relationship is probably with marketing. Indeed, physical distribution is often thought of as a marketing subject, not as part of materials management or logistics.

The "marketing mix" is made up of product, promotion, price, and place, and the latter is created by physical distribution. Marketing is responsible for transferring ownership. This is accomplished by such methods as personal selling, advertising, sales promotion, merchandising, and pricing. Physical distribution is responsible for giving the customer possession of the goods and does so by operating distribution centers, transportation systems, inventories, and order processing systems. It has the responsibility of meeting the customer service levels established by marketing and the senior management of the firm.

Physical distribution contributes to creating demand. Prompt delivery, product availability, and accurate order filling are important competitive tools in promoting a firm's products. The distribution system is a cost, so its efficiency and effectiveness influence the company's ability to price competitively. All of these affect company profits.

Production

Physical supply establishes the flow of material into the production process. The service level must usually be very high because the cost of interrupted production schedules caused by raw material shortage is usually enormous.

There are many factors involved in selecting a site for a factory, but an important one is the cost and availability of transportation for raw materials to the factory and the movement of finished goods to the marketplace. Sometimes, the location of factories is decided largely by the sources and transportation links of raw materials. This is particularly true where the raw materials are bulky and of relatively low value compared to the finished product. The location of steel mills on the Great Lakes is a good example. The basic raw material, iron ore, is bulky, heavy, and of low unit value. Transportation costs must be kept low to make a steel mill profitable. Iron ore from mines in either northern Quebec or Minnesota is transported to the mills by boat, the least costly mode of transportation. In other cases, the availability of low-cost transportation makes it possible to locate in areas remote from markets, but where labor is inexpensive.

Unless a firm is delivering finished goods directly to a customer, demand on the factory is created by the distribution center orders and not directly by the final customer. As noted in Chapter 11, this can have severe implications on the demand pattern at the factory. Although the demand from customers may be relatively uniform, the factory reacts to the demand from the distribution centers for replenishment stock. If the distribution centers are using an order point system, the demand on the factory will not be uniform and will be dependent rather than independent. The distribution system is the factory's customer, and the way that the distribution system interfaces with the factory will influence the efficiency of factory operations.

TRANSPORTATION

Transportation is an essential ingredient in the economic development of any area. It brings together raw materials for production of marketable commodities and distributes the products of industry to the marketplace. As such, it is a major contributor to the economic and social fabric of a society and aids economic development of regional areas.

The carriers of transportation can be divided into five basic modes:

1. Rail.
2. Road, including trucks, buses, and automobiles.
3. Air.
4. Water, including ocean-going, inland, and coastal ships.
5. Pipeline.

Each mode has different cost and service characteristics. These determine which method is appropriate for the types of goods to be moved. Certain types of traffic are simply more logically moved within one mode than they are in another. For example, trucks are best suited to moving small quantities to widely dispersed markets, but trains are best suited to moving large quantities of bulky cargo such as grain.

Costs of Carriage

To provide transportation service, any carrier, whatever mode, must have certain basic physical elements. These elements are ways, terminals, and vehicles. Each results in a cost to the carrier and, depending on the mode and the carrier, may be either capital (fixed) or operating (variable) costs. **Fixed costs** are costs that do not change with the volume of goods carried. The purchase cost of a truck owned by the carrier is a fixed cost. No matter how much it is used, the cost of the vehicle does not change. However, many costs of operation, such as fuel, maintenance, and driver's wages, depend on the use made of the truck. These are **variable costs.**

Ways are the paths over which the carrier operates. They include the right of way (land area being used), plus any roadbed, tracks, or other physical facilities needed on the right of way. The nature of the way and how it is paid for vary with the mode. They may be owned and operated by the government or by the carrier or provided by nature.

Terminals are places where carriers load and unload goods to and from vehicles and make connections between local pickup and delivery service and line-haul service. Other functions performed at terminals are weighing; connections with other routes and carriers; vehicle routing, dispatching, and maintenance; and administration and paperwork. The nature, size, and complexity of the terminal varies with the mode and size of the firm and the types of goods carried. Terminals are generally owned and operated by the carrier but, in some special circumstances, may be publicly owned and operated.

Vehicles of various types are used in all modes except pipelines. They serve as carrying and power units to move the goods over the ways. The carrier usually owns or leases the vehicles, although sometimes the shipper owns or leases them.

Besides ways, terminals, and vehicles, a carrier will have other costs such as maintenance, labor, fuel, and administration. These are generally part of operating costs and may be fixed or variable.

Rail

Railways provide their own ways, terminals, and vehicles, all of which represent a large capital investment. This means that most of the total cost of operating a railway is fixed. Thus, railways must have a high volume of traffic to absorb the fixed costs. They will not want to install and operate rail lines unless there is a large enough volume of traffic. Trains move goods by train loads composed of perhaps a hundred cars each with a carrying capacity in the order of 160,000 pounds.

Therefore, railways are best able to move large volumes of bulky goods over long distances. Their frequency of departure will be less than trucks, which can move when one truck is loaded. Rail speed is good over long distances, the service is generally reliable, and trains are flexible about the goods they can carry. Train service is cheaper than road for large quantities of bulky commodities such as coal, grain, potash, and containers moved over long distances.

Road

Trucks do not provide their own ways (roads and highways) but pay a fee to the government as license, gasoline, and other taxes and tolls for the use of roads. Terminals are usually owned and operated by the carrier but may be either privately owned or owned by the government. Vehicles are owned, or leased, and operated by the carrier. If owned, they are a major capital expense. However, in comparison to other modes, the cost of a vehicle is small. This means that for road carriers most of their costs are operating (variable) in nature.

Trucks can provide door-to-door service as long as there is a suitable surface on which to drive. In the United States and Canada, the road network is superb. The unit of movement is a truckload, which can be up to about 100,000 pounds. These two factors—the excellent road system and the relatively small unit of movement—mean that trucks can provide fast flexible service almost anywhere in North America. Trucks are particularly suited to distribution of relatively small-volume goods to a dispersed market.

Air

Air transport does not have ways in the sense of fixed physical roadbeds, but it does require an airway system that includes air traffic control and navigation systems. These systems are usually provided by the government. Carriers pay a user charge that is a variable cost to them. Terminals include all of the airport facilities, most of which are provided by the government. However, carriers are usually responsible for providing their own cargo terminals and maintenance facilities, either by owning or renting the space. The carrier provides the aircraft either through ownership or leasing. The aircraft are expensive and are the single most important cost element for the airline. Since operating costs are high, airlines' costs are mainly variable.

The main advantage of air transport is speed of service, especially over long distances. Most cargo travels in passenger aircraft, and thus many delivery schedules are tied to those of passenger service. The service is flexible about destination provided there is a suitable landing strip. Transportation cost for air cargo is higher than for other modes. For these reasons, air transport is most often suitable for high-value, low-weight cargo or for emergency items.

Water

Waterways are provided by nature or by nature with the assistance of the government. The St. Lawrence Seaway system is an example of this. The carrier thus has no capital cost in providing the ways but may have to pay a fee for using the waterway.

Terminals may be provided by the government but are increasingly privately owned. In either case, the carrier will pay a fee to use them. Thus, terminals are mainly a variable cost. Vehicles (ships) are either owned or leased by the carrier and represent the major capital or fixed cost to the carrier.

The main advantage of water transport is cost. Operating costs are low, and since the ships have a relatively large capacity, the fixed costs can be absorbed over large volumes. Ships are slow and are door to door only if the shipper and the consignee are on a waterway. Therefore, water transportation is most useful for moving low-value, bulky cargo over relatively long distances where waterways are available.

Pipelines

Pipelines are unique among the modes of transportation in that they move only gas, oil, and refined products on a widespread basis. As such, they are of little interest to most users of transportation. Capital costs for ways and pipelines are high and are borne by the carrier, but operating costs are very low.

LEGAL TYPES OF CARRIAGE

Carriers are legally classified as public (for hire) or private (not for hire). In the latter, individuals or firms own or lease their vehicles and use them to move their own goods. Public transport, on the other hand, is in the business of hauling for others for pay. All modes of transport have public and for-hire carriers.

For-hire carriers are subject to economic regulation by federal, state, or municipal governments. Depending on the jurisdiction, economic regulation may be more or less severe, and in recent years, there has been a strong move by government to reduce regulations. Economic regulation has centered on three areas:

1. Regulation of rates
2. Control of routes and service levels
3. Control of market entry and exit

Private carriers are not subject to economic regulation but, like public carriers, are regulated in such matters as public safety, license fees, and taxes.

For Hire

A for-hire carrier may carry goods for the public as a common carrier or under contract to a specified shipper.

Common carriers make a standing offer to serve the public. This means that whatever products they offer to carry will be carried for anyone wanting their service. With some minor exceptions, they can carry only those commodities they are licensed to carry. For instance, a household mover cannot carry gravel or fresh vegetables. Common carriers provide the following:

* Service available to the public
* Service to designated points or in designated areas
* Scheduled service
* Service of a given class of movement or commodity

Contract carriers haul only for those with whom they have a specific formal contract of service, not the general public. Contract carriers offer a service according to a contractual agreement signed with a specific shipper. The contract specifies the character of the service, performance, and charges.

Private

Private carriers own or lease their equipment and operate it themselves. This means investment in equipment, insurance, and maintenance expense. A company normally only considers operating its own fleet if the volume of transport is high enough to justify the capital expense.

Service Capability

Service capability depends on the availability of transportation service, which in turn depends on the control that the shipper has over the transportation agency. The shipper must go to the marketplace to hire a common carrier and is subject to the schedules and regulations of that carrier. Least control is exercised over common carriers. Shippers can exercise most control over their own vehicles and have the highest service capability with private carriage.

Other Transportation Agencies

There are several transportation agencies that use the various modes or combinations of the modes. Some of these are the post office, freight forwarders, couriers, and shippers. They all provide a transportation service, usually as a common carrier. They may own the vehicles, or they may contract with carriers to move their goods. Usually, they consolidate small shipments into large shipments to make economic loads.

TRANSPORTATION COST ELEMENTS

There are four basic cost elements in transportation. Knowledge of these costs enables a shipper to get a better price by selecting the right shipping mode. The four basic costs are as follows:

1. Line haul
2. Pickup and delivery
3. Terminal handling
4. Billing and collecting

We will use motor transport as an example, but the principles are the same for all modes.

Goods move either directly from the shipper to the consignee or through a terminal. In the latter, they are picked up in some vehicle suitable for short-haul local travel. They are then delivered to a terminal where they are sorted according to destination and loaded onto highway vehicles for travel to a destination terminal. There, they are again sorted, loaded on local delivery trucks, and taken to the consignee. Figure 13.3 shows this pattern schematically.

Line-Haul Costs

When goods are shipped, they are sent in a moving container that has a weight and volume capacity. The carrier, private or for hire, has basic costs to move this container, which exist whether the container is full or not. For a truck, these include such items as gasoline, the driver's wages, and depreciation due to usage. These costs vary with the distance traveled, not the weight carried. The carrier has essentially the same basic costs whether the truck moves full or empty. If it is half full, the basic costs must be spread over only those goods in the truck.

Figure 13.3 Shipping patterns.

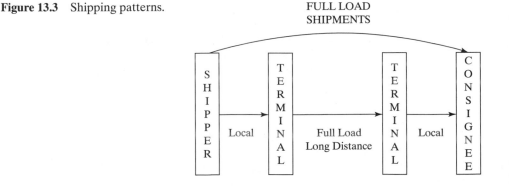

Therefore, total line-haul costs vary directly with the distance shipped, not on the weight shipped. For example, if for a given commodity, the line-haul cost is $3 per mile and the distance is 100 miles, the total line-haul cost is $300. If the shipper sends 50,000 pounds, the total line-haul cost is the same as if 10,000 pounds is sent. However, the line-haul costs (LHC) per hundred weight (cwt.) is different.

$$\text{LHC/cwt.} = \frac{300}{500}$$
$$= \$0.60 \text{ per cwt. [for 50,000 lb. (500 cwt.)]}$$
$$\text{LHC/cwt.} = \frac{300}{100}$$
$$= \$3 \text{ per cwt. [for 10,000 lb. (100 cwt.)]}$$

Thus, the **total line-haul cost** varies with (a) the cost per mile and (b) the distance moved. However, the **line-haul cost per hundred weight** varies with (a) the cost per mile, (b) the distance moved, and (c) the weight moved.

▼ ───────────────────────────────────────

EXAMPLE PROBLEM

For a particular commodity, the line-haul cost is $2.50 per mile. For a trip of 500 miles and a shipment of 600 cwt., what is the cost of shipping per cwt.? If the shipment is increased to 1000 cwt., what is the saving in cost per cwt.?

Answer

Total line-haul cost = $2.5 × 500 = $1250

Cost per cwt. = $1250 ÷ 600 = $2.083

If 1000 cwt. is shipped:

Cost per cwt. = $1250 ÷ 1000 = $1.25

Saving per cwt. = $2.08 – $1.25 = $0.83

───

The carrier has two limitations or capacity restrictions on how much can be moved on any one trip: the weight limitation and the cubic volume limitation of the vehicle. With some commodities, their density is such that the volume limitation is reached before the weight limitation. If the shipper wants to ship more, a method of increasing the density of the goods must be found. This is one reason that some lightweight products are made so they nest (for example, disposable cups) and bicycles and wheelbarrows are shipped in an unassembled state. This is not to frustrate us poor mortals who try to assemble them but to increase the density of the product so more weight can be shipped in a given vehicle. The same principle applies to goods stored in distribution centers. The more compact they are, the more can be stored in a given space. Therefore, if shippers want to reduce transportation cost, they should (a) increase the weight shipped and (b) maximize density.

▼

EXAMPLE PROBLEM

A company ships barbecues fully assembled. The average line-haul cost per shipment is $12.50 per mile, and the truck carries 100 assembled barbecues. The company decides to ship the barbecues unassembled and figures they can ship 500 barbecues in a truck. Calculate the line-haul cost per barbecue assembled and unassembled. If the average trip is 300 miles, calculate the saving per barbecue.

Answer

Line-haul cost assembled = $12.50 ÷ 100 = $0.125 per barbecue per mile

Line-haul cost unassembled = $12.50 ÷ 500 = $0.025 per barbecue per mile

Saving per mile = $0.125 – 0.025 = $0.10

Trip saving = 300 × $0.10 = $30.00 per barbecue

Pickup and Delivery Costs

Pickup and delivery costs are similar to line-haul costs except that the cost depends more on the time spent than on the distance traveled. The carrier will charge for each pickup and the weight picked up. If a shipper is making several shipments, it will be less expensive if they are consolidated and picked up on one trip.

Terminal Handling

Terminal-handling costs depend on the number of times a shipment must be loaded, handled, and unloaded. If full truckloads are shipped, the goods do not need to be handled in the terminal but can go directly to the consignee. If part loads are shipped, they must be taken to the terminal, unloaded, sorted, and loaded onto a highway vehicle. At the destination, the goods must be unloaded, sorted, and loaded onto a local delivery vehicle.

Each individual parcel must be handled. A shipper who has many customers, each ordering small quantities, will expect the terminal-handling costs to be high because there will be a handling charge for each package.

The basic rule for reducing terminal-handling costs is to reduce handling effort by consolidating shipments into fewer parcels.

Billing and Collecting

Every time a shipment is made, paperwork must be done and an invoice made out. Billing and collecting costs can be reduced by consolidating shipments and reducing the pickup frequency.

Total Transportation Costs

The total cost of transportation consists of line-haul, pickup and delivery, terminal-handling, and billing and collecting costs. To reduce shipping costs, the shipper needs to do the following:

- Decrease line-haul costs by increasing the weight shipped.
- Decrease pickup and delivery cost by reducing the number of pickups. This can be done by consolidating and increasing the weight per pickup.
- Decrease terminal-handling costs by decreasing the number of parcels by consolidating shipments.
- Decrease billing and collecting costs by consolidating shipments.

For any given shipment, the line-haul costs vary with the distance shipped. However, the pickup and delivery, terminal-handling, and billing costs are fixed. The total cost for any given shipment thus has a fixed cost and a variable cost associated with it. This relationship is shown in Figure 13.4. The carrier will consider this relationship and either charge a fixed cost plus so much per mile or offer a tapered rate. In the latter, the cost per mile for short distances far exceeds that for longer distances.

The rate charged by a carrier will also vary with the commodity shipped and will depend upon the following:

- *Value.* A carrier's liability for damage will be greater the more valuable the item.
- *Density.* The more dense the item, the greater the weight that can be carried in a given vehicle.

Figure 13.4 Distance versus cost of carriage.

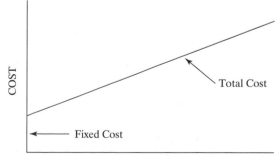

- *Perishability.* Perishable goods often require special equipment and methods of handling.
- *Packaging.* The method of packaging influences the risk of damage and breakage.

In addition, carriers have two rate structures, one based on full loads called truckload (TL) or carload (CL) and one based on less than truckload (LTL) and less than carload (LCL). For any given commodity, the LTL rates can be up to 100% higher than the TL rates. The basic reason for this differential lies in the extra pickup and delivery, terminal-handling and billing, and collection costs. Truckers, airlines, and water carriers accept less than full loads, but usually the railways do not accept LCL shipments.

WAREHOUSING

The last chapter discussed the management of warehouses. This section is concerned with the role of warehouses in a physical distribution system.

Warehouses include plant warehouses, regional warehouses, and local warehouses. They may be owned and operated by the supplier or intermediaries such as wholesalers, or may be public warehouses. The latter offer a general service to their public that includes providing storage space and warehouse services. Some warehouses specialize in the kinds of services they offer and the goods they store. A freezer storage is an example. The service functions warehouses perform can be classified into two kinds:

1. The **general warehouse** where goods are stored for long periods and where the prime purpose is to protect goods until they are needed. There is minimal handling, movement, and relationship to transportation. Furniture storage or a depository for documents are examples of this type of storage. It is also the type used for inventories accumulated in anticipation of seasonal sales.

2. The **distribution warehouse** has a dynamic purpose of movement and mixing. Goods are received in large-volume uniform lots, stored briefly, and then broken down into small individual orders of different items required by the customer in the marketplace. The emphasis is on movement and handling rather than on storage. This type of warehouse is widely used in distribution systems. The size of the warehouse is not so much its physical size as it is the **throughput,** or volume of traffic handled.

As discussed in the last chapter, warehouses, or distribution centers, are places where raw materials, semi-finished, or finished goods are stored. They represent an interruption in the flow of material and thus add cost to the system. Items should be warehoused only if there is an offsetting benefit gained from storing them.

Role of Warehouses

Warehouses serve three important roles: transportation consolidation, product mixing, and service.

Transportation consolidation. As shown in the last section, transportation costs can be reduced by using warehouses. This is accomplished by consolidating small (LTL) shipments into large (TL) shipments.

Consolidation can occur in both the supply and distribution systems. In physical supply, LTL shipments from several suppliers can be consolidated at a warehouse before being shipped TL to the factory. In physical distribution, TL shipments can be made to a distant warehouse and LTL shipments made to local users. Figure 13.5 shows the two situations graphically. Transportation consolidation in physical distribution is sometimes called **break-bulk,** which means the bulk (TL) shipments from factories to distribution centers are broken down into small shipments going to local markets.

Product mixing. While transportation consolidation is concerned with reduction of transportation costs, product mixing deals with the grouping of different items into an order and the economies that warehouses can provide in doing this. When customers place orders, they often want a mix of products that are produced in different locations.

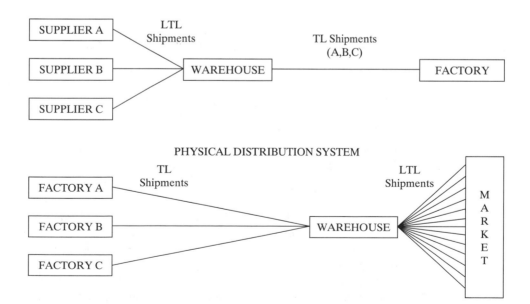

Figure 13.5 Transportation consolidation.

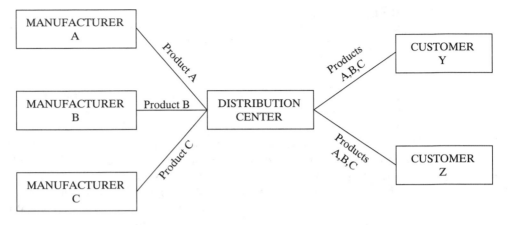

Figure 13.6   Product mixing.

Without a distribution center, customers would have to order from each source and pay for LTL transport from each source. Using a distribution center, orders can be placed and delivered from a central location. Figure 13.6 illustrates the concept.

Service.   Distribution centers improve customer service by providing place utility. Goods are positioned close to markets so the markets can be served more quickly.

Warehousing and Transportation Costs

Any distribution system should try to provide the highest service level (the number of orders delivered in a specified time) at the lowest possible cost. The particular shipping pattern will depend largely upon the following:

- Number of customers
- Geographic distribution of the customers
- Customer order size
- Number and location of plants and distribution centers

Suppliers have little or no control over the first three but do have some control over the last. They can establish local distribution centers in their markets. With respect to transportation, it then becomes a question of the cost of serving customers direct from the central distribution center or from the regional distribution center. If truckload shipments are made, the cost is less from the central distribution center, but if LTL shipments are made, it may be cheaper to serve the customer from the local distribution center.

EXAMPLE PROBLEM

Suppose a company with a plant located in Toronto is serving a market in the northeastern United States with many customers located in Boston. If they ship direct to customers from the Toronto plant, most shipments will be less than truckload. However, if they locate a distribution center in Boston, they can ship truckload (TL) to Boston and distribute by local cartage (LTL) to customers in that area. Whether this is economical or not depends on the total cost of shipping direct compared with shipping via the distribution center. Assume the following figures represent the average shipments to the Boston area:

> Plant to customer LTL: $100/cwt.
>
> Plant to distribution center TL: $50/cwt.
>
> Inventory-carrying cost (distribution center): $10/cwt.
>
> Distribution center to customer LTL: $20/cwt.

Is it more economical to establish the distribution center in Boston? If the annual shipped volume is 10,000 cwt., what will be the annual saving?

Answer

Costs if a distribution center is used:

$$
\begin{aligned}
\text{TL Toronto to Boston} &= \$50 \text{ per cwt.} \\
\text{Distribution center costs} &= \$10 \text{ per cwt.} \\
\text{LTL in Boston area} &= \underline{\$20 \text{ per cwt.}} \\
\text{Total cost} &= \$80 \text{ per cwt.} \\
\text{Saving per cwt.} &= \$100 - \$80 = \$20 \\
\text{Annual saving} &= \$20 \times 10,000 = \$200,000
\end{aligned}
$$

Market Boundaries

Continuing with the previous example problem, the company can now supply customers in other locations directly from the factory in Toronto or through the distribution center in Boston. The question is to decide which locations should be supplied from each source. The answer, of course, is the source that can service the location at least cost.

Laid-down cost (LDC) is the delivered cost of a product to a particular geographic point. The delivered cost includes all costs of moving the goods from A to B. In the previous example problem, the laid-down cost of delivering from

Toronto would be the transportation cost per mile × the miles to a particular destination. The LDC from Boston would include all costs of getting the goods to Boston, inventory costs in the Boston distribution center, and the transportation costs in getting to a particular destination.

$$LDC = P + TX$$

Where

P = product costs

T = transportation costs per mile

X = distance

The product cost includes all costs in getting the product to the supply location and storing it there. In the previous example, the product cost at Boston includes the TL cost of delivery to Boston and the inventory cost at Boston.

EXAMPLE PROBLEM

Syracuse is 300 miles from Toronto. The product cost for an item is $10 per cwt., and the transportation cost per mile per cwt. is $0.20. What is the laid-down cost per cwt.?

Answer

LDC = Product cost + (transportation cost per mile)(distance)

= $10 + ($0.20 × 300) = $70 per cwt.

Market boundary. The market boundary is the line between two or more supply sources where the laid-down cost is the same. Consider Figure 13.7. There are two sources of supply: A and B. The market boundary occurs at Y where the LDC from A is the same as B.

In the example shown in Figure 13.7, the distance between A and B is 100 miles. If we let the distance from A to Y be X miles, then the distance from B to Y is $(100 - X)$ miles. Assume supply A is the factory and supply B is a distribution center. Assume the product cost at A is $100 and product cost from B is $100 plus TL transportation from A to B and inventory costs at B. For this example, assume the

Figure 13.7   Market boundary.

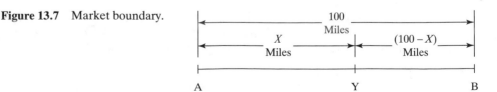

TL transportation and inventory carrying costs are $10 per unit so the product cost from B is $110. Transportation costs from either A or B are $0.40 per unit per mile.

Point Y occurs where:

$$LDC_A = LDC_B$$
$$100 + 0.40X = 110 + 0.40(100 - X)$$
$$X = 62.5$$

Thus a point Y, 62.5 miles from A, marks the market boundary between A and B.

EXAMPLE PROBLEM

The distance between Toronto and Boston is about 500 miles. Given the cost structure in the previous example problems and an LTL transportation cost of $0.20 per cwt., calculate the location of the market boundary between Toronto and Boston. Assume the product cost at Toronto is $10 per cwt.

Answer

The product cost at Boston is the sum of the product cost at Toronto, plus the cost of TL shipment from Toronto to Boston, plus the handling costs at Boston.

$$
\begin{aligned}
\text{Product cost at Boston} &= \text{product cost at A} + \text{TL transportation} + \text{handling costs} \\
&= \$10 + \$50 + \$10 \\
&= \$70
\end{aligned}
$$

The market boundary occurs where

$$LDC_T = LDC_B$$
$$\$10 + \$0.20X = \$70 + \$0.20(500 - X)$$
$$0.4X = 160$$
$$X = 400$$

The market boundary is 400 miles from Toronto or 100 miles from Boston.

Effect on Transportation Costs of Adding More Warehouses

We have seen from the previous example that establishing a distribution center in Boston reduces total transportation costs. Similarly, if a second distribution center is established, perhaps in Cleveland, we expect total transportation costs to be reduced further.

Generally, as more distribution centers are added to the system, we can expect the following:

- The cost of truckload (and carload) shipments to the distribution centers to increase.
- The cost of LTL shipments to customers to decrease.
- The total cost of transportation to decrease.

As expected, the major savings is from the addition of the first few distribution centers. Eventually, as more distribution centers are added, the savings decrease. The first distribution center added to the system is located to serve the largest market; the second distribution center, the second largest market, and so on. The number of customers served by additional distribution centers decreases, and the volume that can be shipped TL to the additional distribution centers is less than to the first distribution centers. Figure 13.8 shows the relationship that exists between transportation costs and the number of distribution centers in a system.

Figure 13.8 Transportation cost versus number of warehouses.

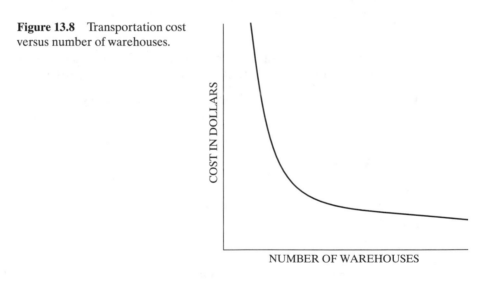

PACKAGING

The basic role of packaging in any industrial organization is to carry the goods safely through a distribution system to the customer. The package must do the following:

- Identify the product
- Contain and protect the product
- Contribute to physical distribution efficiency

For consumer products, the package may also be an important part of the marketing program.

Physical distribution must not only move and store products but also identify them. The package serves as a means of identifying the product in a way not possible from its outward appearance. When shoes are offered in ten sizes, the package becomes an important identifier.

Packaging must contain and protect the product, often against a wide range of hazards such as shock, compression, vibration, moisture, heat, solar radiation, oxidation, and infestation by animals, insects, birds, mold, or bacteria. Packages are subject to distribution hazards in loading and off-loading, in movement, in transportation, and in warehousing and storage. The package must be robust enough to protect and contain the product through all phases of distribution.

Packaging is a pure cost that must be offset by the increased physical distribution efficiency that the package can provide.

There are usually at least three levels of packaging required in a distribution system. First is a primary package that holds the product—the box of cornflakes. Next, for small packages, a shipping container such as a corrugated box is needed. Finally, there is a third level of packaging where several primary or secondary packages are assembled into a unit load.

Unitization

Unitization is the consolidation of several units into large units, called **unit loads,** so there is less handling. A unit load is a load made up of a number of items, or bulky material, arranged or constrained so the mass can be picked up or moved as a single unit too large for manual handling. Material handling costs decrease as the size of the unit load increases. It is more economical to move the product by cartons rather than individually and still more economical to move several cartons in one unit load.

This principle is used when we go shopping and put a number of articles into bags and then put the bags into the trunk of the car. In industry, unit loads are used instead of shopping bags.

There are a number of unit-load devices such as sheets, racks, or containers. One of the most common is the **pallet.**

The pallet is a platform usually measuring $48'' \times 40'' \times 4''$ and designed so that it can be lifted and moved by a forklift industrial truck. Packages are arranged on it so that several packages may be moved at one time. Loaded with packages, it forms a cube that is a unit load.

Unitization can be successive. Shippers place their products into primary packages, the packages into shipping cartons, the cartons onto pallets, and the pallets into warehouses, trucks, or other vehicles.

To use the capacity of pallets, trucks (or other vehicles), and warehouses, there should be some relationship between the dimensions of the product, the primary package, the shipping cartons, the pallet, the truck, and the warehouse space. The

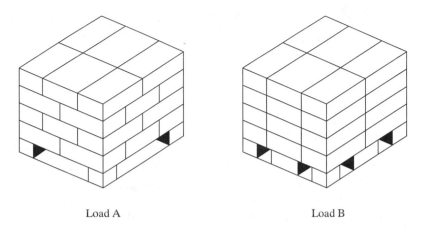

Load A   Load B

Figure 13.9   Stable and unstable pallet loads.

packages should be designed so space on the pallet is fully utilized and so the cartons interlock to form a stable load. Figure 13.9 shows two unit loads each using the total space of the pallet. However, load B does not interlock and is not stable.

Pallets fit into trucks and railway cars. The dimensions mentioned earlier were selected so pallets would fit into nominal 50′ railway cars and 40′ truck trailers with a minimum of lost space. Figure 13.10 shows the layout in railcars and trailers.

Thus to get the highest cube utilization, consideration must be given to the dimensions of the product, the carton, the pallet, the vehicle, and the warehouse.

MATERIALS HANDLING

Materials handling is the short-distance movement that takes place in or around a building such as a plant or distribution center. For a distribution center, this means the unloading and loading of transport vehicles and the dispatch and recall of goods to and from storage. In addition, the racking systems used in distribution centers are usually considered as part of materials handling.

Some objectives of materials handling are as follows:

1. To increase cube utilization by using the height of the building and by reducing the need for aisle space as much as possible.
2. To improve operating efficiency by reducing handling. Increasing the load per move will result in fewer moves.
3. To improve the service level by increasing the speed of response to customer needs.

There are many types of materials handling equipment. For convenience, they can be grouped into three categories: conveyors, industrial trucks, and cranes and hoists.

Figure 13.10 Railcar and trailer pallet position plan.

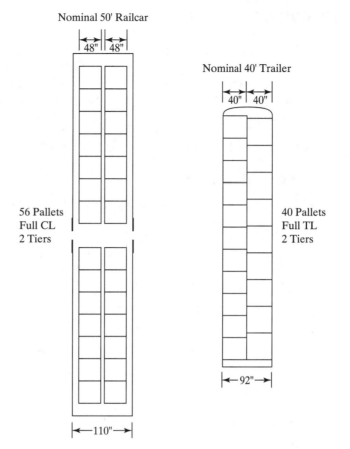

Nominal 50' Railcar

Nominal 40' Trailer

56 Pallets
Full CL
2 Tiers

40 Pallets
Full TL
2 Tiers

Conveyors are devices that move material (or people) horizontally or vertically between two fixed points. They are expensive, create a fixed route, and occupy space continuously. As a result, they are used only where there is sufficient throughput between fixed points to justify their cost.

Industrial trucks are vehicles powered by hand, electricity, or propane. Diesel and gasoline are not used indoors because they are noxious and lethal. Industrial trucks are more flexible than conveyors in that they can move anywhere there is a suitable surface and no obstructions. They do not occupy space continuously. For these reasons, they are the most often-used form of materials handling in distribution centers and in manufacturing.

Cranes and hoists can move materials vertically and horizontally to any point within their area of operation. They use overhead space and are used to move heavy or large items. Within their area of operation, they are very flexible.

MULTI-WAREHOUSE SYSTEMS

This section will look at the result of adding more distribution centers to the system. As might be expected, there is an effect on the cost of warehousing, materials handling, inventories, packaging, and transportation. Our purpose will be to look at how all of these costs and the total system cost behave. We also want to know what happens to the service level as more distribution centers are added to the system. To make valid comparisons, we must freeze the sales volume. We can then compare costs as we add distribution centers to the system.

Transportation Costs

In the section on transportation, we saw that if shipments to customers are in less-than-full vehicle lots, the total transportation cost is reduced by establishing a distribution center in a market area. This is because more weight can be shipped for greater distances by truck or carload and the LTL shipments can be made over relatively short distances. Generally, then, as more distribution centers are added to a system, we expect the following:

- The cost of TL shipments increases.
- The cost of LTL shipments decreases.
- The total cost of transportation decreases.

The major savings are made with the addition of the first distribution centers. Eventually, as more distribution centers are added, the marginal savings decrease.

Inventory-Carrying Cost

The average inventory carried depends on the order quantity and the safety stock. The average order quantity inventory in the system should remain the same since it depends on demand, the cost of ordering, and the cost of carrying inventory.

The total safety stock will be affected by the number of warehouses in the system. Safety stock is carried to protect against fluctuations in demand during the lead time and depends, in part, on the number of units sold. In Chapter 11, it was shown that the standard deviation varies as the square root of the ratio of the forecast and lead-time intervals. Similarly, for the same SKU, the standard deviation varies approximately as the square root of the ratio of the different annual demands. Suppose that the average demand is 1000 units and, for a service level of 90%, the safety stock is 100 units. If the 1000 units is divided between two distribution centers each having a demand of 500 units, the safety stock in each is:

$$SS = 100 \sqrt{\frac{500}{1000}} = 71 \text{ units}$$

With two distribution centers and the same total sales, the total safety stock increases to 142 from 100. Thus, with a constant sales volume, as the number of distribution centers increases, the demand on each decreases. This causes an increase in the total safety stock in all distribution centers.

Warehousing Costs

The fixed costs associated with distribution centers are space and materials handling. The space needed depends on the amount of inventory carried. As we have seen, as more distribution centers are added to the system, more inventory has to be carried, which requires more space.

In addition, there will be some duplication of nonstorage space such as washrooms and offices. So as the number of distribution centers increases, there will be a gradual increase in distribution center space costs.

Operating costs also increase as the number of distribution centers increases. Operating costs depend largely on the number of units handled. Since there is no increase in sales, the total number of units handled remains the same, as does the cost of handling. However, the nondirect supervision and clerical costs increase.

Materials Handling Costs

Materials handling costs depend upon the number of units handled. Since the sales volume remains constant, the number of units handled should also remain constant. There will be little change in materials handling costs as long as the firm can ship unit loads to the distribution center. However, if the number of distribution centers increases to the point that some nonunitized loads are shipped, materials handling costs increase.

Packaging Costs

Per-unit packaging costs will remain the same, but since there will be more inventory, total packaging costs will rise with inventory.

Total System Cost

We have assumed that total system sales remain the same. Figure 13.11 shows graphically how the costs of transportation, warehousing, materials handling inventory, and packaging behave as distribution centers are added to the system. Up to a point, total costs decrease and then start to increase. It is the objective of logistics to determine this least-cost point.

System Service Capability

The service capability of the system must also be evaluated. One way of assessing this is by estimating the percentage of the market served within a given period. Figure 13.12 represents such an estimate.

As expected, the service level increases as the number of distribution centers increases. It increases rapidly from one to two distribution centers and much less rapidly as the number is further increased. The first distribution center is built to serve the best market, the next to serve the second best market, and so on. Let us assume that a study has been made of a system of one to ten distribution centers and the costs are as shown in Figure 13.13.

Figure 13.11   Total system cost.

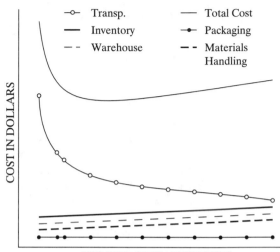

Figure 13.12   Estimate of market reached versus number of warehouses.

Number of Warehouses	Percentage of Market Reached in 1 Day
1	30
2	70
3	87
10	95

Figure 13.13   Cost versus number of warehouses.

Cost ($1000)	Number of Locations			
	1	2	3	10
Transportation	$8000	$6000	$5000	$4500
Warehousing	500	600	700	900
Material Handling	1000	1000	1100	1400
Inventory	400	425	460	700
Packaging	100	100	100	100
Total Cost	$10,000	$8125	$7300	$7800

A three-distribution center system would provide the least total cost. Figure 13.13 shows that by moving from three to ten distribution centers, the one-day service level increases by 8%. Management must decide which system to select. The decision must be based on adequate analysis of the choices available and a comparison of the increase in costs and service level.

QUESTIONS

1. Name and describe the three functions in the flow of materials from supplier to consumer. What are the differences between physical supply and physical distribution?

2. What is the primary function of the transaction channel and the distribution channels?

3. The particular way that goods move depends in part on four factors. What are they?

4. What are the objectives of a physical distribution system?

5. Name and describe each of the six system activities in a physical distribution system.

6. What are the cost-tradeoff and total-cost concepts? Why are they important?

7. Describe the relationship between marketing and physical distribution. How does physical distribution contribute to creating demand?

8. Why is the demand placed on a central distribution center or a factory by distribution centers considered dependent?

9. What are the five basic modes of transportation?

10. What are the three physical elements in a transportation system? For each of the five modes, describe who provides them and how they are funded.

11. Describe why train service is cheaper than road transport for large quantities of bulky commodities moving over long distances.

12. Why can trucks provide a fast, flexible service for the distribution of small volumes of goods to a dispersed market?

13. What are the major characteristics of water and air transport?

14. What are the major legal types of carriage? What are the three areas of economic regulation? To which legal type of carriage do they apply?

15. Compare common and contract carriage. How do they differ from private carriage? Which will give the highest level of service?

16. On what do total line-haul costs and line-haul costs per hundred weight depend? What two ways can shippers reduce line-haul costs?

17. Describe how a shipper can reduce the following:
    a. Pickup and delivery costs
    b. Terminal-handling costs
    c. Billing and collecting costs

18. The rates charged by a shipper vary with the commodity shipped. Name and describe four factors that affect the rates.

19. Why are LTL rates more expensive than TL rates?

20. Name and describe the two basic types of warehouses.

21. Name and describe the three important roles warehouses serve.

22. Name four factors that affect shipping patterns. Which can a supplier control?

23. What is the laid-down cost? What is a market boundary? Why are laid-down costs important in determining market boundaries?

24. As more distribution centers are added to a system, what happens to the cost of truckload, less than truckload, and total transportation costs?

25. What are the three roles of packaging in a distribution system? Describe why each is important.

26. What is unitization? Why is it important in physical distribution? Why is it successive?

27. What are three prime objectives of materials handling? Describe the characteristics of conveyors, industrial trucks, cranes, and hoists.

28. As more warehouses are added to the system, what would we expect to happen to the following:

   a. Transportation costs

   b. Inventory costs

   c. Materials handling costs

   d. Packaging costs

   e. Total system costs

   f. System service capability

PROBLEMS

13.1 A company normally ships to a customer by rail at a cost of $500 per load. The transit time is 14 days. The goods can be shipped by truck for $700 per load and a transit time of four days. If transit inventory cost is $25 per day, what does it cost to ship each way?

   Answer. Rail, $850; truck, $800

13.2 A company manufactures component parts for machine tools in North America and ships them to southeast Asia for assembly and sale in the local market. The components are shipped by sea, transit time averages six weeks, and the shipping costs $1000 per shipment. The company is considering moving the parts by air at an estimated cost of $7500, the shipment taking two days to get there. If inventory in transit for the shipment costs $150 per day, should they ship by air?

   In Chapter 8, it was said that forecasts are more accurate for nearer periods of time. Should this be considered? What activities are affected by the shorter lead time?

13.3 For a given commodity, the line-haul cost is $10 per mile. For a trip of 200 miles and a shipment of 300 cwt., what is the cost per hundred weight? If the shipment is increased to 500 cwt., what is the saving in cost per hundred weight?

   Answer. $2.67 per cwt.

13.4 A company ships a particular product to a market located 1000 miles from the plant at a cost of $4 per mile. Normally it ships 500 units at a time. What is the line-haul cost per unit?

13.5 In problem 13.4, if the company can ship the units unassembled, it can ship 800 units in a truck. What is the line-haul cost per unit now?

   Answer. Line-haul cost per unit = $5

13.6 A company processes feathers and ships them loose in a covered truck. The line-haul cost for an average shipment is $250, and the truck carries 2000 pounds of feathers. A bright new graduate has just been hired and has suggested that they should bale the feathers into 500-pound bales. This would make them easier to handle and also allow them to be compressed into about one-tenth of the space they now occupy. How many pounds of feathers can the truck now carry? What is the present line-haul cost per pound? What will it be if the proposal is adopted?

13.7 A company in Calgary serves a market in the northwestern United States. Now it ships LTL at an average cost of $25 per unit. If the company establishes a distribution center in the market, it estimates the TL cost will be $12 per unit, inventory-carrying costs will be $5 per unit, and the local LTL cost will average $6 per unit. If the company forecasts annual demand at 100,000 units, how much will they save annually?

   Answer. Annual saving = $200,000

13.8 A company ships LTL to customers in a market in the Midwest at an average cost of $40 per cwt. It proposes establishing a distribution center in this market. If TL shipment costs $20 per cwt., the estimated inventory-carrying costs are $5 per cwt., and the local cartage (LTL) cost is estimated at $6 per cwt. If the annual shipped volume is 100,000 cwt., what will the annual savings be by establishing the distribution center?

13.9 A company has a central supply facility and a distribution center located 500 miles away. The central supply product cost is $20, TL transportation rates from central supply to the DC are $50 per unit, and inventory-carrying costs are $4 per unit. Calculate the market boundary location and the laid-down cost at the market boundary. LTL rates are $1 per unit per mile.

   Answer. Market boundary is 277 miles from central supply. LDC = $297.

13.10 Suppose the company in problem 13.8 had another market area located between the parent plant and the proposed distribution center. The LTL costs from the plant to that market are $35 per cwt. The company estimates that LTL shipments from the distribution center will cost $4 per cwt. Should it supply this market from the distribution center or central supply?

13.11 A company can ship LTL direct to customers in city A or use a public warehouse located in city B. It has determined the following data.

   Cost per cwt. for shipping LTL to city A is $0.70 + $0.30 per mile.

   Cost per cwt. for shipping TL to warehouse is $0.40 + $0.15 per mile.

   Warehouse handling costs are $0.30 per cwt.

   Distances: Plant to city A = 115 miles

   Plant to city B = 135 miles

   From city B to city A = 30 miles

   a. What is the total cost per cwt. to ship from the plant direct to customers in city A?

   b. What is the total cost per cwt. to ship via the warehouse in city B?

   c. In this problem, the cost per cwt. has a fixed and a variable component. Why?

14

Products and Processes

INTRODUCTION

The effect and the efficiency of operations management, Just-in-Time manufacturing, and total quality management all depend on the way products are designed and the processes selected. The way products are designed determines the processes that are available to make them. The product design and the process determine the quality and cost of the product. Quality and cost determine the profitability of the company. This chapter studies the relationship between product design and process design and the costs associated with different types of processes. Finally, the chapter looks at the improvement of existing processes.

PRODUCT DEVELOPMENT PRINCIPLES

A few organizations supply a single product, but most supply a range of similar or related products. There are two conflicting factors to be considered in establishing the range of products to supply.

- If the product line is too narrow, customers may be lost.
- If the product line is too wide, customers may be satisfied, but operating costs will increase because of the lack of specialization.

Sales organizations are responsible for increasing sales and revenue. They want to offer product variety to consumers. Often this means the organization must offer a variety of products, many of which sell in small volumes.

Operations, on the other hand, would like to produce as few products as possible and make them in long runs. In this way they could reduce the number of setups (and cost) and probably lower run costs by using special machinery. They would fulfill their mandate to produce at the lowest cost.

Somehow the needs of sales and the economics of production must be balanced. Usually this balance can be obtained with good programs of

- Product simplification
- Product standardization
- Product specialization

Simplification

Simplification is the process of making something easier to do or make. It seeks to cut out waste by getting rid of needless product varieties, sizes, and types. The emphasis is not in cutting out products simply to reduce variety, but to remove unnecessary products and variations.

As well as reducing the variety of parts, product design can often be simplified to reduce operations and material costs. For example, the use of a snap-on plastic cap instead of a screw cap reduces the cost of both materials and labor.

Standardization

In product design, a standard is a carefully established specification covering the product's material, configuration, measurements, and so on. Thus, all products made to a given specification will be alike and interchangeable. Light bulbs are a good example of standardization. The sockets and wattage are standardized and the light bulbs are interchangeable.

A range of standard specifications can be established so it covers most uses for the item. Men's shirts are made in a range of standard collar sizes and sleeve lengths so nearly everyone can be fitted. Most shirt manufacturers also use the same standards so the consumer can get the same size shirt from any manufacturer and expect it to fit.

Because product standardization allows parts to be interchangeable, as long as the range of standard specifications has been well-chosen, a smaller variety of components is needed. Using the example of light bulbs, the wattages are standardized at 40, 60, and 100 watts. This range allows users to pick wattages that satisfy their needs and reduce the number of different bulbs, thus reducing inventory.

Another aspect of standardization is the way parts fit together. If the designs of assemblies are standardized so various models or products are assembled in the same way, then mass production is possible. The automotive industry designs automobiles so many different models can be assembled on the same assembly line. For example, several different engines can be mounted in a chassis because the engines are mounted in the same way and designed so they will all fit into the engine compartment.

Modularization.   Standardization does not necessarily reduce the range of choice for the customer. By standardizing on *component parts,* a manufacturer can make a variety of finished goods, one of which will probably satisfy the customer. Automobile manufacturing is a prime example of this. Cars are usually made from a few standard components and a series of standard options so the consumer has a selection from which to choose. For example, the Mazda Miata contains 80% standard parts, which enables Mazda to produce the car quickly and at low cost thus making a profit even though sales are comparatively small. Chrysler uses one platform—the basic frame of the vehicle—for all models of its minivan. Thus the company has only one set of frame costs for all minivans.

Specialization

Specialization is concentration of effort in a particular area or occupation. Electricians, doctors, and lawyers specialize in their chosen fields. In product specialization, a firm may produce and market only one or a limited range of similar products. This leads to process and labor specialization, which increases productivity and decreases costs.

    With a limited range of products, productivity can be increased and costs reduced by

- Allowing the development of machinery and equipment specially designed to make the limited range of products quickly and cheaply.
- Reducing the number of setups because of fewer task changes.
- Allowing labor to develop speed and dexterity because of fewer task changes.

Specialization is sometimes called focus and can be based either on product and market or on process.

Product and market focus   can be based on characteristics such as customer grouping (serving similar customers), demand characteristics (volume), or degree of customization. For example, one company may specialize in a limited range of high-volume products, while another may specialize in providing a wider range of low-volume products with a high level of customization.

Process focus   is based on the similarity of process. For example, automobile manufacturers specialize in assembling automobiles. Other factories and companies supply the assemblers with components and the assembler specializes in assembly operations.

Focused factory. Currently there is a trend toward more specialization in manufacturing whereby a factory specializes in a narrow product mix for a niche market. Generally focused factories are thought to produce more effectively and economically than more complex factories, the reason being that repetition and concentration in one area allows the workforce and management to gain the advantages of specialization. The focused factory may be a "factory within a factory," an area in an existing factory set aside to specialize in a narrow product mix.

Specialization has the disadvantage of inflexibility. Often it is difficult to use highly specialized labor and equipment for tasks other than those for which they were trained or built.

In summary, the three ideas of simplification, standardization, and specialization are different but interrelated. Simplification is the elimination of the superfluous and is the first step toward standardization. Standardization is establishing a range of standards that will meet most needs. Finally, specialization would not be possible without standardization. Specialization is concentration in a particular area and therefore implies repetition, which cannot be arrived at without standard products or procedures.

A program of product simplification, standardization, and specialization allows a firm to concentrate on the things it does best, provides the customers with what they want, and allows operations to perform with a high level of productivity. Reducing part variety will create savings in raw material, work-in-process, and finished goods inventory. It will also allow longer production runs, improve quality because there are fewer parts, and improve opportunities for automation and mechanization. Such a program contributes significantly to reducing cost.

PRODUCT SPECIFICATION AND DESIGN

Product design is responsible for producing a set of specifications that manufacturing can use to make the product. Products must be designed to be

- Functional
- Capable of low-cost processing

Functional means that the product will be designed to perform as specified in the marketplace. The marketing department produces a market specification laying down the expected performance, sales volume, selling price, and appearance values of the product. Product design engineers design the product to meet the market specifications. Engineers establish the dimensions, configurations, and specifications so the item, if properly manufactured, will perform as expected in the marketplace.

Low-cost processing. The product must be designed so it can be made at least cost. The product designer specifies materials, tolerances, basic shapes, methods of joining parts, and so on and, through these specifications, sets the minimum product cost. Usually, many different designs will satisfy functional and appearance specifications. The job then is to pick the design that will minimize manufacturing cost.

Poor design can add cost to processing in several ways:

- The product and its components may be designed to make using the most economical methods impossible.
- Parts may be designed so excessive material has to be removed.
- Parts may be designed so operations are difficult.
- Lack of standardized components may mean batches of work have to be small. Using standard parts across a range of products reduces the number of parts in inventory, tooling, and operator training and permits the use of special-purpose machinery. All this reduces product cost.
- Finally, product design can influence indirect costs such as production planning, purchasing, inventory management, and inspection. For example, one design may call for 20 different nonstandard parts, while another uses 15 standard parts. The effort required to plan and control the flow of materials and the operations will be greater, and more costly, in the first case than the second.

Low cost has to be designed into the product.

Simultaneous Engineering

To design products for low-cost manufacture requires close coordination between product design and process design. If the two groups can work together, they have a better chance of designing a product that will function well in the marketplace and can be manufactured at least cost. This relationship between product design and process design can spell the success or failure of a product. If a product cannot be produced at a cost that will allow a profit to be made, then it is a failure for the firm.

The traditional approach to product and process design has been a little like a relay race. When product design finished, it would pass the work to process design and let that department figure out how to make it. This system has proved time consuming and expensive. Figure 14.1 shows, with some humor, what can happen without strong communication and interaction between all parties in the product development cycle.

Today, many organizations concurrently develop the design for the product and the process. Often a team is made up of people from product design, process design, quality assurance, production planning and inventory control, purchasing, marketing, field service, and others who contribute to, or are affected by, the delivery of the product to the customer. This group works together to develop the product design so it meets the needs of the customer and can be made and delivered to the customer at low cost.

Figure 14.1 Communication is essential.

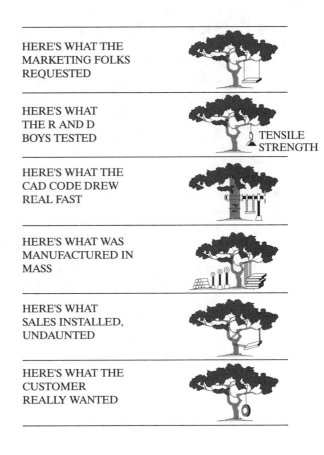

HERE'S WHAT THE
MARKETING FOLKS
REQUESTED

HERE'S WHAT
THE R AND D
BOYS TESTED

TENSILE
STRENGTH

HERE'S WHAT THE
CAD CODE DREW
REAL FAST

HERE'S WHAT WAS
MANUFACTURED IN
MASS

HERE'S WHAT
SALES INSTALLED,
UNDAUNTED

HERE'S WHAT THE
CUSTOMER
REALLY WANTED

There are several advantages to this approach:

- Time to market is reduced. The organization that gets its products to market before the competition gains a strong competitive advantage.

- Cost is reduced. Involving all stakeholders early in the process means less need for costly design changes later.

- Better quality. Because the product is designed for ease of manufacture, and ease of maintaining quality in manufacture, the number of rejects will be reduced. Because quality is improved, the need for after-sales service is reduced.

- Lower total system cost. Because all groups affected by the product design are consulted, all concerns are addressed. For example, field service might need a product that is designed so it is easy to service in the field, thus reducing servicing costs.

PROCESS DESIGN

Operations management is responsible for producing the products and services the customer wants, when wanted, with the required quality, at minimum cost and maximum effectiveness and productivity. Processes are the means by which operations management reaches these objectives.

> *A process is a method of doing something, generally involving a number of steps or operations. Process design is the developing and designing of the steps.*

Everything we do involves a process of some description. When we go to the bank to deposit or withdraw money, prepare a meal, or go on a trip we are involved in a process. Sometimes, as consumers, we are personally involved in the process. Most of us have waited at a check-out counter in a store and wondered why management has not devised a better process for serving customers.

Nesting.   Another way of looking at the hierarchy of processes is the concept of nesting. Small processes are linked to form a larger process. Consider Figure 14.2. Level zero shows a series of steps, each of which may have its own series of steps. One of the operations on level zero is expanded into its component parts and shown on level one. The nesting can continue to further levels of detail.

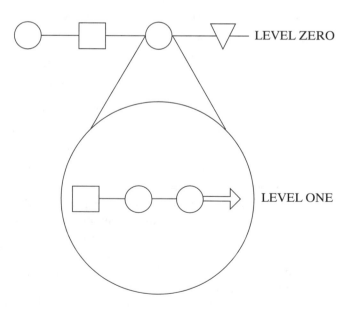

Figure 14.2   Nesting concept.

FACTORS INFLUENCING PROCESS DESIGN

Five basic factors must be considered when designing a process.

Product design and quality level.   The product's design determines the basic processes needed to convert the raw materials and components into the finished product. For example, if a steak is to be barbecued, then the process must include a barbecue operation. The process designer can usually select from a variety of different machines or operations to do the job. The type of machine or operation selected will depend upon the quantity to be produced, the available equipment, and the quality level needed.

The desired quality level affects the process design, because the process must be capable of achieving that quality level and doing it repeatedly. If the process cannot do that, operations will not be able to produce what is wanted except with expensive inspection and rework. The process designer must be aware of the capabilities of machines and processes and select those that will meet the quality level at least cost.

Demand patterns and flexibility needed.   If there is variation in demand for a product, the process must be flexible enough to respond to these changes quickly. For example, if a full-service restaurant sells a variety of foods, the process must be flexible enough to switch from broiling hamburgers to making pizzas. Conversely, if a pizza parlor sells nothing but pizzas, the process need not be designed to cook any other type of food. Flexible processes require flexible equipment and personnel capable of doing a number of different jobs.

Quantity/capacity considerations.   Product design, the quantity to produce, and process design are closely related. Both product and process design depend on the quantity needed. For example, if only one of an item is to be made, the design and the process used will be different than if the volume is 100,000 units. The quantity needed and the process design determine the capacity needed. Figure 14.3 shows this relationship. Note that all three are directly connected to the customer.

Figure 14.3   Product design, process design, and capacity are closely related.

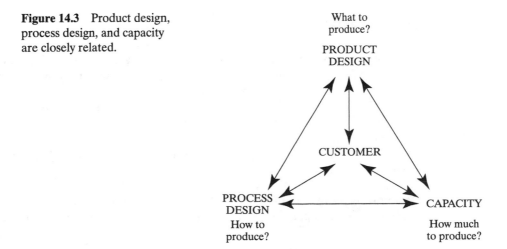

Customer involvement.   Chapter 1 discussed four manufacturing strategies— engineer-to-order, make-to-order, assemble-to-order, and make-to-stock—and the extent of customer involvement in each. Process design will depend on which strategy is chosen.

Make or buy decision.   A manufacturer has the alternative of making parts in-house or of buying them from an outside supplier. Few companies make everything or buy everything they need. Indeed, on the average, North American manufacturers purchase more than 50% of the cost of goods manufactured. A decision has to be made about which items to make and which to buy. While cost is the main determinant, other factors such as the following are usually considered:

Reasons to Make In-House
- Can produce for less cost than a supplier.
- To utilize existing equipment to fullest extent.
- To keep confidential processes within control of the firm.
- To maintain quality.
- To maintain workforce.

Reasons to Buy out
- Requires less capital investment.
- Uses specialized expertise of suppliers.
- Allows the firm to concentrate on its own area of specialization.
- Provides known and competitive prices.

The decision to make or buy is clear for many items such as nuts and bolts, motors or components that the firm does not normally manufacture. For other items that are in the firm's specialty area, a specific decision will have to be made.

PROCESSING EQUIPMENT

Processing equipment can be classified in several ways, but for our purposes one of the most convenient is the degree of specialization of machinery and equipment.

General-purpose machinery   can be used for a variety of operations or can do work on a variety of products within its machine classification. For example, a home sewing machine can sew a variety of materials, stitches, and patterns within its basic capability. Different auxiliary tools can be used to create other stitches or for particular sewing operations.

Special-purpose machinery   is designed to perform specific operations on one work piece or a small number of similar work pieces. For example, a sewing machine built or equipped to sew shirt collars would be a special-purpose machine capable of sewing collars on any size shirt of any color but not capable of performing other sewing operations unless it was modified extensively.

General-purpose machinery is generally less costly than special-purpose machinery. However, its run time is slow, and because it is operated by humans, the quality level tends to be lower than when using special-purpose machinery. Special-purpose machinery is less flexible but can produce parts much quicker than general-purpose machinery.

PROCESS SYSTEMS

Depending on the product design, volume, and available equipment, the process engineer must design the system to make the product. Based on material flow, processes can be organized in three ways:

- Flow
- Intermittent
- Project (fixed position)

The system used will depend on the demand for the item, range of products, and the ease or difficulty of moving material. All three systems can be used to make discrete units such as automobiles or textbooks, or to make nondiscrete products such as gasoline, paint, or fertilizer.

Flow Processes

Workstations needed to make the product, or family of similar products, are grouped together in one department, and are laid out in the sequence needed to make the product. Examples are assembly lines, cafeterias, oil refineries, and steel rolling mills. Work flows from one workstation to another at a nearly constant rate and with no delays. There is some form of mechanical method of moving goods between workstations. If the units are discrete, such as automobiles, flow manufacturing is called *repetitive manufacturing.* If not, for example gasoline, it is called *continuous manufacturing.* The typical flow pattern is shown in Figure 14.4.

WORKSTATIONS

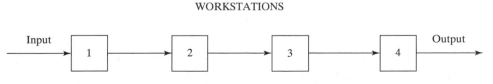

Figure 14.4 Material flow: flow process.

Flow systems produce only a limited range of similar products. For example, an assembly line produces a certain type of refrigerator and cannot be used to assemble washing machines. The operations used to make one are different and in a different sequence than those used for making the other. Demand for the family of products has to be large enough to justify setting up the line economically. If sufficient demand exists, flow systems are extremely efficient because:

- Workstations are designed to produce a limited range of similar products, so machinery and tooling can be specialized.
- Because material flows from one workstation to the next, there is very little build-up of work-in-process inventory.
- Because of the flow system and the low work-in-process inventory, lead times are short.
- In most cases, flow systems substitute capital for labor and standardize what labor there is into routine tasks.

Because it is so cost-effective, this system of processing should be used wherever and to whatever extent it can be.

Intermittent Processes

In intermittent manufacturing, goods are not made continuously as in a flow system but are made at intervals in lots or batches. Workstations must be capable of processing many different parts. Thus it is necessary to use general-purpose workstations and machinery that can perform a variety of tasks.

General-purpose workstations do not produce goods as quickly as special-purpose workstations used in flow manufacturing. Usually, workstations are organized into departments based on similar types of skills or equipment. For example, all welding and fabrication operations are located in one department, machine tools in another, and assembly in yet another department. Work moves only to those workstations needed to make the product and skips the rest. This results in the jumbled flow pattern shown in Figure 14.5.

Intermittent processes are flexible. They can change from one part or task to another more quickly than can flow processes. This is because they use general-purpose machinery and skilled flexible labor that can perform the variety of operations needed.

Control of work flow is managed through individual work orders for each lot or batch being made. Because of this and the jumbled pattern of work flow, manufacturing planning and control problems are severe. Often, many work orders exist, each of which can be processed in different ways.

Provided the volume of work exists to justify it, flow manufacturing is less costly than intermittent manufacturing. There are several reasons for this:

- Setup costs are low. Once the line is established, changeovers are needed infrequently to run another product.
- Since work centers are designed for specific products, run costs are low.

Figure 14.5 Material flow: intermittent process.

WORKSTATIONS

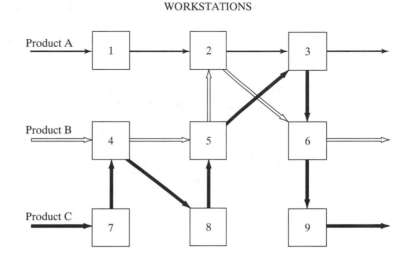

- Because products move continuously from one workstation to the next, work-in-process inventory will be low.
- Costs associated with controlling production are low because work flows through the process in a fixed sequence.

But the volume of specific parts must be enough to use the capacity of the line and justify the capital cost.

Project (Fixed Position) Processes

Project, or fixed position, manufacturing is mostly used for large complex projects such as locomotives, ships, or buildings. The product may remain in one location for its full assembly period, as with a ship, or it may move from location to location after considerable work and time are spent on it. Large aircraft are made this way. Project manufacturing has little advantage except it avoids the very high costs of moving the product from one workstation to another.

There are many variations and combinations of these three basic types of processes. Companies will try to find the best combination to make their particular products. In any one company it is not unusual to see examples of all three being used.

SELECTING THE PROCESS

Generally, the larger the volume (quantity) to be produced, the greater the opportunity to use special-purpose processes. The more special purpose an operation, the faster it will produce. Often the capital cost for such machinery or for special tools or fixtures is high. Capital costs are called fixed costs and the production, or run, costs are called variable costs.

Fixed costs   do not vary with the volume being produced. Purchase costs of machinery and tools and setup costs are considered fixed costs. No matter what volume is produced, these costs remain the same. If it costs $200 to set up a process, this cost will not change no matter how much is produced.

Variable costs   vary with the quantity produced. Direct labor (labor used directly in the making of the product) and direct material (material used directly in the product) are the major variable costs. If the run time for a product is 12 minutes per unit, the labor cost $10 per hour, and the material cost $5 per unit, then:

$$\text{Variable cost} = \frac{12}{60} \times \$10.00 + \$5.00 = \$7.00 \text{ per unit}$$

Let:

$$
\begin{aligned}
FC &= \text{Fixed cost} \\
VC &= \text{Variable cost per unit} \\
x &= \text{Number of units to be produced} \\
TC &= \text{Total cost} \\
UC &= \text{Unit (average) cost per unit} \\
\text{Total cost} &= \text{Fixed cost} + (\text{Variable cost per unit})(\text{number of units produced})
\end{aligned}
$$

Then:

$$TC = FC + VCx$$

$$\text{Unit cost} = \frac{\text{Total cost}}{\text{number of units produced}} = \frac{TC}{x}$$

EXAMPLE

A process designer has a choice of two methods for making an item. Method A has a fixed cost of $2000 for tooling and jigs and a variable cost of $3 per unit. Method B requires a special machine costing $20,000 and the variable costs are $1 per piece. Let x be the number of units produced.

	Method A	Method B
Fixed cost	$2000.00	$20,000.00
Variable unit cost	$3.00	$1.00
Total cost	$2000.00 + 3x	$20,000.00 + 1x
Unit (average) cost	$\dfrac{\$2000.00 + 3x}{x}$	$\dfrac{\$20,000.00 + 1x}{x}$

Table 14.1 shows what happens to the total cost as quantities produced are increased. The total cost data in this table are shown graphically in Figure 14.6. From Table 14.1 and Figure 14.6, we can see that initially the total cost and unit cost of method A are less than method B. This is because the fixed cost for method B is greater and has to be absorbed over a small number of units. While the total cost for both methods increases as more units are produced, the total cost for method A increases faster until, at some quantity between 8,000 and 10,000 units, the total cost for method B becomes less than for method A.

Similarly, the unit cost for both methods decreases as more units are produced. However, the unit cost for method B decreases at a faster rate until, at some quantity between 8,000 and 10,000 units, they are equal.

Volume	Total Cost (Dollars)		Unit Cost (Dollars)	
(Units)	Method A	Method B	Method A	Method B
2000	$8,000	$22,000	$4.00	$11.00
4000	14000	24000	3.5	6
6000	20000	26000	3.33	4.33
8000	26000	28000	3.25	3.5
10000	32000	30000	3.2	3
12000	38000	32000	3.17	2.67
14000	44000	34000	3.14	2.43
16000	50000	36000	3.13	2.25

Table 14.1   Total and average cost versus quantity produced.

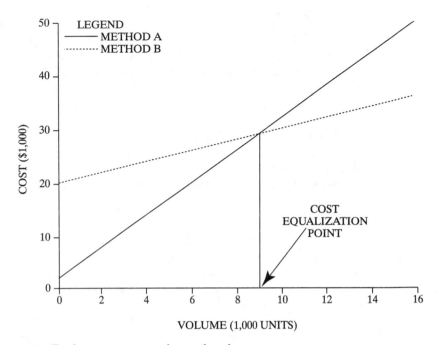

Figure 14.6   Total cost versus quantity produced.

Cost Equalization Point (CEP)

Knowing the quantity beyond which the cost of using method B becomes less than for method A enables us to decide easily which process to use to minimize the total cost (and the unit cost).

This quantity is called the cost equalization point (CEP) and is the volume for which the total cost (and unit cost) of using one method is the same as another. For our example, the total cost calculations are as follows.

$$
\begin{aligned}
TC_A &= TC_B \\
FC_A + VC_A x &= FC_B + VC_B x \\
\$2,000 + \$3x &= \$20,000 + \$1x \\
3x - 1x &= 20,000 - 2,000 \\
2x &= 18,000 \\
x &= 9,000 \text{ units}
\end{aligned}
$$

The CEP is 9,000 units. At this quantity the total cost of using method A will be the same as for method B.

$$
\begin{aligned}
TC_A &= FC_A + VC_A x = 2,000 + (3 \times 9,000) = \$29,000 \\
TC_B &= FC_B + VC_B x = 20,000 + (1 \times 9,000) = \$29,000
\end{aligned}
$$

We can get the same results by using unit costs instead of total costs.

From the preceding calculations we can say:

- If the volume (quantity to produce) is less than the CEP, the method with the lower fixed cost will cost less.
- If the volume is greater than the CEP, the method with the greater fixed cost will cost less.

For example, if the volume were 5,000 units, method A would cost less and if the volume were 10,000 units, method B would be least cost.

Variable costs, mostly direct labor and material, can be reduced by substituting machinery and equipment (capital) for direct labor. This increases the fixed costs and decreases the variable costs. But to justify this economically the volume must be high enough to reduce the total or unit cost of production.

Increasing Volume

The obvious way to increase volume is to increase sales. However, a finished product is made up of several purchased or manufactured components. If the volume of these components can be increased, then the unit cost of the components, and the final product, will be reduced.

The volume of components can be increased without increasing sales by a program of simplification and standardization, discussed earlier in this chapter.

If a subassembly or component part can be standardized for use in more than one final product, then the volume of the subassembly or part is increased without an increase in the total sales volume. Thus more specialized and faster-running processes can be justified and the cost of operations reduced.

Standardization of parts is a major characteristic of modern mass production. At the turn of the century, Henry Ford revolutionized manufacturing by standardizing the finished product—one model of car. The joke often heard then was that you could have any color you wanted as long as it was black. Today a vast range of models are made, but if each model was exploded into its subassemblies and component parts, we would find specific components common to a great number of models. In this way, modern manufacturers can provide the consumer with a wide choice of finished products made from standard parts and components.

CONTINUOUS PROCESS IMPROVEMENT (CPI)

People have always been concerned with how best to do a job and the time it should take to do it. Process improvement is concerned with improving the effective use of human and other resources. Continuous implies an ongoing activity; improvement implies an increase in the productivity or value of quality or condition. Hence the name Continuous Process Improvement.

Continuous process improvement consists of a logical set of steps and techniques used to analyze processes and to improve them.

Improving productivity.   Productivity can be improved by spending money (capital) on better and faster machines and equipment. However, with any given amount of capital, a method must be designed to use the machinery and equipment most productively. A workstation might consist of highly sophisticated machinery and equipment worth $1 million or more. Its productivity and return on investment depend on how the equipment is used and how the operator manages it. CPI determines how equipment is used and managed.

Continuous process improvement is a low-cost method of designing or improving work methods to maximize productivity. The aim is to increase productivity by better use of existing resources. Continuous process improvement is concerned with removing work content, not with spending money on better and faster machines.

Peter Drucker has said "efficiency is doing things right; effectiveness is doing the right things." CPI aims to do the right things and to do them efficiently.

People involvement.   Today management recognizes the need to maximize the potential of flexible, motivated workers. People are capable of thinking, learning, problem solving and contributing to productivity. With existing processes and equipment, people are the primary source of improvement because they are the experts in the things they do.

Process improvement is not solely the responsibility of industrial engineers. Everyone in the workforce must be given the opportunity to improve the processes they work with. Techniques that help to analyze and improve work are not complicated and can be learned. Indeed, the idea of continuous improvement is based on the participation of operators and improvement in methods requiring relatively little capital.

Workers have two jobs:

- Their "as defined" job.
- To improve their "as defined" job.

Teams.   One of the features of CPI is team involvement. A team is a group of people working together to achieve common goals and objectives. The members of the team should be all those who are involved with the process. Teams are successful because of the emphasis placed on people. Not all problems can be solved by teams, nor are all people suited to teams. However, they are often effective. Often problems cross functional lines and thus multi-functional teams are common.

Continuous process improvement can still be effectively carried out by the individuals.

The Six Steps in Continuous Process Improvement

The CPI system is based on the *scientific method.* This general method is used to solve many kinds of problems. The six steps are as follows:

1. Select the process to be studied.
2. Record the existing method to collect the necessary data in a useful form.
3. Analyze the recorded data to generate alternative improved methods.
4. Evaluate the alternatives to develop the best method of doing the work.
5. Install the method as standard practice by training the operator.
6. Maintain the new method.

Select the Process

The first step is to decide what to study. This depends on the ability to recognize situations that have good potential for improvement. Observation of existing methods comes first.

Observe. The important feature in observation is a questioning attitude. Questions such as why, when, and how must be asked whenever we observe something. This attitude needs development because we tend to assume that the familiar method is the only one. Often we hear "We have always done it this way!" "This way" is not necessarily the only, most productive, or most effective way.

Any situation can be improved but some have better potential than others. Indicators that show areas most needing improvement include:

- High scrap, reprocessing, rework, and repair costs.
- Back tracking of material flow caused by poor plant layout.
- Bottlenecks.
- Excessive overtime.
- Excessive manual handling of materials, both from workplace to workplace and at the workplace.
- Employee grievances without true assignable causes.

Select. The purpose of continuous process improvement is to improve productivity to reduce operating, product, or service costs. In selecting jobs or operations for method improvement, there are two major considerations: economic and human.

Economic considerations. The cost of the improvement must be justified. The cost of making the study and installing the improvement must be recovered from the savings in a reasonable time. One to two years is a commonly used period.

The job size must justify the study. Almost anything can be improved, but the improvement must be worthwhile. Suppose a method improvement saves one hour on a job taking five hours, performed once a month or 12 times a year. The reduction in time is 20% and the total time saved in a year is 12 hours. Another method improvement saves one minute on a job taking 10 minutes, performed 200 times per week. The time saved in this case is only 10% but will be $(1 \times 200 \times 52) \div 60 = 173.3$ hours per year, a much higher rate of return on the investment made in the study.

The human factor governs the success of method improvement. The resistance to change, by both management and worker, must always be remembered. Working situations characterized by high fatigue, accident hazards, absenteeism, and dirty and unpleasant conditions should be identified and improved. Sometimes it is difficult to give specific economic justification for such improvements, but the intangible benefits are extensive and should weigh heavily in selecting studies.

Pareto diagrams. Pareto analysis can be used to select problems with the greatest economic impact. The theory of Pareto analysis is the same as that used in the ABC analysis discussed in Chapter 9. This theory says that a few items (usually about 20%) account for most of the cost or problems. It separates the "vital few" from the "trivial many." Examples of the "vital few" are

- A few processes account for the bulk of scrap.
- A few suppliers account for most rejected parts.
- A few problems account for most process downtime.

The steps in making a Pareto analysis are as follows:

1. Determine the method of classifying the data: by problem, cause, nonconformity, and so forth.
2. Select the unit of measure. This is usually dollars but may be the frequency of occurrence.
3. Collect data for an appropriate time interval, usually long enough to include all likely conditions.
4. Summarize the data by ranking the items in descending order according to the selected unit of measure.
5. Calculate the total cost.

6. Calculate the percentage for each item.
7. Construct a bar graph showing the percentage for each item and a line graph of the cumulative percentage.

EXAMPLE PROBLEM

A product has experienced a number of failures in the field. Data is collected according to the type of failure with the following results. Type A—11, type B—8, type C—5, type D—60, type E—100, type F—4, other—12. Construct a table summarizing the data in descending order of magnitude. From this table, construct a Pareto diagram.

Answer

Type of Failure	Number of Failures	Percent	Cumulative Percentage
E	100	50.0	50.0
D	60	30.0	80.0
A	11	5.5	85.5
B	8	4.0	89.5
C	5	2.5	92.0
F	4	2.0	94.0
O (Other)	12	6.0	100.0
Total	200	100.0	

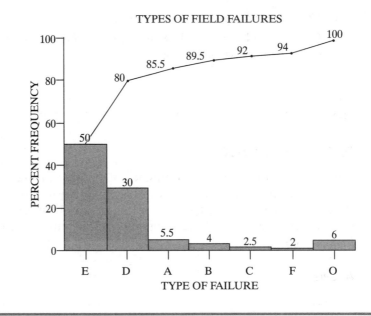

Note that Pareto analysis does not report what the problems are, only where they seem to occur. In the previous example, further investigation into the causes of failure types E and D will give the best return for the effort spent. It is important to select the categories carefully. For example, in the last problem, if the location of failures were recorded rather than type of failure, the results would be quite different and perhaps not significant.

Cause-and-effect diagram.   Sometimes called a fishbone or an Ishikawa diagram, the cause-and-effect diagram is a very useful tool for identifying root causes. Figure 14.7 shows such a diagram.

The fishbone diagram is best used by a group or team working together. It can be constructed by discussion and brainstorming. The steps in developing a fishbone diagram are:

1. Identify the problem to be studied and state it in a few words. For example, the reject rate on machine A is 20%.

2. Generate some ideas about the main causes of the problem. Usually all probable root causes can be classified into six categories.

   • Materials. For example, from consistent to inconsistent raw materials.

   • Machines. For example, a well-maintained machine versus a poorly maintained one.

   • People. For example, a poorly trained operator instead of a well-trained one.

   • Methods. For example, changing the speed on a machine.

Figure 14.7 Cause-and-effect diagram.

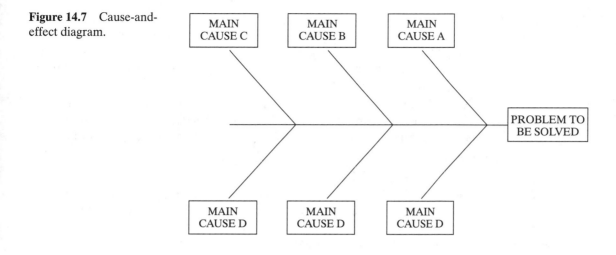

- Measurement. For example, measuring parts with an inaccurate gauge.
- Environment. For example, increased dust or humidity.

3. Brainstorm all possible causes for each of the main causes.
4. Once all the causes have been listed, try to identify the most likely root causes and work on these.

Record

The next step is to record all the facts relating to the existing process. To be able to understand what to record, it is necessary to define the process being studied. Recording defines the process. The following must be determined to properly define the process.

- **The process boundaries.** All processes, big or small, begin and end somewhere. Starting and ending points form the boundaries of the process. For example, the starting point in the process of getting to work in the morning might be getting out of bed. The ending point might be arrival at the desk or classroom.
- **Process flow.** This is a description of what happens between the starting and ending points. Usually this is a listing of the steps taken between the start and finish of the process. There are several recording techniques to help perform this step. Some of these will be discussed later in this chapter.
- **Process inputs and outputs.** All processes change something. The things that are changed are called inputs and they may be physical, such as raw materials, or informational, such as data. Outputs are the result of what goes on in the process. For example, raw materials are converted into something more useful or data is manipulated to produce reports.

- **Components** are the resources used in changing inputs to outputs. They are composed of people, methods, and equipment. Unlike process inputs, components do not become part of the output, but are part of the process. For example, in producing a report the word processor, graphics program, computer, printer, et cetera, are all components.
- **Customer.** Processes exist to serve customers and customers ultimately define what a process is supposed to do. If customer needs are not considered, there is a risk of improving things that do not matter to the users of the output.
- **Suppliers** are those who provide the inputs. They may be internal to the organization or external.
- **Environment.** The process is controlled or regulated by external and internal factors. The external factors are beyond control and include customers' acceptance of the process output, competitors, and government regulation. Internal factors are in the organization and can be controlled.

Figure 14.8 shows a schematic of the system.

   The next step is to record all facts relating to the existing method. A record is necessary because it is difficult to record and maintain a large mass of detail in our heads for the duration of the analysis. Recording helps us consider all elements of the problem in a logical sequence and makes sure we do consider all the steps in the process. The record of the present method also provides the basis for both the critical examination and the development of an improved method.

Figure 14.8   Schematic of a process.

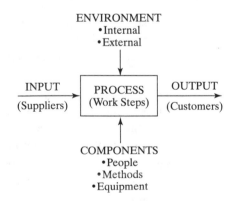

Classes of activity. Before discussing some of the charts used, we should look at the kinds of activities recorded. All activity can generally be classified into one of six types. As a method of shorthand, there are six universally used symbols for these activities. The activities and symbols are shown in Figure 14.9. One of the activities shown, decision, may not always be used. If not, a decision is usually considered an operation.

Following are descriptions of some of the various charting techniques.

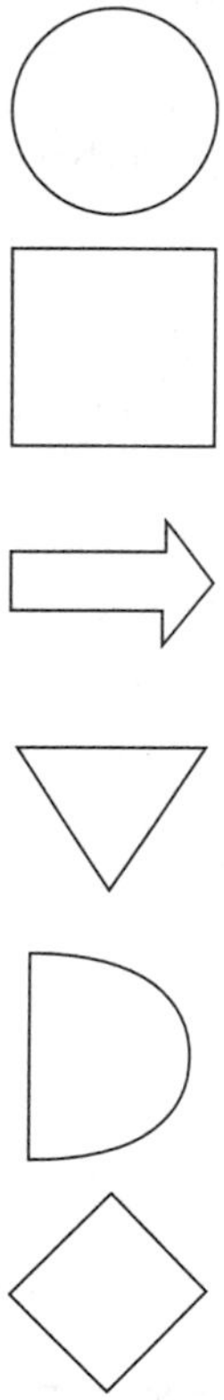

Operation. The main step in a process, method, or procedure. Usually the part, material, or product is modified during the operation.

Inspection. An inspection for quality or a check for quantity. Where process can be measured, regulated, and controlled.

Movement. The movement of workers, material, equipment, or information from place to place.

Storage. A controlled storage from which material is issued or received.

Delay. A delay in the sequence of events; for example, material waiting to be worked on.

Decision. Where a decision is made.

Figure 14.9 Classes of activity.

Operations Process Charts. These charts record in sequence only the main operation and inspections. They are useful for preliminary investigation and give a bird's-eye view of the process. Figure 14.10 shows such a chart.

The description, and sometimes the times, for each operation are also shown. An operations process chart would be used to record product movement.

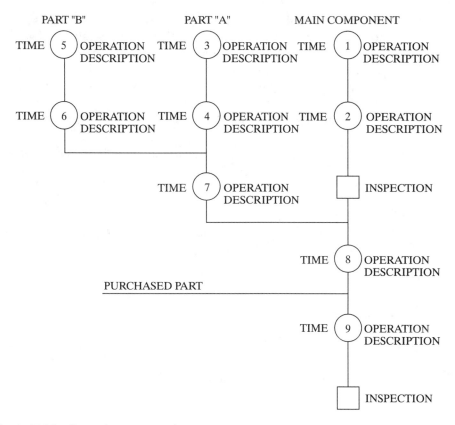

Figure 14.10 Operations process chart.

Process flow diagram. A process flow diagram shows graphically and sequentially the various steps, events, and operations that make up a process. It provides a picture, in the form of a diagram, of what actually happens when a product is made or a service performed. In addition to the six symbols shown in Figure 14.9, others may be used to show such things as rework and documentation. Figure 14.11 shows an example of a process flow diagram. In this example, the process starts when the goods are received and ends when a check is sent to the supplier.

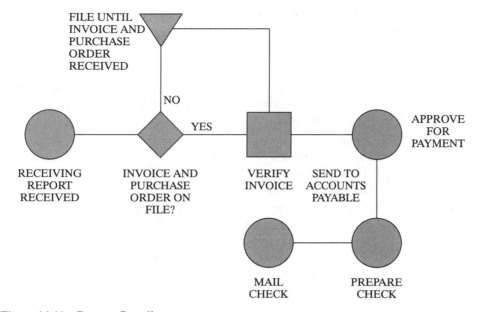

Figure 14.11 Process flow diagram.

Analyze

Examination and analysis are the key steps in continuous process improvement. Although all the other steps are significant, they either lead up to, or result from, the critical analysis. This step involves analyzing every aspect of the present method and evaluating all proposed possible methods.

Find the root cause. Frequently it is difficult to separate symptoms from the root causes of problems. Often symptoms are what we see and it is difficult to trace back to the root cause. To find root causes you must have a questioning attitude. For the analyst, "why" is the most important word. Every aspect of the existing method should be questioned. In "The Elephant's Child," from *The Just-So Stories,* Rudyard Kipling wrote:

> *I keep six honest serving-men*
> *(They taught me all I know);*
> *Their names are What and Why and When*
> *And How and Where and Who.*

Kipling personifies six words normally used as questions; they are also the "serving men" of the methods analyst.

A rule of thumb common to many problem-solving methods says it is necessary to ask (and answer) the "why" question up to five times before one reaches the root cause of a problem.

Three approaches can help in examining.

- A questioning attitude. This implies an open mind, examining the facts as they are, not as they seem, avoiding preconceived ideas, and avoiding hasty judgments.
- Examining the total process to define what is accomplished, how, and why. The answers to these questions will determine the effectiveness of the total process. The results may show that the process is not even needed.
- Examining the parts of the process. Activities can be divided into two major categories: Those in which something is happening to the product (worked on, inspected, or moved) and those in which nothing constructive is happening to the product (delay or storage, for example). In the first category, value is added only when the part is being worked on. Setup, put away, and move, while necessary, add cost to the product but do not increase its value. They must be minimized. Value will be added when the product is being worked on but, again, the goal is to maximize the productivity of these operations.

Develop

When developing possible solutions, there are four approaches to take to help develop a better method.

- *Eliminate* all unnecessary work. Question why the work is being done in the first place and if it can be eliminated.
- *Combine* operations wherever possible. Thus material handling will be reduced, space saved, and the throughput time reduced. This is a major thrust of just-in-time manufacturing.
- *Rearrange the sequence* of operations for more effective results. This is an extension of the previous approach. If sequences are changed, then possibly they can be combined.
- *Simplify* wherever possible by making the necessary operations less complex. If the questioning attitude is used, then complexity should be reduced. Usually the best solutions are the simple ones.

Principles of motion economy.   There are several principles of motion economy, among which are the following.

1. Locate materials, tools, and workplace within normal working areas and pre-position tools and materials.
2. Locate the work done most frequently in the normal working areas and everything else within the maximum grasp areas.

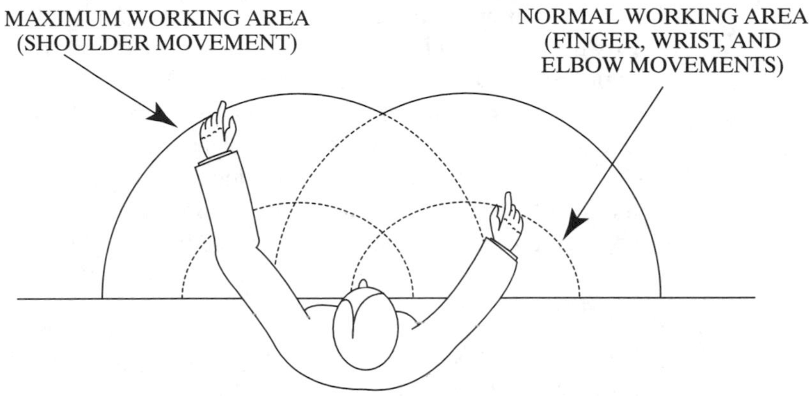

Figure 14.12 Working areas.

MAXIMUM WORKING AREA
(SHOULDER MOVEMENT)

NORMAL WORKING AREA
(FINGER, WRIST, AND
ELBOW MOVEMENTS)

3. Arrange work so motions of hands, arms, legs, and so on are balanced by being made simultaneously, in opposite directions, and over symmetrical paths. Both hands should be working together and should start and finish at the same time. The end of one cycle should be located near the start of the next cycle.

4. Conditions contributing to operator fatigue must be reduced to a minimum. Provide good lighting, keep tools and materials within maximum working areas (see Figure 14.12), provide for alternate sitting and standing at work, and design workplaces of proper height to eliminate stooping.

Human and environmental factors.   In addition to the principles of motion economy, other important matters that influence human and environmental conditions must be considered. These include safety, comfort, cleanliness, and personal care; so provision must be made for lighting, ventilation and heating, noise reduction, seating, and stimulation.

Of these, stimulation may be less obvious. In highly repetitive work, workers may become bored and dissatisfied, which may lead to emotional problems. A pleasant environment created by attractive color schemes in plants or offices, location of windows, and music during working hours, can do much to reduce stress and absenteeism.

Job design. Methods improvement is based on the concepts of scientific management. It concentrates on the task and ways of removing work content (waste) from tasks. It gives little consideration to the human being and higher-level needs, such as self-esteem and self-fulfilment, and work can become repetitive and boring. Job design is an attempt to provide more satisfying meaningful jobs and to use the worker's mental and interpersonal skills. These improvements include the following.

Job enlargement. A worker's job is expanded by clustering similar or related tasks into one job. For example, a job might be expanded to include a sequence of activities instead of only one activity. This is called *horizontal enlargement.*

Job enrichment adds more meaningful, satisfying, and fulfilling tasks. The job not only includes production operations but many setup, scheduling, maintenance, and control responsibilities.

Job rotation trains workers to do several jobs so they can be moved from one job to another. This is called **cross training.**

All these factors help to produce a more motivated and flexible workforce. In modern manufacturing, where quick response to customers' needs is essential, these characteristics can mean the difference between business success and failure.

Install

So far, the work done by the analyst has been planning. Now the plans must be put into action by installing the new method.

In planning the installation, consideration must be given to the best time to install, the method of installing, and the people involved. Then the analyst needs to be sure that equipment, tooling, information, and the people are all available. At installation time, a dry run will show whether all equipment and tooling are working properly.

Training the operator is the most important part of the installation. If the operator has been involved in designing the method change, this should not be difficult. The worker will be familiar and comfortable with the change and will probably feel some sense of ownership.

Figure 14.13   Learning curve.

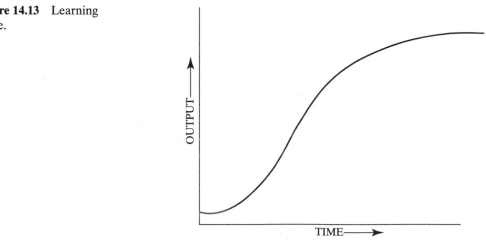

Learning curve.   Over time, as the operator does the tasks repetitively, speed will increase and errors decrease. This process is known as the learning curve and is illustrated in Figure 14.13. Note there is no time scale shown. Depending on the task, a worker may progress through the learning curve in a few minutes or, for high-skill jobs, several months or years.

Maintain

Maintaining is a follow-up activity that has two parts. The first is to be sure that the new method is being done as it should be. This is most critical for the first few days and close supervision may be necessary. The second is to evaluate the change to be sure that the planned benefits are accomplished. If not, the method must be changed.

QUESTIONS

1. Describe simplification, standardization, and specialization. Why are they important and why are they interrelated?
2. What are the advantages and disadvantages of standardization?
3. What are the advantages and disadvantages of specialization?
4. What are product focus and process focus based on? What is a focused factory?
5. What is the advantage to modular design?
6. How are production costs affected by:
   a. Standard sizes?
   b. Universal fit parts?
7. What are the two criteria for designing a product?
8. Why is product design important to operations costs?
9. Why is product design important to quality?
10. What is simultaneous engineering and what are some of its advantages?
11. What is a process? What is process nesting?
12. What are the five basic factors that must be considered when designing a process?
13. Why are product design, quantity to produce, and process design intimately related?
14. Give four reasons why companies will make a product in-house and why they will buy out.
15. Describe general-purpose and special-purpose machinery. Compare each for flexibility of use, operator involvement, run time per piece, setup time, quality, capital cost, and application.
16. What is flow processing and what are its advantages and disadvantages?
17. What is intermittent processing and when is it used? Contrast it with flow processing.
18. When is project (fixed position) processing used?
19. Define fixed and variable costs and give examples of each in manufacturing. What is total cost and what is the equation for it?

20. What is the cost equalization point?

21. How can the variable cost be reduced? What does this do to the fixed costs and what is needed to economically justify this course of action?

26. What are the six steps in continuous process improvement?

27. Name and describe the two considerations in selecting a job to be studied.

28. What is a Pareto diagram and why is it useful?

29. What is a cause-and-effect diagram and why is it useful?

30. Why is it necessary to record?

31. Describe each of the following as they relate to recording:

    a. Process boundaries

    b. Process flow

    c. Process inputs and outputs

    d. Process components

    e. Suppliers

    f. Process control

32. What are the six symbols used in method analysis? Are there other symbols that can be used?

33. Describe each of the following:

    a. Operations process chart

    b. Process flow diagram

34. What is the purpose of the analysis step? What is the basic question?

35. What are the four approaches that should be taken when developing a better method?

36. What are the four principles of motion economy?

37. Describe job design.

38. What is the learning curve?

PROBLEMS

14.1 Select a product with which you are familiar. How do you think it might be redesigned to make it easier to manufacture and possibly more useful to the user?

14.2 Given the following fixed and variable costs and the volumes, calculate the total and unit costs.

Fixed Cost	Variable cost	Volume (units)	Total cost	Unit cost
$100.00	$10.00	100		
$200.00	$5.00	1000		
$50.00	$15.00	10		
$1,000.00	$1.00	1000		
$500.00	$20.00	500		

14.3 A process costs $200 to set up. The run time is five minutes per piece and the run cost is $30 per hour. Determine:

a. The fixed cost

b. The variable cost

c. The total cost and unit cost for a lot of 500.

d. The total cost and unit cost for a lot of 1000.

Answer.	a. Fixed cost	=	$200.00
	b. Variable cost	=	$2.50
	c. Total cost	=	$1450.00
	Unit cost	=	$2.90
	d. Total cost	=	$2750.00
	Unit cost	=	$2.75

14.4   A manufacturer has a choice of purchasing and installing a heat-treating oven or having the heat treating done by an outside supplier. The manufacturer has developed the following cost estimates:

	Heat treat in-house	Purchase services
Fixed cost	$25,000.00	$0.00
Variable cost	$10.00	$17.50

  a. What is the cost equalization point?

  b. Should the company have the heat treating done outside if the annual volume is:

  3,000 units

  5,000 units

  c. What would be the unit (average) cost for the selected process for each of the volumes in b above?

  Answer.   CEP is at 3333.33 units

  3000 units. Purchase. Unit cost   = $17.50

  5,000 units. In house. Unit cost   = $15.00

14.5   Bananas are on sale at the Cross Towne store for 29¢ per pound. They normally sell for 49¢ per pound at your corner store. If round trip bus fare costs $1.80 to the Cross Towne store, is it worth going for? Discuss other ways of taking advantage of this bargain.

14.6   Given the following costs, which process should be used for an order of 400 pieces of a given part? What will be the unit cost for the process selected?

	Buy	Process A	Process B
Setup		$20.00	$150.00
Tooling		$10.00	$20.00
Labor/unit		$4.00	$3.75
Material/unit		$2.00	$2.00
Purchase cost	$6.10		

14.7   The Light Company is planning on producing a new type of light shade, the parts for which may be made or bought. If purchased, they will cost $2 per unit. Making the parts on a semi-automatic machine will involve a $5,000 fixed cost for tooling and $1.30 per unit variable cost. The alternative is to make the parts on an automatic machine. The tooling costs are $15,000 but the variable cost is reduced to 60¢ per unit.

  a. Calculate the cost equalization point between buying and the semi-automatic machine.

  b. Calculate the cost equalization point between the semi-automatic and automatic machines.

    c. Which method should be used for expected sales of:

i)	5,000	units
ii)	6,000	units
iii)	8,000	units
iv)	10,000	units
v)	20,000	units

    d. What is the unit cost for the selected process for each of the sales in *c* above?

Answer.	i)	Unit cost	=	$2.00
	ii)	Unit cost	=	$2.00
	iii)	Unit cost	=	$1.925
	iv)	Unit cost	=	$1.80
	v)	Unit cost	=	$1.20

14.8 A major mail order house collected data on the reasons for return shipments over a three-month period with the following results. Wrong size 50,000, order canceled 15,000, wrong address 3000, other 15,000. Construct a Pareto diagram.

Reason	Number	Percent	Cumulative Percent
Total			

14.9　A firm experienced abnormal scrap and collected data to see which parts were causing the problem with the following results. Part A—$5720, part B—$10,250, Part C—$820, Part D—$1130, Part F—$700. Complete the following table listing the errors in descending order of importance. Construct a Pareto diagram.

Part	Number	Percent	Cumulative Percent
Total			

14.10　Draw an operation process chart for the assembly of a ballpoint pen. The pen is made from three subassemblies (see Figure 14.14):

　　1. upper barrel

　　2. cartridge

　　3. lower barrel

Operation　1　attach clip to upper barrel

　　　　　　2　insert button into upper barrel

　　　　　　3　insert rotor into upper barrel

　　　　　　4　press ball into tip of cartridge

　　　　　　5　insert tip into cartridge

　　　　　　6　fill cartridge with ink

Inspection　1　test cartridge

Operation　7　insert cartridge into upper barrel

　　　　　　8　slip spring over cartridge

　　　　　　9　screw lower barrel and upper barrel together

Inspection　2　test operation of pen

　　　　　　3　final inspection

Figure 14.14 Ballpoint pen assembly.

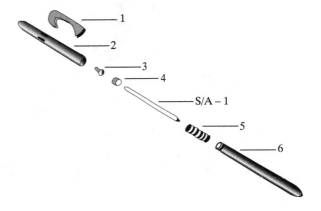

14.11 Draw a process flow diagram for your activities from the time you wake up until you arrive at work. Can you think of ways to improve the process?

15

Just-in-Time Manufacturing

INTRODUCTION

In the past 20 years, manufacturing has become much more competitive and the global economy is now a forceful reality. Countries such as Japan and others on the Pacific Rim can produce goods of consistently superior quality and deliver them to North American markets at a competitive price and schedule. They have responded to changing market needs and often have detected those needs before the consumer. The Walkman, developed by Sony, is an example of Japanese market awareness. Because of such competition, North America has lost the edge in the manufacture of such goods as radios, televisions, cameras, and ships.

How have the Japanese been able to do this? It is not because of their culture, geography, government assistance, new equipment, or cheap labor, but because they practice just-in-time manufacturing. **Just-in-time manufacturing (JIT)** is a philosophy that relates to the way a manufacturing company organizes and operates its business. It is not a magic formula or a set of new techniques that suddenly makes a manufacturer more productive. Rather, it is the very skillful application of existing industrial and manufacturing engineering principles. The Japanese have not taught us new tricks but have forced us to examine some of our basic assumptions and approach manufacturing with a different philosophy.

JUST-IN-TIME PHILOSOPHY

Just-in-time manufacturing is defined in many ways, but the most popular is the *elimination of all waste and continuous improvement of productivity.* Waste means anything other than the minimum amount of equipment, parts, space, material, and workers' time absolutely necessary to add value to the product. This means there should be no surplus, there should be no safety stocks, and lead times should be minimal: "If you can't use it now, don't make it now."

The long-term result of eliminating waste is a cost-efficient, quality-oriented, fast-response organization that is responsive to customer needs. Such an organization has a huge competitive advantage in the marketplace.

Adding Value

What constitutes value to the user? It is having the right parts and quantities at the right time and place. It is having a product or service that does what the customer wants, does it well and consistently, and is available when the customer wants it. Value satisfies the actual and perceived needs of the customer and does it at a price the customer can afford and considers reasonable. Another word for this is *quality.* Quality is meeting and exceeding customers' expectations.

Value starts in the marketplace when marketing must decide what the customer wants. Design engineering must design the product so it will provide the required value to the customer. Manufacturing engineering must first design a process to make the product and then build the product according to certain specifications. The loop is complete when the product is delivered to the customer. Figure 15.1 shows this loop schematically. If any part of the chain does not add value for the customer, there is waste.

Adding value to a product does not mean adding cost. Users are not concerned with the manufacturer's cost but only with the price they must pay and the value they receive. Many activities increase cost without adding value and, as much as possible, these activities should be eliminated.

Figure 15.1   Product cycle.

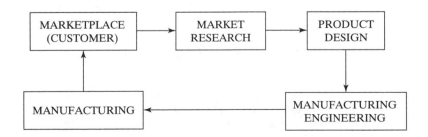

WASTE

Anything in the product cycle that does not add value to the product is waste. This section will look at the causes of waste in each element of the product cycle.

Waste Caused by Poor Product Specification and Design

Waste making starts with the policies set by management in responding to the needs of the marketplace. Management is responsible for establishing policies for the market segments the company wishes to serve and for deciding how broad or specialized the product line is to be. These policies affect the costs of manufacturing. For example, if the range and variety of product are large, production runs will be short, and machines must be changed over frequently. There will be little opportunity to use specialized machinery and fixtures. On the other hand, a company with a limited product range can probably produce goods on an assembly-line basis and take advantage of special-purpose machinery. In addition, the greater the diversity of products, the more complex the manufacturing process becomes, and the more difficult it is to plan and control.

Component standardization.   As stated in Chapter 14, companies can specialize in the products they make and still offer customers a wide range of options. If companies standardize on the component parts used in the different models they make, they can supply the customer with a variety of models and options made from standard components. Parts standardization has many advantages in manufacturing. It creates larger quantities of specific components that allow longer production runs. This, in turn, makes it more economical to use more specialized machinery, fixtures, and assembly methods. Standardization reduces the planning and control effort needed, the number of items required, and the inventory that has to be carried.

The "ideal" product is one that meets or exceeds customer expectations, makes the best use of material, and can be manufactured with a minimum of waste (at least cost). As well as satisfying the customer, the product's design determines both the basic manufacturing processes that have to be used and the cost and quality of the product. The product should be designed so it can be made by the most productive process with the smallest number of operations, motions, and parts and includes all of the features that are important to the customer.

Chapter 14 discussed the principles of product design in more detail.

Waste Caused in Manufacturing

Manufacturing takes the design and specifications of the product and, using the manufacturing resources, converts them into useful products. First, however, manufacturing engineering must design a system capable of making the product. They do so by selecting the manufacturing steps, machinery, and equipment and by designing the plant layout and work methods. Manufacturing must then plan and control the operation to produce the goods. This involves manufacturing planning and control, quality management, maintenance, and labor relations.

Toyota has identified seven important sources of waste in manufacturing. The first four relate to the design of the manufacturing system and the last three to the operation and management of the system:

1. *The process.* The best process is one that has the capability to consistently make the product with an absolute minimum of scrap, in the quantities needed, and with the least cost added. Waste, or cost, is added to the process if the wrong type or size of machines are used, if the process is not being operated correctly, or if the wrong tools and fixtures are used.

2. *Methods.* Waste is added if the methods of performing tasks by the operators cause wasted movement, time, or effort. Activities that do not add value to the product should be eliminated. Searching for tools, walking, or unnecessary motions are all examples of waste.

3. *Movement.* Moving and storing components adds cost but not value. For example, goods received may be stored and then issued to production. This requires labor to put away, find, and deliver to production. Records must be kept, and a storage system maintained. Poorly planned layouts may make it necessary to move products over long distances, thus increasing the movement cost and possibly storage and record-keeping costs.

4. *Product defects.* Defects interrupt the smooth flow of work. If the scrap is not identified, the next workstation receiving it will waste time trying to use the defective parts or waiting for good material. Schedules must be adjusted. If the next step is the customer, then the cost will be even higher. Sorting out or reworking defects are also waste.

5. *Waiting time.* There are two kinds of waiting time: that of the operator and that of material. If the operator has no productive work to do or there are delays in getting material or instructions, there will be waste. Ideally, material passes from one work center to the next and is processed without waiting in queue.

6. *Overproduction.* Overproduction is producing products beyond those needed for immediate use. When this occurs, raw materials and labor are consumed for parts not needed, resulting in unnecessary inventories. Considering the costs of carrying inventory, this can be very expensive. Overproduction causes extra handling of material, extra planning and control effort, and quality problems. Because of the extra inventory and work-in-process, overproduction adds confusion, tends to bury problems in inventory, and often leads to producing components that are not needed instead of those that are. Overproduction is not necessary as long as market demand is met. Machines and operators do not always need to be fully utilized.

7. *Inventory.* As we saw in Chapter 9, inventory costs money to carry, and excess inventory adds extra cost to the product. However, there are other costs in carrying excess inventory.

To remain competitive, a manufacturing organization must produce better products at lower cost while responding quickly to the marketplace. Let us look at the role inventory plays in each of these steps.

A better product suggests one that has features and quality superior to others. The ability to take advantage of product improvement opportunities depends on the speed with which engineering changes and improvements can be implemented. If there are large quantities of inventory to work through the system, it takes longer and is more costly before the engineering changes reach the marketplace. Lower inventories improve quality. Suppose that a component is made in batches of 1000 and that a defect occurs on the first operation. Eventually, the defect will be caught, very often after several more operations have been completed. Thus, all 1000 pieces have to be inspected. Because much time has elapsed since the first operation when the defect occurred, it is also difficult to pinpoint the cause of the problem. If the batch size had been 100 instead of 1000, it would have moved through the system more quickly and been detected earlier—and there would only be 100 to inspect.

Companies can offer better prices if their costs are low. Lower inventories reduce cost. Also, if work-in-process inventory is reduced, less space is needed in manufacturing, resulting in cost savings.

Responsiveness to the marketplace depends on being able to provide shorter lead times and better due date performance. In Chapter 6, we saw that manufacturing lead time depends on queue and queue depends on the number and the batch size of the orders in process. If batch size is reduced, the queue and lead time will be reduced. Chapter 8 noted that forecasts were more accurate for nearer periods of time. Reducing lead time improves forecast accuracy and provides better order promising and due-date performance.

JUST-IN-TIME ENVIRONMENT

Many elements are characteristic of a JIT environment. They may not all exist in a particular manufacturing situation, but in general they provide some principles to help in the development of a JIT system. These can be grouped under the following headings:

- Flow manufacturing
- Process flexibility
- Total quality management
- Total productive maintenance
- Uninterrupted flow
- Continuous process improvement
- Supplier partnerships
- Total employee involvement

Flow Manufacturing

The just-in-time concept was developed by companies such as Toyota and some major appliance and consumer electronic manufacturers. These companies manufacture goods in a repetitive manufacturing environment.

Repetitive manufacturing is the production of discrete units on a flow basis. In this type of system, the workstations required to make the product, or family of products, are located close together and in the sequence needed to make the product. Work flows from one workstation to the next at a relatively constant rate and often with some materials handling system to move the product. Figure 15.2 shows a schematic of flow manufacturing.

Figure 15.2   Flow manufacturing.   WORKSTATIONS

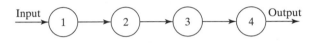

These systems are discussed in Chapter 14. They are suitable for a limited range of similar products such as automobiles, televisions, or microwave ovens. Because work centers are arranged in the sequence needed to make the product, the system is not suitable for making a variety of different products. Therefore, the demand for the family of products must be large enough to justify economically setting up the line. Flow systems are usually very cost effective.

Work cells.   Many companies do not have a product line that lends itself to flow manufacturing. For example, many companies do not have sufficient volume of specific parts to justify setting up a line. Companies with this kind of product line usually organize their production on a functional basis by grouping together similar or identical operations. Lathes will be placed together as will milling machines, drills, and welding equipment. Figure 15.3 shows a schematic of this kind of layout including routing for a hypothetical product (saw, lathe, grinder, lathe, drill). Product moves from one workstation to the other in lots or batches. This type of

Figure 15.3   Functional layout.

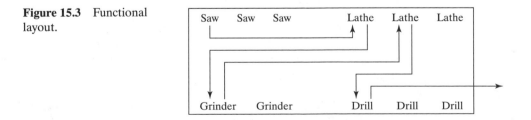

production produces long queues, high work-in-process inventory, long lead times, and considerable materials handling.

Usually this kind of layout can be improved. It depends on the ability to detect product flows. This can be done by grouping products together into product families. Products will be in the same family if they use common work flow or routing, materials, tooling, setup procedures, and cycle times. Workstations can then be set up in miniature flow lines or work cells. For example, suppose the product flow shown in Figure 15.3 represents the flow for a family of products. The work centers required to make this family can be laid out according to the steps to make that family. Figure 15.4 is a schematic of such a layout.

Figure 15.4   Work cell layout.

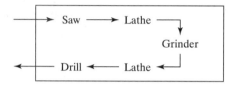

Parts can now pass one by one, or in very small lots, from one workstation to the next. This has several benefits:

- Queue and lead times going through the cell are reduced drastically.
- Production activity control and scheduling are simplified. The cell has only one work center to control as opposed to five in a conventional system.
- Floor space needed is reduced.
- Feedback to preceding operations is immediate. If there is a quality problem, it will be found out immediately.

Work cells permit high-variety, low-volume manufacturing to be repetitive. For work cells to be really effective, product design and process design must work together so parts are designed for manufacture in work cells. Component standardization becomes even more important.

Process Flexibility

Process flexibility is desirable so the company can react swiftly to changes in the volume and mix of their products. To achieve this, operators and machinery must be flexible, and the system must be able to be changed over quickly from one product to another.

Machine flexibility.   It often makes more sense to have two relatively inexpensive general-purpose machines than one large, expensive special-purpose piece of equipment. Smaller general-purpose machines can be adapted to particular jobs

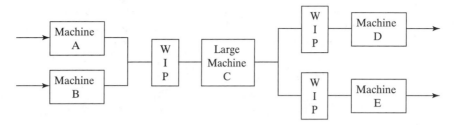

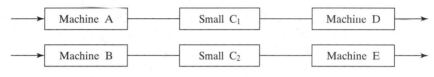

Flow with Small-Capacity Dedicated Equipment

Figure 15.5 Large versatile equipment versus small dedicated equipment.

with appropriate tooling. Having two instead of one makes it easier to dedicate one to a work cell. Ideally, the machinery should be low-cost and movable. Figure 15.5 illustrates the concept.

Quick changeover.   Quick changeover requires short setup times. Short setup times also have the following advantages:

- *Reduced economic-order quantity.* The economic lot size depends on the setup cost. If the setup time can be reduced, the lot size can be reduced. For example, if the economic-order quantity is 100 units and the setup can be reduced to 25%, the economic-order quantity decreases to 50. Inventory is cut in half, and queue and lead times are reduced. Reductions in setup of even greater magnitude are possible. The general opinion is that setup can be cut 50% by organizing the work and having the right tools and fixtures available when needed. For example, in one instance, a changeover on a die press was videotaped. The operator doing the changeover was not in view for more than 50% of the time! He was away from the machine getting tools, dies, and so on. One system for setup reduction, called the "four-step method," claims that reductions of 90% can be achieved without major capital expense. This is accomplished by organizing the preparation, streamlining the setup, and eliminating adjustments.

- *Reduced queue and manufacturing lead time.* Manufacturing lead time depends mostly on the queue. In turn, queue depends on the order quantity and scheduling. Reducing setup time reduces the order quantity and queue and lead times.

- *Reduced work-in-process (WIP) inventory.* Work-in-process inventory depends on the number of orders in process and their size. If the order quantity is reduced, the WIP is reduced. This frees up more floor space, allowing work centers to be moved closer together, thus reducing handling costs and promoting the creation of work cells.
- *Improved quality.* When order quantities are small, defects have less time to be buried. Because they are more quickly and easily exposed, their cause will more likely be detected and corrected.
- *Improved process and material flow.* Inventory acts as a buffer, burying problems in processes and in scheduling. Reducing inventory reduces this buffer and exposes problems in the production process and in the materials control system. This gives an opportunity to correct the problems and improve the process.

Operator flexibility.   Flexible machinery and flexible processes need flexible people to operate them. Not only should people be trained in their own jobs, they should also be cross-trained in other skills and in problem-solving techniques. Only with well-trained people can the benefits of process flexibility be realized.

Total Quality Management

Total quality management (TQM) is the subject of Chapter 16. This section will focus on quality from the point of view of manufacturing.

Quality is important for two reasons. If quality is not present in what is supplied to the customer and the product is defective, the customer will be dissatisfied. If a process produces scrap, it creates disrupted schedules that delay supplying the customer, increases inventory or causes shortages, wastes time and effort on work centers, and increases the cost of the product.

Who is the user? Ultimately, it is the company's customer, but the user is also the next operation in the process. Quality at any one work center should meet or exceed the expectations (needs) of the next step in the process. This is important in maintaining the uninterrupted flow of material. If defects occur at one work center and are not detected until subsequent operations, time will be wasted, and the quantity needed will not be supplied.

For manufacturing, quality does not mean inspecting the product to segregate good from bad parts. Manufacturing must ensure that the process is capable of producing the required quality consistently and with as close to zero defects as possible. Manufacturing must do all it can to improve the process to achieve this and then monitor the process to make sure it remains in control. Daily monitoring can best be done by the operator. If defects are discovered, the process should be stopped, and the cause of the defects corrected.

The benefits of a good-quality program are less scrap, less rework, less inventory (inventory just in case there is a problem), better on-time production, timely deliveries, and more satisfied customers.

Quality at the source.   Quality at the source means doing it right the first time and, if something does go wrong, stopping the process and fixing it. People become their own inspectors, personally responsible for the quality of what they produce.

Total Productive Maintenance

Traditional maintenance might be called "breakdown maintenance," meaning maintenance is done only when a machine breaks down. The motto of breakdown maintenance is "If it ain't broke, don't fix it." Unfortunately, breakdowns occur only when a machine is in operation, resulting in disrupted schedules, excess inventory, and delayed deliveries. In addition, lack of proper maintenance results in wear and poor performance. For example, if a car is not properly maintained, it will break down, not start, or perform poorly on the road.

For a process to continue to produce the required quality, machinery must be maintained in excellent condition. This can best be achieved through a program of preventive maintenance. This is important for more reasons than quality. Low work-in-process inventories mean there is little buffer available. If a machine breaks down, it will quickly affect other work centers. Preventive maintenance starts with daily inspections, lubrication, and cleanup. Since operators usually understand how their equipment should "feel" better than anyone else, it makes more sense to have them handle this type of regular maintenance.

Total productive maintenance takes the ideas of total preventive maintenance one step further. According to the eighth edition of the APICS Dictionary, total productive maintenance is "preventive maintenance plus continuing efforts to adapt, modify, and refine equipment to increase flexibility, reduce material handling, and promote continuous flow." As such, it emphasizes the JIT principle of eliminating waste.

Uninterrupted Flow

Ideally, material should flow smoothly from one operation to the next with no delays. This is most likely to occur in repetitive manufacturing where the product line is limited in variety. However, the concept should be the goal in any manufacturing environment. Several conditions are needed to achieve uninterrupted flow of materials: uniform plant loading, a pull system, valid schedules, and linearity.

Uniform plant loading.   The work done at each workstation should take about the same time. In repetitive manufacturing, this is called **balancing the line,** which means that the time taken to perform tasks at each workstation on the line is the same or very nearly so. The result will be no bottlenecks and no buildup of work-in-process inventory.

Pull system.   Demand on a workstation should come from the next workstation. The pull system starts at the end of the line and pulls product from the preceding operation as needed. The preceding operation does not produce anything unless a signal is sent from the following operation to do so. The system for signaling demand depends on the physical layout and conditions in the plant. The most well-known system is the Kanban system. The details vary, but it is basically a two-bin, fixed-order-quantity, order-point system. A small inventory of parts is held at the user operation—for example, two containers (bins) of parts. When one bin is used up, it is sent back to the supplier operation and is the signal for the supplier operation to make a container of parts. The containers are a standard size and hold a fixed number of parts (order quantity). This system also makes the counting and control of WIP inventory much easier.

Valid schedules.   There should be a well-planned valid schedule. The schedule sets the flow of materials coming into the factory and the flow of work through manufacturing. To maintain an even flow, the schedule must be level. In other words, the same amount should be produced each day. Furthermore, the mix of products should be the same each day. For example, suppose a company makes a line of dog clippers composed of three models: economy, standard, and deluxe. The demand for each is 500, 600, and 400 per week, respectively, and the capacity of the assembly line is 1500 per week. The company can develop the schedule shown in Figure 15.6. This will satisfy demand and will be level based on capacity. However, inventory will build up and, if there is no safety stock, a variation in demand will create a shortage. For example, if there is a surge in demand for the deluxe model in week 1, none may be available for sale in week 2.

Figure 15.6   Master production schedule.

Week	On Hand	1	2	3
Economy	0	1500		
Standard	600		1500	300
Deluxe	800			1200
Total	1400	1500	1500	1500

An alternate schedule is shown in Figure 15.7. With this schedule, inventory is reduced, and the ability to respond to changes in model demand increases. The number of setups increases, but this is not a problem if setup times are small. The idea can be carried further by making some of each model each day. Now it would mean producing 100, 120, and 80 of each model each day. If the line has complete flexibility, these can be produced in the following mixed sequence of 15. This is repeated 20 times during the day for a total output of 300:

E: Economy

S: Standard

D: Deluxe Sequence: ESD, ESD, ESD, ESD, SES

Figure 15.7 MPS leveled by week.

Week	On Hand	1	2	3
Economy	250	500	500	500
Standard	300	600	600	600
Deluxe	200	400	400	400
Total	750	1500	1500	1500

The company makes some of everything each day in the proportions to meet demand. Inventories are at a minimum. If demand shifts between models, the assembly line can respond daily. This is called **mixed-model scheduling.** The schedule is leveled, not only for capacity, but also for material.

Linearity. The emphasis in JIT is on achieving the plan—no more, no less. This concept is called **linearity** and is usually reached by scheduling to less than full capacity. If an assembly line can produce 100 units per hour, it can be scheduled for perhaps 700 units for an eight-hour shift. If there are problems during the shift, there is extra time so the 700-unit schedule can be maintained. If there is time left over after the 700 units are produced, it can be spent on jobs such as cleanup, lubricating machinery, getting ready for the next shift, or solving problems.

Continuous Process Improvement

This topic is an element in both JIT and TQM and was discussed in Chapter 14. Elimination of waste depends on improving processes continuously. Thus continuous improvement is a major feature of just-in-time manufacturing.

Supplier Partnerships

If good schedules are to be maintained and the company is to develop a just-in-time environment, it is vital to have good, reliable suppliers. They establish the flow of materials into the factory.

Partnering implies a long-term commitment between two or more organizations to achieve specific goals. Just-in-time philosophy places much emphasis not only on supplier performance but also on supplier relations. Suppliers are looked on as co-producers, not as adversaries. The relationship with them should be one of mutual trust and cooperation.

There are three key factors in partnering.

1. *Long-term commitment.* This is necessary to achieve the benefits of partnering. It takes time to solve problems, improve processes, and build the relationship need.

2. *Trust.* Trust is needed to eliminate an adversarial relationship. Both partners must be willing to share information and form a strong working relationship. Open and frequent communications are necessary. In many cases the parties have access to each other's business plans and technical information.

3. *Shared vision.* All partners must understand the need to satisfy the customer. Goals and objectives should be shared so that there is a common direction.

If properly done, partnering should be a win-win situation. The benefits to the buyer include the following:

- The ability to supply the quality needed all the time so there will be no need for inbound inspection. This implies that the supplier will have, or develop, an excellent process quality improvement program.

- The ability to make frequent deliveries on a just-in-time basis. This implies that the supplier will become a just-in-time manufacturer.

- The ability to work with the buyer to improve performance, quality, and cost. For a supplier to become a just-in-time supplier, a long-term relationship must be established. Suppliers need to have that assurance so they can plan their capacity and make the necessary commitment to a single customer.

In return, the supplier has the following benefits.

- A greater share of the business with long-term security.
- Ability to plan more effectively.
- More competitive as a just-in-time supplier.

Supplier selection.   Chapter 7 noted that the factors to be considered when selecting suppliers were technical ability, manufacturing capability, reliability, after-sales service, and supplier location. In a partnership there are other considerations based on the partnership relationship. They include the following:

1. The supplier has a stable management system and is sincere in implementing the partnership agreement.
2. There is no danger of the supplier breaching the organization's secrets.
3. The supplier has an effective quality system.
4. The supplier shares the vision of customer satisfaction and delighting the customer.

Supplier certification.   Once the supplier is selected, the next step is a certification process that begins after the supplier has started to ship product. Organizations can set up their own criteria for certification or can use one such as developed by the *American Society for Quality Control.* This emphasizes the absence of defects both in product and nonproduct categories (e.g., billing errors), and a good documentation system, such as the ISO 9000 system.

Total Employee Involvement

A successful JIT environment can be achieved only with the cooperation and involvement of everyone in the organization. The ideas of elimination of waste and continuous improvement that are central to the JIT philosophy can be accomplished only through people cooperating.

Instead of being receivers of orders, operators must take responsibility for improving processes, controlling equipment, correcting deviations, and becoming vehicles for continuous improvement. Their jobs include not only direct labor but also a variety of traditionally indirect jobs such as preventive maintenance, some setup, data recording, and problem solving. As discussed earlier in this chapter, employees must be flexible in the tasks they do. Just as machines must be flexible and capable of quick changeover, so must the people who run them.

The role of management must change. Traditionally, management has been responsible for planning, organizing, and supervising operations. Many of their traditional duties are now done by line workers. In a JIT environment, more emphasis is placed on the leadership role. Managers and supervisors must become coaches and trainers, develop the capability of employees, and provide coordination and leadership for improvements.

Traditionally, staff have been responsible for such things as quality control, maintenance, and record keeping. Under JIT, line workers do many of these duties. Staff responsibilities then become those of training and assisting line workers to do the staff duties assigned to them.

MANUFACTURING PLANNING AND CONTROL IN A JIT ENVIRONMENT

The philosophy and techniques of just-in-time manufacturing discussed in this chapter are related to how processes and methods of manufacture are designed. The major responsibility for designing processes and methods lies with manufacturing and industrial engineering. Manufacturing planning and control is responsible for managing the flow of material and work through the manufacturing process, not designing the process. However, manufacturing planning and control is governed by, and must work with, the manufacturing environment, whatever it is. Figure 15.8 shows the relationship.

Figure 15.8   JIT manufacturing.

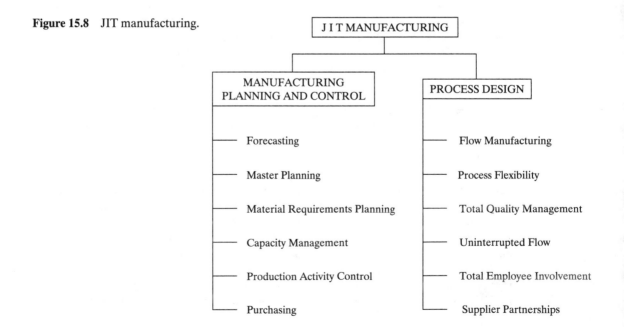

This section will study the effect a JIT environment has on manufacturing planning and control. No matter what planning and control system is used, these four basic questions have to be answered:

1. What are we going to make?
2. What do we need to make it?
3. What do we have?
4. What must we get?

The logic of these questions always applies, whether we are going to cook a meal or make a modern jet aircraft. Systems for planning and control vary. The manufacturing planning and control system (manufacturing resource planning) discussed in this text has proved effective in any manufacturing environment. The complexity of the manufacturing process, the number of finished items and parts, the levels in the bill of material, and the lead times have made the planning and control problems either simple or complex. If anything can be done to simplify these factors, the planning and control system will be simpler. In general, the JIT philosophy simplifies these factors, thus making the planning and control problems easier.

The sections that follow will look at how the various parts of the manufacturing planning and control system relate to a JIT environment. In general, JIT manufacturing does not make the manufacturing planning and control system obsolete but, in some ways, does change the focus. The JIT process is not primarily a planning and control system. It is a philosophy and a set of techniques for designing and operating a manufacturing plant. Planning and control are still needed in JIT manufacturing.

Forecasting

The major effect that JIT has on forecasting is shortened lead time. This does not affect forecasting for business planning or production planning, but it does for master production scheduling. If lead times are short enough that production rates can be matched to sales rates, forecasting for the master production schedule becomes less important.

Production Planning

The JIT system emphasizes relationships with suppliers. One purpose of production planning is to arrange for long-lead-time purchases. The JIT process has the potential for reducing those lead times, but more importantly, it provides an environment in which the supplier and buyer can work together to plan the flow of material.

Master Production Scheduling

Several scheduling factors are influenced by JIT manufacturing:

1. Master scheduling tries to level capacity, and JIT tries to level the schedule based on capacity and material flow. Figure 15.9 illustrates the difference.

2. The shorter lead times reduce time fences and make the master production schedule more responsive to customer demand. The ideal lead time is so short that the company can respond to actual sales, not to forecast. Where the company builds to a seasonal demand or to satisfy promotion, a forecast is still necessary. Planning horizons can also be reduced.

3. The JIT system requires a stable schedule to operate. This principle is supported by using time fences. These are established based on lead times and the commitment of materials and resources. If lead times can be reduced through JIT practices, the time fences can be reduced.

Figure 15.9   MPS leveled by capacity and by material.

Level Capacity Schedule

Week	1	2	3
Model A	900		800
Model B	300	1600	800
Model C		300	1200
Total	1200	1200	1200

Level Material and Capacity Schedule

Week	1	2	3
Model A	900		800
Model B	300	1600	800
Model C		300	1200
Total	1200	1200	1200

4. Traditionally, weekly time buckets are used. This gives manufacturing an organizational buffer to plan and organize actual work flow. Because of reduced lead times and schedule stability, it is possible to use daily time buckets in a JIT environment.

Material Requirements Planning

Material requirements planning (MRP) plans the material flow based on the bill of material, lead times, and available inventory. Just-in-time practices will modify this approach in several ways:

- The MRP time buckets are usually one week. As lead times are reduced and the flow of material improved, these can be reduced to daily buckets.

- The MRP netting logic is based on generating order quantities calculated using the planned order releases of the parent, the inventory on hand, and any order quantity logic used. In a pure JIT environment, there is no inventory on hand, and the order quantity logic is to make exactly what is needed. Therefore, there is no netting required. If the lead times are short enough, component production occurs in the same time bucket as the gross requirement, and no offsetting is required.

- Bills of material can frequently be flattened in a JIT environment. With the use of work cells and the elimination of many inventory transactions, some levels in bills of material become unnecessary.

Both MRP and JIT are based on establishing a material flow. In a repetitive manufacturing environment, this is set by the model mix and the flow rate. The product to be made is decided by the need of the following workstation, which is ultimately the assembly line.

However, many production situations do not lend themselves to level scheduling and the pull system. Some examples are as follows:

- Where the demand pattern is unstable.
- Where custom engineering is required
- Where quality is unpredictable
- Where volumes are low and occur infrequently

Capacity Management

Capacity planning's function is to determine the need for labor, equipment, and material to meet the priority plans. Leveling schedules should make the job easier. Capacity control focuses on adjusting capacity daily to meet demand. Leveling should make this task easier, but so will the JIT emphasis on cutting out waste and problems that cause ineffective use of capacity. Linearity, the practice of scheduling extra capacity, will improve the ability to meet priority schedules.

Inventory Management

Because the JIT system reduces the inventory in the system, in some respects this should make inventory management easier. However, if order quantities are reduced and annual demand remains the same, more work orders and more paperwork must be tracked, and more transactions recorded. The challenge then is to reduce the number of transactions that have to be recorded. One system used is called **backflushing,** or **post-deduct.**

Material flows from raw material to work-in-process to finished goods. In a post-deduct system, raw materials are recorded into work-in-process. When work is completed and becomes finished goods, the work-in-process inventory is relieved by multiplying the number of units completed by the number of parts in the bill of material. The system works if the bills of material are accurate and if the manufacturing lead times are short.

SUMMARY

The just-in-time philosophy and techniques that seek the elimination of waste and continuous improvement were developed for repetitive manufacturing and are perhaps most applicable there. However, many basic concepts are appropriate to any form of manufacturing organization.

As noted in Chapter 14, in intermittent manufacturing, material is processed at intervals in lots or batches, not continuously. It is characterized by large variation in product design, process requirements, and order quantities. In the extreme, every job is made to customer specification, and there is no commonality in product design. General-purpose equipment is used to make products. A contract machine shop is an example. However, many companies make a variety of standard products in small volumes manufactured either to stock or to customer order. Many of these have families of products, and if these can be identified, work cells can be set up.

No matter what the characteristics of the intermittent shop, several JIT principles can be applied:

- Employee involvement
- Workplace layout
- Total quality control
- Total productive maintenance
- Setup time reduction
- Supplier relations
- Inventory reduction

The JIT manufacturing environment requires a planning and control system. The manufacturing resource planning system is complementary to JIT manufacturing. The way in which some functions are used changes to reflect the differences in the manufacturing environment, but, in general, the functions performed in the MRP II system are those that have to be done in a JIT planning and control system.

QUESTIONS

1. What is the definition of waste as it is used in this text?
2. What is value to the user? How is it related to quality?
3. What are the elements in a product cycle loop? For what is each responsible?
4. Why is product specialization important? Who is responsible for setting the level of product specialization?
5. What is meant by component standardization? Why is it important in eliminating waste?
6. Why is product design important to manufacturing? How can the design add waste in manufacturing?
7. After a product is designed, what is the responsibility of manufacturing engineering?
8. Explain why each of the following are sources of waste:
   a. The process
   b. Methods
   c. Movement
   d. Product defects
   e. Waiting time
   f. Overproduction
9. Explain how inventory affects product improvement, quality, prices, and the ability to respond quickly to the marketplace.
10. What is repetitive manufacturing? What are its advantages? What are its limitations?
11. What is a work cell? How does it operate? What conditions are necessary to establish one? What are its advantages?
12. Why is process flexibility desirable? What two conditions are required?
13. Name and describe five advantages of low setup time.
14. What are the two reasons why quality is important?
15. What is quality for manufacturing? How is it obtained?
16. Why is productive maintenance important?
17. What are the four conditions needed for uninterrupted flow? Describe each.
18. What is the difference between leveling based on capacity and leveling based on material flow?

19. Why would a JIT manufacturer schedule seven hours of work in an eight-hour shift?

20. Why are supplier relations particularly important in a JIT environment?

21. Why is employee involvement important in a JIT environment?

22. What are the differences in a master production schedule in a JIT environment?

23. What effect does a JIT environment have on MRP?

24. Describe the backflush or post-deduct system of inventory record keeping.

PROBLEMS

15.1 A company carries ten items in stock, each with an economic-order quantity of $20,000. Through a program of component standardization, the ten items are reduced to five. The total annual demand is the same, but the annual demand for each item is twice what it was before. In Chapter 10, we learned that the economic-order quantity varies as the square root of the annual demand. Since the annual demand for each item is now doubled, calculate:

a. The new EOQ

b. The total average inventory before standardization

c. The total average inventory after standardization

15.2 In problem 15.1, if the annual carrying cost is 20% per unit, what will be the annual savings in carrying cost?

15.3 A company has an annual demand for a product of 1000 units, a carrying cost of $20 per unit per year, and a setup cost of $100. Through a program of setup reduction, the setup cost is reduced to $10. Run costs are $2 per unit. Calculate:

a. The EOQ before setup reduction

b. The EOQ after setup reduction

c. The total and unit cost before and after setup reduction

15.4 A company produces a line of golf putters composed of three models. The demand for model A is 500 per month, for B is 400, and for C is 300. What would be the mixed model sequence if some of each were made each day?

Answer. ABC, ABC, ABC, ABA

15.5 The following MPS summary schedule is leveled for capacity. Using the table below, level the schedule for material as well.

Week	1	2	3	4	5
Model A			800	1600	1600
Model B	600	1600	800		
Model C	1000				
Total	1600	1600	1600	1600	1600

Answer.

Week	1	2	3	4	5
Model A					
Model B					
Model C					
Total					

16

Total Quality Management

WHAT IS QUALITY?

We all know, or think we know, what quality is. But often it means different things to different people. When asked to define quality, people's responses are influenced by personal opinion and perception. Answers are often general or vague, for example, "It's the best there is," "Something that lasts a long time and gives good service," and "Something with style."

While the definitions of quality vary, the one we will use contains the most commonly accepted ideas in business today.

Quality means user satisfaction: that goods or services satisfy the needs and expectations of the user.

To achieve quality according to this definition, we must consider quality and product policy, product design, manufacturing, and final use of the product.

Quality and product policy.   Product planning involves decisions about the products and services that a firm will market. A product or service is a combination of tangible and intangible characteristics that a company hopes the customer will accept and be willing to pay a price for. Product planning must decide the market segment to be served, the level of performance expected, and the price to be charged, and must estimate the expected sales volume. The basic quality level of a product is thus specified by senior management according to its understanding of the wants and needs of the market segment.

Quality and product design.   A firm's studies of the marketplace should yield a general specification of the product, outlining the expected performance, appearance, price, and volume. Product designers must then build into the product the quality level described in the general specification. They determine the materials to be used, dimensions, tolerances, product capability, and service requirements. If product designers do not do this properly, the product or service will be unsuccessful in the marketplace because it may not adequately satisfy the needs and expectation of the user.

Quality and manufacturing.   At the least, manufacturing is responsible for meeting the minimum specifications of the product design. Tolerances establish the acceptable limits and are usually expressed as the amount of allowable variation about the desired amount. For example, the length of a piece of lumber may be expressed as $7' \, 6'' \pm \frac{1}{8}''$. This means that the longest acceptable piece would be $7' \, 6 \frac{1}{8}''$ and the shortest acceptable piece would be $7' \, 5 \frac{7}{8}''$ long. If an item is within tolerance, then the product should perform adequately. If it is not, it is unacceptable. However, the closer an item is to the nominal or target value, the better it will perform and the less chance there is of creating defects.

Quality in manufacturing means that, at a minimum, all production must be within specification limits and the less variation from the nominal the better the quality. Manufacturing must strive to produce excellent—not merely adequate—products. Every product or service produced will have some form of tolerance expressed. For example, the weight of bars of soap, the frequency response of compact disks, or the time spent waiting in line will all have a plus and minus tolerance.

Quality and use.   To the user, quality depends on an expectation of how the product should perform. This is sometimes expressed as "fitness for use." Customers do not care *why* a product is defective, but they care *if* it is defective. If the product has been well conceived, well designed (meets customer needs), well made, well priced, and well serviced, then the quality is satisfactory. If the product exceeds the customer's expectations, that is superb quality.

Figure 16.1 shows the loop formed by product policy, product design, operations, and the user. Quality must be added by each link.

Quality has a number of dimensions, among which are the following:

Performance. The primary operating characteristics, such as the power of an engine. Performance implies that the product or service is ready for the customer's use at the time of sale. The phrase "fitness for use"—that the product does what it is supposed to do—is often used to describe this. Three dimensions to performance are important: reliability, durability, and maintainability.

Figure 16.1 Product development cycle.

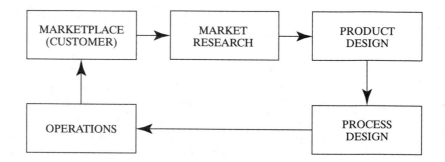

Reliability means consistency of performance. It is measured by the length of time a product can be used before it fails.

Durability refers to the ability of a product to continue to function even when subjected to hard wear and frequent use.

Maintainability refers to being able to return a product to operating condition after it has failed.

Features. These are secondary characteristics—little extras, such as remote control on a VCR.

Conformance. Meeting established standards or specifications. This is manufacturing's responsibility.

Warranty. An organization's public promise to back up its products with a guarantee of customer satisfaction.

Service. Service is an intangible generally made up of a number of things such as availability, speed of service, courtesy, and competence of personnel.

Aesthetics. Aesthetics means pleasing to the senses; for example, the exterior finish or the appearance of a product.

Perceived quality. Total customer satisfaction is based on the complete experience with an organization, not just the product. Many intangibles such as a company's reputation or past performance influence perceived quality.

Price. Customers pay for value in what they buy. Value is the sum of the benefits the customer receives and can be more than the product itself. All the dimensions listed above are elements of value.

These dimensions are not necessarily interrelated. A product can be superb in one or a few dimensions and average or poor in others.

TOTAL QUALITY MANAGEMENT (TQM)

TQM is an approach to improving both customer satisfaction and the way organizations do business. TQM brings together all of the quality and customer-related process improvement ideas. It is people oriented. According to the eighth edition of the

APICS dictionary, "it is based on the participation of all members of an organization in improving processes, products, services, and the culture they work in."

The objective of TQM is to provide a quality product to customers at a lower price. By increasing quality and decreasing price, profit and growth will increase, which will increase job security and employment.

TQM is both a philosophy and a set of guiding principles that lead to a continuously improving organization.

Basic Concepts.   There are six basic concepts in TQM.

1. *A committed and involved management* directing and participating in the quality program. TQM is a continuous process that must become part of the organization's culture. This requires senior management commitment.

2. *Focus on the customer.* This means listening to the customer so goods and services meet customer needs at a low cost. It means improving design and processes to reduce defects and cost.

3. *Involvement of the total work force.* Total quality management is the responsibility of everyone in the organization. It means training all personnel in the techniques of product and process improvement and creating a new culture. It means empowering people.

4. *Continuous process improvement.* Processes can and must be improved to reduce cost and increase quality. (This topic was discussed in Chapter 14 and will not be covered in this chapter.)

5. *Supplier partnering.* A partnering rather than adversarial relationship must be established.

6. *Performance measures.* Improvement is not possible unless there is some way to measure the results.

These basic concepts will be discussed in more detail in the following sections.

Management Commitment

If senior management is not committed and involved, then TQM will fail. These managers must start the process and should be the first to be educated in the TQM philosophy and concepts. The chief executive officer and senior management should form a quality council whose purpose is to establish a clear vision of what is to be done, develop a set of long-term goals, and direct the program.

The quality council must establish core values that help define the culture of the organization. Core values include such principles as customer-driven quality, continuous improvement, employee participation, and fast response. As well, the council must establish quality statements that include a vision, mission, and quality policy statements. The vision statement describes what the organization should become five to ten years in the future. The mission statement describes the function of the organization: who we are, who our customers are, what the organization does, and how it does it. The quality policy statement is a guide for all in the organization about how products and services should be provided. Finally, the quality council must establish a strategic plan that expresses the TQM goals and objectives of the organization and how it hopes to achieve them.

Customer Focus

Total quality management implies an organization that is dedicated to delighting the customer by meeting or exceeding customer expectations. It means not only understanding present customer needs but also anticipating customers' future needs.

A customer is a person or organization who receives products or services. There are two types of customers, external and internal. *External customers* exist outside the organization and purchase goods or services from the organization. *Internal customers* are persons or departments who receive the output from another person or department in an organization. Each person or operation in a process is considered a customer of the preceding operation. If an organization is dedicated to delighting the customer, internal suppliers must be dedicated to delighting internal customers.

Customers have six requirements of their suppliers.

1. High quality level.
2. High flexibility to change such things as volume, specifications, and delivery.
3. High service level.
4. Short lead times.
5. Low variability in meeting targets.
6. Low cost.

Customers expect improvement in all requirements. These requirements are not necessarily in conflict. Low cost and high flexibility, for example, do not have to be tradeoffs if the process is designed to provide them.

Employee Involvement

TQM is organization wide and is everyone's responsibility. In a TQM environment, people come to work not only to do their jobs but also to work at improving their jobs. To gain employee commitment to the organization and TQM requires the following:

1. *Training.* People should be trained in their own job skills and, where possible, cross trained in other related jobs. As well, they should be trained to use the tools of continuous improvement, problem solving, and statistical process control. Training provides the tools for continual people-driven improvement.
2. *Organization.* The organization must be designed to put people in close contact with their suppliers and customers, internal or external. One way is to organize into customer-, product-, or service-focused cells or teams.
3. *Local ownership.* People should feel ownership of the processes they work with. This results in a commitment to make their processes better and to continuous improvement. They should be empowered.

  Empowerment means giving people the authority to make decisions and take action in their work areas without getting prior approval. For example, a customer service representative can respond to a customer's complaint on the spot rather than getting approval or passing the complaint on to a supervisor. Giving people the authority to make decisions motivates them to accomplish the goals and objectives of the organization and to improve their jobs.

Teams. A team is a group of people working together to achieve common goals or objectives. Good teams can move beyond the contribution of individual members so that the sum of their total effort is greater than their individual efforts. Working in a team requires skill and training, and to work in teams is part of total quality management.

Continuous Process Improvement

This topic, an element in both JIT and TQM, was discussed in Chapter 14. Quality requires continuous process improvement. If a product is excellent in one dimension such as performance, then improved quality in another dimension should be sought.

Supplier Partnerships

Supplier partnerships are very important in both just-in-time manufacturing and total quality management. This topic was discussed in some detail in Chapter 15.

Performance Measures

To determine how well an organization is performing, its progress must be measured. Performance measures can be used to:

- Discover which process needs improvement.
- Evaluate alternative processes.
- Compare actual performance with targets so corrective action can be taken.
- Evaluate employee performance.
- Show trends.

  It is really not a question of whether performance measures are necessary, but of selecting appropriate measures. There is no point in measuring something that does not give valid and useful feedback on the process being measured. There are many basic characteristics that can be used to measure the performance of a particular process or activity, such as the following:

- *Quantity.* For example, how many units a process produces in a period of time. Time standards measure this dimension.
- *Cost.* The amount of resources needed to produce a given output.
- *Time/delivery.* Measurements of the ability to deliver a service or product on time.

- *Quality.* There are three dimensions to quality measurements:

    Function. Does the product perform as specified?

    Aesthetics. Does the product or service appeal to customers? For example, the percentages of people who like certain features of a product.

    Accuracy. This measures the number of nonconformances produced. For example, the number of defects or rejects.

Performance measures should be simple, easy for users to understand, relevant to the user, visible to the user, preferably developed by the user, designed to promote improvement, and few in number.

Measurement is needed for all types of process. Some of the areas and possible measurements are as follows:

- Customer. Number of complaints, on-time delivery, dealer or customer satisfaction.
- Production. Inventory turns, scrap or rework, process yield, cost per unit, time to perform operations.
- Suppliers. On-time delivery, rating, quality performance, billing accuracy.
- Sales. Sales expense to revenue, new customers, gained or lost accounts, sales per square foot.

QUALITY COST CONCEPTS

Quality costs fall into two broad categories: the cost of failure to control quality and the cost of controlling quality.

Costs of Failure

The costs of failing to control quality are the costs of producing material that does not meet specification. They can be broken down into:

- *Internal failure costs.* The costs of correcting problems that occur while the goods are still in the production facility. Such costs are scrap, rework, and spoilage. These costs would disappear if no defects existed in the product before shipment.
- *External failure costs.* The costs of correcting problems after goods or services have been delivered to the customer. They include warranty costs, field servicing of customer goods and all the other costs associated with trying to satisfy customer complaints. External failure costs can be the most costly of all if the customer loses interest in a company's product. These costs would also disappear if there were no defects.

Costs of Controlling Quality

The costs of controlling quality can be broken down into:

- *Prevention costs.* The costs of avoiding trouble by doing the job right in the first place. (Remember the old adage, "An ounce of prevention is better than a pound of cure".) They include training, statistical process control, machine maintenance, and quality planning costs.

- *Appraisal costs.* The costs associated with checking and auditing quality in the organization. They include product inspection, quality audits, testing, and calibration.

Investment in prevention will improve productivity by reducing the cost of failure and appraisal. Figure 16.2 shows the typical pattern of quality costs before and after a quality improvement program. Investing in prevention will increase total costs in the short run, but in the long run prevention will eliminate the causes of failure and reduce total quality costs.

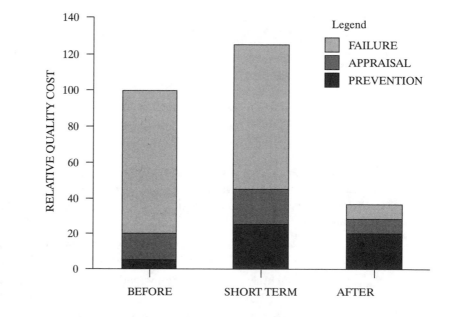

Figure 16.2 Impact of quality improvement on quality-related costs.

VARIATION AS A WAY OF LIFE

Variability exists in everything—people, machines, or nature. People do not perform the same task in exactly the same way each time, nor can machines be relied upon to perform exactly the same way each time. No two leaves are alike. It is really a question of how much variability there is.

Suppose that a lathe made 100 shafts with a nominal diameter of 1″. If we measured those shafts we would find that, while the diameters tended to cluster about 1″, there were some smaller and some bigger. If we plotted the number of shafts of each diameter, we would probably get a distribution as shown in Figure 16.3. Note this forms a histogram.

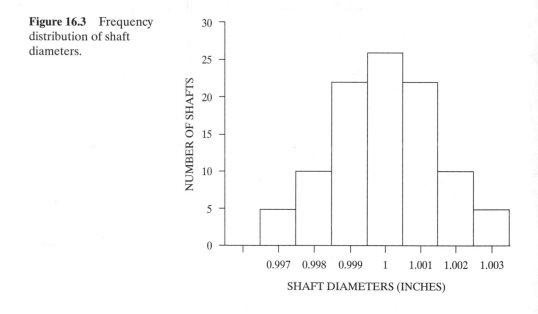

Figure 16.3 Frequency distribution of shaft diameters.

Chance variation.   In nature or any manufacturing process, we can expect to find a certain amount of variation that is inherent in the process. This variation comes from everything influencing the process, but is usually broken into the following six categories.

People	Poorly trained operators tend to be more inconsistent compared with well-trained operators.
Machine	Well-maintained machines tend to give more consistent output than a poorly maintained, sloppy machine.
Material	Consistent raw materials give better results than poor quality, inconsistent, ungraded materials.

Method	Changes in the method of doing a job will alter the quality.
Environment	Changes in temperature, humidity, dust and so on can affect some processes.
Measurement	Measuring tools that may be in error can cause incorrect adjustments and poor process performance.

Breaking all possible variation into these six smaller categories makes it easier to identify the source of variation occurring in a process. If a connection can be found between variation in the product and variation in one of its six sources, then improvements in quality are possible.

There is no way to alter chance variation except to change the process. If the process produces too many defects, then it must be changed.

Assignable variation.   Chance is not the only cause of variation. A tool may shift, a gauge may move, a machine wear, or an operator make a mistake. There is a specific reason for these causes of variation, which is called assignable variation.

Statistical control.   As long as only chance variation exists, the system is said to be in *statistical control*. If there is an assignable cause for variation, the process is not in control and is unlikely to produce good product. As will be shown later in this chapter, the objective of statistical process control is to detect the presence of assignable causes of variation. Statistical process control, then, has two objectives:

- To help select processes capable of producing the required quality with minimum defects.
- To monitor a process to be sure it continues to produce the required quality and no assignable cause for variation exists.

Patterns of Variability

The output of every process has a unique pattern that can be described by its shape, center, and spread.

Shape.   Suppose, instead of measuring the diameters of 100 shafts, we measured the diameters of 10,000 shafts. If we plotted the distribution of the diameters of the 10,000 shafts, the results in Figure 16.3 would be smoothed out and we would have a curve as shown in Figure 16.4. This bell-shaped curve is called a *normal curve* and is commonly encountered in manufacturing processes that are running under controlled conditions. This curve exists in virtually all natural processes from the length of grass on a lawn, to the heights of people, to student grades.

Figure 16.4 Normal
distribution.

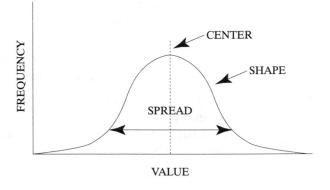

Center. We can see from Figure 16.4 that the normal distribution has most results clustered near one central point, with progressively fewer results occurring away from this center. The center of the distribution can be calculated as follows:

Let

$$\Sigma x = \text{sum of all observations}$$
$$N = \text{number of observations}$$
$$\mu = \text{arithmetic mean (average or center)}$$

Then

$$\mu = \frac{\Sigma x}{N}$$

In the example we are using:

$$\Sigma x = 10{,}000$$
$$N = 10{,}000$$
$$\mu = \frac{10{,}000}{10{,}000} = 1 \text{ inch}$$

Spread.   To evaluate a process, we must know not only what the center is, but also something about the spread or variation. In statistical process control there are two methods of measuring this variation, the range and the standard deviation.

- *Range.* The range is simply the difference between the largest and smallest values in the sample. In the example shown in Figure 16.3, the largest value was 1.003″ and the smallest was .997″. The range would be 1.003″ − .997″ = .006″.

- *Standard deviation.* The standard deviation (represented by the Greek letter σ or sigma) may be thought of as the "average spread" around the center. A distribution with a high standard deviation is "fatter" than one with a low standard deviation. Higher quality products have little variation (a low standard deviation). The measurement of the standard deviation was discussed in Chapter 11 in the section on determining safety stock. We know that:

$$\mu \pm 1\sigma = 68.3 \text{ percent of observations}$$
$$\mu \pm 2\sigma = 95.4 \text{ percent of observations}$$
$$\mu \pm 3\sigma = 99.7 \text{ percent of observations}$$

where:

$$\mu = \text{mean or average}$$
$$\sigma = \text{standard deviation}$$

We can use the standard deviation to estimate the amount of variation (quality) in a product.

Suppose in our example of the shafts that:

$$\mu = 1.000 \text{ inches}$$
$$\sigma = 0.0016 \text{ inches}$$

Applying standard deviation to the previous example we know that:

68.3% of the shafts will have a diameter of $1'' \pm .0016''$ (1σ)

95.4% of the shafts will have a diameter of $1'' \pm .0032''$ (2σ)

99.7% of the shafts will have a diameter of $1'' \pm .0048''$ (3σ)

PROCESS CAPABILITY

Tolerances are the limits of deviation from perfection and are established by the product design engineers to meet a particular design function. For example a shaft might be specified as having a diameter of $1'' \pm .005''$. Thus, any shaft having a diameter from 0.995″ to 1.005″ would be within tolerance. In statistical process control 0.995″ is called the lower specification limit (LSL) and is the minimum acceptable level of output. Similarly 1.005″ is called the upper specification limit (USL) and is defined as the maximum acceptable level of output. *Both the USL and the LSL are related to the product specification and are independent of any process.* Figure 16.5 illustrates this. One process, having a narrow spread (low sigma) will produce product within the specification limits. The other process, having a wide spread (high sigma), will produce defects. The first process is said to be capable, the second is not.

Besides spread, there is another way a process can produce defects. If there is a shift in the mean (average), defects will be produced. Figure 16.6 illustrates the concept.

Figure 16.5   Effect of process spreads.

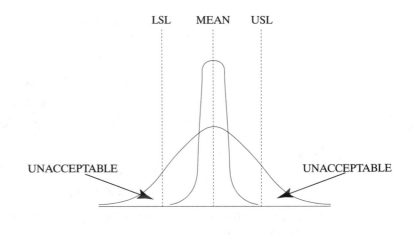

Figure 16.6   Effect of shift in the mean.

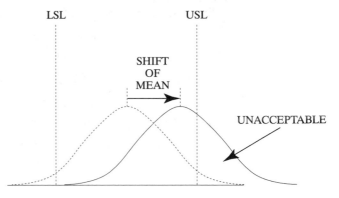

In summary:

- The capability of the process is not related to the product specification tolerance.
- A process must be selected that can meet the specifications.
- Processes can produce defects in two ways, by having too big a spread (sigma) or by a shift in the mean (average).

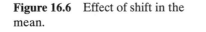

EXAMPLE PROBLEM

In the previous example the process had a standard deviation of .0016″ and a mean of 1″. If the specifications called for a diameter of 1″ ± .005″

    a. Approximately what percentage of the shafts will be within tolerance?

    b. If the tolerances were changed to 1″ ± .002″, approximately what percentage of the shafts will be within tolerance?

Answer

a. Number of standard deviations within tolerance = .005 ÷ .0016 = 3.125.

Sigma is approximately 3.

Approximately 99.7% of shafts will have a diameter of 1″ ± .0048″ and be within tolerance.

b. Number of standard deviations within tolerance = .002 ÷ .0016 = 1.25.

Sigma is approximately 1.

Approximately 68.3% of shafts will have a diameter of 1″ ± .0016″ and be within tolerance.

Process Capability Index (C_p)

The process capability index, combines the process capability and the tolerance into one index. It assumes the process is centered—that there has been no shift of the mean. As well, the index assumes if a process produces 99.7% good parts, it is capable. Thus the factor ± 3σ or 6σ is used in the calculation.

$$C_p = \frac{USL - LSL}{6\sigma}$$

In the previous example:

$$C_p = \frac{1.005 - .995}{6 \times .0016} = 1.04$$

If the capability index is greater than 1.00, the process is capable of producing 99.7% of parts within tolerance and is said to be capable. If C_p is less than 1.00, the process is said to be not capable. Because processes tend to shift back and forwards, a C_p of 1.33 has become a standard of process capability. Some organizations use a higher value such as 2. The larger the capability index the fewer the rejects and the greater the quality. Figure 16.7 shows the concept of the capability index.

Figure 16.7 C_p index greater than 1.

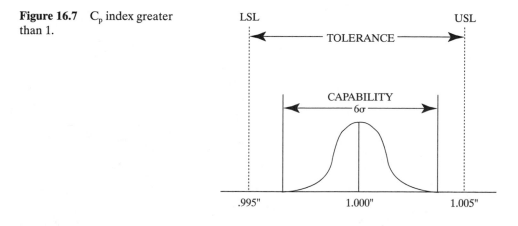

EXAMPLE PROBLEM

The specifications for the weight of a chemical in a compound is 10 grams ± .05 grams. If the standard deviation of the weighing scales is .02, is the process considered capable?

Answer

$$C_p = \frac{10.05 - 9.95}{6 \times .02} = 0.83$$

The C_p is less than one and the process is not considered capable.

The process capability index indicates whether process variation is satisfactory, but it does not measure whether the process is centered properly. Thus it does not protect against out-of-specification product resulting from poor centering. In some cases this is important to know.

C_{pk} Index

This index measures the effect of both center and variation at the same time. The philosophy of the C_{pk} index is that if the process distribution is well within specification on the worst case side then it is sure to be acceptable for the other specification limit. Figure 16.8 illustrates the concept.

Figure 16.8 C_{pk} index.

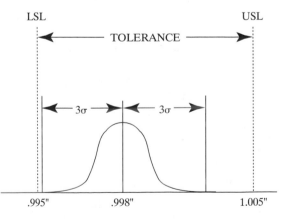

The C_{pk} index is the **lesser** of:

$$\frac{(USL - Mean)}{3\sigma} \qquad \text{or} \qquad \frac{(Mean - LSL)}{3\sigma}$$

The greater the C_{pk}, the further the 3σ limit is from the specification limit and the fewer rejects there will be.

Interpretation of the C_{pk} index is as follows.

C_{pk} Value	Evaluation
Less than + 1	Unacceptable process. Part of process distribution is out of specification.
+ 1 to + 1.33	Marginal process. Process distribution barely within specification.
Greater than 1.33	Acceptable process. Process distribution is well within specification.

EXAMPLE PROBLEM

A company produces shafts with a nominal diameter of 1″ and a tolerance of .005″ on a lathe. The process has a standard deviation of .001″. For each of the following cases calculate the C_{pk} and evaluate the process capability.

   a. A sample has an average diameter of .997.

   b. A sample has an average diameter of .998.

   c. A sample has an average diameter of 1.001.

Answer

   a. $C_{pk} = \dfrac{1.005 - .997}{3 \times .001} = 2.67$ or $= \dfrac{.997 - .995}{.3 \times .001} = 0.67$

   C_{pk} is less than 1. Process is not capable.

   b. $C_{pk} = \dfrac{1.005 - .998}{3 \times .001} = 2.33$ or $= \dfrac{.998 - .995}{3 \times .001} = 1.00$

   C_{pk} is 1. Process is marginal.

   c. $C_{pk} = \dfrac{1.005 - 1.001}{3 \times .001} = 1.33$ or $= \dfrac{1.001 - .995}{3 \times .001} = 2$

   C_{pk} is 1.33. Process is capable.

PROCESS CONTROL

Process control attempts to prevent the production of excessive defects by showing when the probability is high there is an *assignable* cause for variation.

We have seen that variation exists in all processes and that the process must be designed so the spread will be small enough to produce a minimum number of defects. Variation will follow a stable pattern as long as the system of chance causes remains the same and there is no assignable cause of variation. Once a stable process is established, the limits of the resulting pattern of variation can be determined and will serve as a guide for future production. When variation exceeds these limits, it shows a high probability that there is an assignable cause.

We have also seen that a process can produce defects if the spread is too great or if the center or the average are not correct. Some method is needed to measure these two characteristics continually so we can compare what is happening to the product specification. This is done using the $\bar{X}$ (X bar) and R control chart.

Control Charts

Run charts.   Suppose a manufacturer was filling bottles and wanted to check the process to be sure the proper amount of liquid was going into each. Samples are taken every half hour and measured. The average of the samples is then plotted on a chart as shown in Figure 16.9. This is called a run chart. While it gives a visual description of what is happening with the process, it does not distinguish between system (chance) variation and assignable cause variation.

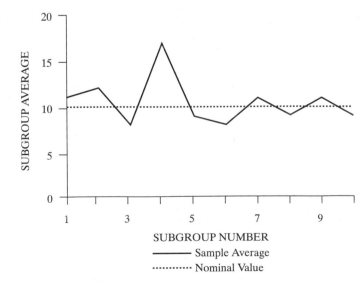

Figure 16.9   Run chart.

$\bar{X}$ (X bar) and R Chart.   A control chart for averages and ranges ($\bar{X}$ and R chart) tracks the two critical characteristics of a frequency distribution—the center and the spread. Small samples (typically three to nine pieces) are taken on a regular basis over time and the sample averages and range plotted. The range is used rather than the standard deviation because it is easier to calculate. Samples are used in control charts rather than individual observations because average values will indicate a change in variation much faster. Figure 16.10 shows an example of an $\bar{X}$ and R chart.

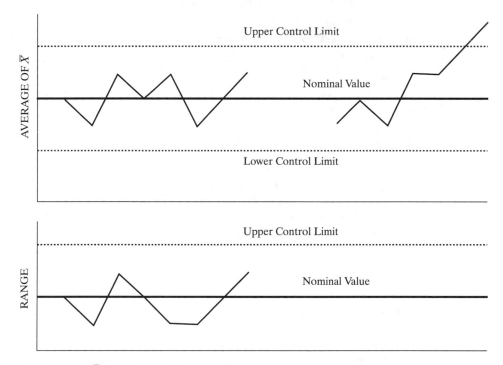

Figure 16.10 $\bar{X}$ and R control chart.

Control Limits

Figure 16.10 also shows two dotted lines, the upper control limit (UCL) and the lower control limit (LCL). These limits are established to help in assessing the significance of the variation produced by the process. *They should not be confused with specification limits,* which are permissible limits of each unit of product (tolerances) and have nothing to do with the process.

Control limits are set so that there is a 99.7% probability that if the process is in control, the sample value will fall within the control limits. When this situation occurs, the process is considered to be in statistical control and there is no assignable cause of variation. The process is stable and predictable. This is shown on the left portion of the chart in Figure 16.10. All the points lie within the limits and the process is in statistical control. Only chance variation is affecting the process. However, when assignable causes of variation exist, the variation will be excessive and the process is said to be out of control. This is shown on the right portion of the chart, which indicates something has caused the process to go out of control.

As explained earlier, two types of changes can occur in a process:

- A shift in the mean or average. This might be caused by a worn tool or a guide that has moved. This will show up on the $\bar{X}$ portion of the chart.

- A change in the spread of the distribution. If the range increases but the sample averages remain the same, we might expect this kind of problem. It might be caused by a gauge or tool becoming loose, or by some part of the machine becoming worn. This will show up on the R portion of the chart.

The $\bar{X}$ and R chart is used to measure variables, such as the diameter of a shaft, which can be measured on a continual scale.

Control Charts for Attributes

An attribute refers to quality characteristics that either conform to specification or do not; for example, visual inspection for color, missing parts, and scratches. A go-no-go gauge is a good example. Either the part is within tolerance or it is not. These characteristics cannot be measured but they can be counted. Attributes are usually plotted using a proportion defective or p-chart.

SAMPLE INSPECTION

Statistical process control monitors the process and detects when the process goes out of control, thus minimizing the production of defective parts. Traditional inspection inspects the batch of parts after they are made, and on the basis of the inspection, accepts or rejects the batch. There are two inspection procedures, 100% and acceptance sampling.

100% inspection means testing every unit in the lot. This is appropriate when the cost of inspection is less than the cost of any loss resulting from failure of the parts; for example, when inspection is very easy to do or the inspection is part of the process. A light bulb manufacturer could easily build a tester to see if every light bulb lights, or a baker could visually inspect all products prior to packaging them. In cases in which the cost of failure is exceptionally high, 100% inspection is vital. Products for the medical and aerospace industry may be checked many times because of the importance of product performance or the high cost of failure.

Acceptance sampling consists of taking a sample of a batch of product and using it to estimate the overall quality of the batch. Based on the results of the inspection a decision is made to reject or accept the entire batch. There is a chance that a good batch will be rejected or a bad batch will be accepted. Sampling inspection is necessary under some conditions.

Reasons for sampling inspection. There are four reasons for using sample inspection.

Testing the product is destructive. The ultimate pull strength of a rope or the sweetness of an apple can be decided only by destroying the product.

There is not enough time to give 100% inspection to a batch of product. On election day newscasters are eager to get coverage. Once a small percentage of the votes are in, a guess is made of the final outcome, usually with some estimate of the error (say 19 times out of 20).

It is too expensive to test all of the batch. Market sampling is an example of this, as are surveys of public attitude.

Human error is estimated to be as high as 3% when performing long-term repetitive testing. There are good reasons to have a representative sample taken of a batch rather than to hazard this high an error.

Conditions necessary for sampling inspection. The use of statistical sampling depends on the following conditions:

All items must be produced under similar or identical conditions. Sampling the incoming produce to a food processing plant would require separate samples for separate farmers or separate fields.

A random sample of the lot must be taken. A random sample implies that every item in the lot has an equal chance of being selected.

The lot to be sampled should be a homogeneous mixture. This means that defects will occur in any part of the batch. (The apples on the top should be the same quality as the apples on the bottom!)

The batches to be inspected should be large. Sampling is rarely performed on small lots and is much more accurate in very large samples.

Sampling Plans

Sampling plans are designed to provide some assurance of the quality of goods while taking costs into consideration. Lots are defined as good if they contain no more than a specified level of defects, called the acceptable quality level (AQL). A plan is designed to have a minimum allowable number (or %) of defects in the sample in order to accept the lot. Above this level of defects, the lot will be rejected.

Selecting a particular plan depends on three factors:

Consumer's risk. The probability of accepting a bad lot is called the consumer's risk. Since sampling inspection does not produce results with 100% accuracy, there is always a risk that a lot containing more than the desired number of defects will be accepted. The consumer will want to be sure that the sampling plan has a low probability of accepting bad lots.

Producer's risk.   The probability of rejecting a good lot is called the producer's risk. Since sampling involves probabilities, there is a chance that a batch of good products will be rejected. The producer will want to ensure that the sampling plan has a low probability of rejecting good lots.

Cost.   Inspection costs money. The objective is to balance the consumer's risk and the producer's risk against the cost of the sampling plan. The larger the sample, the smaller the producer's and consumer's risks and the more the likelihood that good batches will not be rejected and poor batches accepted. However, the larger the sample size, the greater the inspection cost. Thus there is a balance between the producer's and consumer's risks and the cost of inspection.

Example.   For simplicity, a single sampling plan will be considered (there are others). The plan will specify the sample size (n)—the number of randomly selected items to be taken—from a given size of production lot (N). These will be inspected for some known characteristic and the plan will set a maximum allowable number of defective products in the sample (c). If more than this number of defectives is found in the sample, the entire lot is rejected. If the allowable number of defects, or fewer, are found in the sample the lot is accepted. Figure 16.11 illustrates an example of a single sampling plan.

  The larger the sample, the smaller the producer's and consumer's risk and the more the likelihood that good batches will not be rejected and poor batches accepted. However, the larger the sample size, the greater the inspection cost. Thus there is a tradeoff between the producer's and consumer's risks and the cost of inspection.

Figure 16.11   Sampling plan.

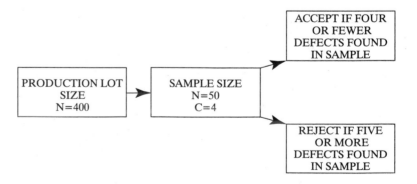

ISO 9000

The International Organization for Standardization (ISO), based in Geneva, Switzerland, developed a series of five standards (ISO 9000, 9001, 9002, 9003, and 9004) for quality systems that were first published in 1987. These were adopted by the European Community (EC) and have become universally accepted standards, particularly in North America. Besides being a requirement for doing business in Europe, customers throughout the world have come to expect a quality standard and to demand ISO 9000 certification.

The standards are intended to prevent nonconformities during all stages of business functions. Originally they were written for contractual situations between suppliers and customers in which the supplier would develop a quality system that conformed to the ISO standards and was satisfactory to the customer. The customer would then audit the system for acceptability. This resulted in multiple audits for different suppliers and different customers. Because of this, a third party registration system was created.

Third party registration system. A third party, called a registrar, assesses the adequacy of the supplier's quality system. When the system conforms to the registrar's interpretation of the applicable ISO 9000 standard, the registrar issues a certificate of registration. The registrar continues to survey the supplier and makes full reassessments every three or four years. The registrar is a "qualified" certifying agent who usually has been accredited by the Registrar Accreditation Board, which is affiliated with the American Society of Quality Control. The third party registration system ensures customers that a supplier has a quality system in place and it is monitored, thus eliminating the need for audits by each customer.

ISO 9000 Series Standards

The ISO 9000 series of standards is generic and can be made to fit any service or manufacturing organization's needs. Its purpose is to standardize quality terms and definitions and use those terms to demonstrate a supplier's capability of controlling its processes. In simple terms, the standards require the supplier to say what it is doing to ensure quality, do what it says, and prove it has done so by documentation.

The ISO 9000 consists of five standards. ISO 9000 explains the basic quality concepts, defines key terms, and provides guidelines for selecting, using, and modifying ISO 9001, 9002, and 9003. ISO 9004 provides guidance in implementing a quality system. ISO 9001 provides a model for quality assurance in design, production, installation and servicing. ISO 9002 provides a model for quality assurance in production and installation. ISO 9003 provides a model for quality assurance in final inspection. Figure 16.12 demonstrates the coverage of each standard.

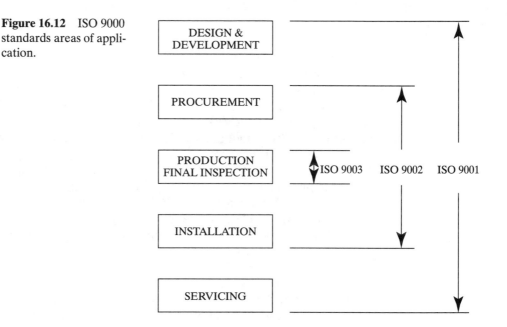

Figure 16.12 ISO 9000 standards areas of application.

ISO 9000 Elements

There are twenty ISO 9000 elements that describe what is required in each of the areas. All twenty are required for ISO 9001, but fewer are required for ISO 9002 and ISO 9003. Table 16.1 briefly describes each of the elements and which standard they apply to.

ISO 9000 places emphasis on all internal processes, especially manufacturing, sales, administration, and technical support. It provides stability in the business system and occurs at the beginning of the quality evolution.

BENCHMARKING

Benchmarking is a systematic method by which organizations can compare their performance in a particular process to that of a "best in class" organization, finding out how that organization achieves those performance levels and applying them to their own organization. Continuous improvement, as discussed in Chapter 14, seeks to make improvement by looking inward and analyzing current practice. Benchmarking looks outward to what competitors and excellent performers outside the industry are doing.

	Quality System Element	ISO 9001	ISO 9002	ISO 9003
1	Management responsibility. The quality process is properly designed and managed.	X	X	X
2	Quality system. The establishing and maintaining of a documented quality system.	X	X	X
3	Contract review. Capability to define clearly and carry out supplier and customer contracts.	X	X	
4	Design control. Capability to ensure that product design meets specifications.	X		
5	Document and data control. Procedures exist to establish and maintain control of all documents and data relating to the quality of a product or service.	X	X	X
6	Purchasing. To ensure purchased items will meet specification.	X	X	
7	Customer-supplied products. Control of customer-supplied products so they are isolated from other use.	X	X	
8	Product identification and traceability. Procedures to identify all products during production, delivery, and installation.	X	X	X
9	Process control. Procedures to ensure processes are operating correctly.	X	X	
10	Inspection and testing. Procedures for receiving, in process, and final inspection.	X	X	X
11	Control of inspection, measuring, and test equipment.	X	X	X

Table 16.1   Standards for quality system elements (continued on next page).

Table 16.1   Standards for quality system elements (continued from previous page).

	Quality System Element	ISO 9001	ISO 9002	ISO 9003
12	Inspection and test status. A product's condition (accepted, rejected, on hold) must be identified throughout production and installation.	X	X	X
13	Control of nonconforming product. Nonconforming product must be removed and isolated so it cannot contaminate good product.	X	X	X
14	Corrective action. There is a procedure to fix known problems.	X	X	
15	Handling, storage, packaging, and delivery. Procedures exist to handle, store, protect, and deliver product throughout the process.	X	X	X
16	Control of quality records. Proper quality records are established and maintained.	X	X	X
17	Internal quality audits. To ensure the quality system is working as planned.	X	X	
18	Training. Adequate personnel training exists.	X	X	X
19	Servicing. Procedures exist for performing and documenting after-delivery service.	X		
20	Statistical techniques. Suitable statistical techniques are implemented.	X	X	X

The steps in benchmarking are:

1. Select the process to benchmark. This is much the same as the first step in the continuous improvement process.

2. Identify an organization that is "best-in-class" in performing the process you want to study. This may not be a company in the same industry. The classic example is Xerox using L. L. Bean, a mail order sales organization, as a benchmark when studying its own order entry system.

3. Study the benchmarked organization. Information may be available internally, be in the public domain, or may require some original research. Original research includes questionnaires, site visits, and focus groups. Questionnaires are useful when information is gathered from many sources. Site visits involve meeting with the "best-in-class" organization. Many companies select a team of workers to be on a benchmark team who meet with their counterparts in the other organization. Focus groups are panels that may be composed of groups such as customers, suppliers, or benchmark partners, brought together to discuss the process.

4. Analyze the data. What are the differences between your process and the benchmark organization? There are two aspects to this. One is comparing the processes and the other is measuring the performance of those processes according to some standard. The measurement of performance requires some unit of measure, referred to as the metrics. Typical performance measures are quality, service response time, cost per order, and so forth.

JIT, TQM, AND MRPII

Although the purposes of MRPII, JIT, and TQM are different, there is a close relationship among them. JIT is a philosophy that seeks to eliminate waste and focuses on decreasing nonvalue-added activities by improving processes and reducing lead time. TQM places emphasis on customer satisfaction and focusing the whole company to that end. While JIT seems to be inward looking (the elimination of waste in the organization and lead time reduction) and TQM outward looking (customer satisfaction), they both have many of the same concepts. Both emphasize management commitment, continuous process improvement, employee involvement, and supplier partnerships. Performance measurement is necessarily a part of process improvement in both JIT and TQM. JIT places emphasis on quality as a means of reducing waste and thus embraces the ideas of TQM. TQM is directed to satisfying the customer, which is also an objective of JIT.

JIT and TQM are mutually reinforcing. They should be considered two sides of the same coin—providing customers what they want at low cost.

MRPII is primarily concerned with managing the flow of materials into, through, and out of an organization. Its objectives are to maximize the use of the organization's resources and provide the required level of customer service. It is a planning and execution process that must work with existing processes, be they good or bad. JIT is directed toward process improvement and lead time reduction, and TQM is directed toward customer satisfaction. Thus both JIT and TQM are part of the environment in which MRPII must work. Improved processes, better quality, employee involvement, and supplier partnerships can only improve the effectiveness of MRPII. Figure 16.13 illustrates the relationship graphically.

Figure 16.13   MRPII, JIT, and TQM.

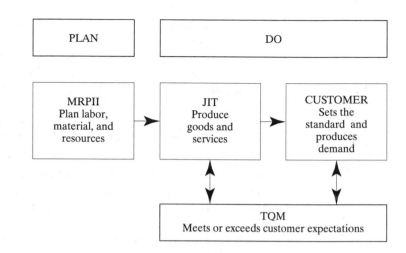

Source: Adapted from *Basics of Supply Chain Management,* APICS—The Ed. Soc. for Res. Mgt., Falls Church, VA, 1997. Reprinted with permission.

QUESTIONS

1. What is the definition of quality?
2. In which four areas must quality be considered? How do they interrelate?
3. Name and describe the eight dimensions to quality.
4. What is total quality management and what are its objectives?
5. What are the six basic concepts of TQM?
6. What does customer focus mean? Who is the customer?
7. What is empowerment and why is it important in TQM?
8. What are the three key factors in supplier partnerships?
9. What is the purpose of performance measurement?
10. Name and describe each of the costs of quality. What is the best way to reduce the costs of quality?
11. What is chance variation? What are the causes of it? How can it be altered?
12. What is assignable variation?
13. What is a normal distribution? Why is it important in quality control?
14. What is the arithmetic mean or average?
15. What is meant by the spread? What are the two measures of it?
16. What percentage of the observations will fall within 1, 2, and 3 standard deviations of the mean?
17. In which two ways can a process create defects?
18. What is tolerance and how does it relate to the USL and LSL?
19. What is the purpose of process capability index (C_p) and the C_{pk}? How do they differ?
20. What is the purpose of process control? What kind of variation does it try to detect?

21. What is a run chart?
22. What is an $\bar{X}$ and R chart?
23. What are upper and lower control limits?
24. What is the difference between variables and attributes?
25. When is it appropriate to use 100% inspection?
26. What is acceptance sampling? When is it appropriate to use?
27. What are the consumer's risk and the producer's risk?
28. Why was the third party registration system established for ISO 9000 certification?
29. What is the difference between ISO 9001, 9002, and 9003?
30. What is benchmarking and how is it different from continuous improvement?

PROBLEMS

16.1 The specification for the length of a shaft is $12'' \pm .001''$. If the process standard deviation is .00033, approximately what percentage of the shafts will be within tolerance?

Answer. There are approximately 99.7% of shafts within tolerance.

16.2 In problem 1, if the tolerance changes to $+ .0007''$, approximately what percentage of the shafts will be within tolerance?

16.3 The specification for the thickness of a piece of steel is $.5'' \pm .05''$. The standard deviation of the band saw is .015. Using C_p, calculate whether the process is capable or not.

Answer. $C_p = 1.11$. Process is barely capable.

16.4 The specification for the weight of a chemical in a compound is 10 grams $\pm$ .05 grams. If the standard deviation of the weighing scales is .02, is the process considered capable?

16.5 The specification for the diameter of a hole is $.75'' \pm .015''$. The standard deviation of the drill press is $.007''$. Using C_p, calculate whether the process is capable or not.

16.6 If, in problem 3, the process is improved so the standard deviation is .0035, is the process capable now?

Answer. Process is capable.

16.7 In problem 6, what is the C_{pk} when:

a. The process is centered on .75? Is the process capable?

b. The process is centered on .74? Is the process capable?

Answer. a. Process centered on 0.75. The process is capable.

b. Process centered on 0.74. The process is not capable

16.8 A company fills plastic bottles with 8 ounces of shampoo. The tolerance is $\pm$ 0.1 ounces. The process has a standard deviation of .02 ounces. For the following situations, calculate the C_{pk} and evaluate the process capability.

a. A sample has an average of quantity of 7.93 ounces

b. A sample has an average of quantity of 7.98 ounces

c. A sample has an average of quantity of 8.04 ounces

Readings

CHAPTER 1

Buffa, E. S., and R. K. Sarin, *Modern Production Operations Management,* 8th ed., Chap. 2. New York: Wiley, 1987.

CHAPTER 2

Fogarty, D. W., and T. R. Hoffmann, *Production and Inventory Management,* Chap. 1. Cincinnati, OH: South-Western, 1983.

Ling, R. C., and W. E. Goddard, *Orchestrating Success.* New York: John Wiley & Sons, 1988.

Plossl, G. W.,. *Production and Inventory Control, Principles and Techniques,* 2nd ed., Chap. 7. Englewood Cliffs, NJ: Prentice Hall, 1985.

Vollmann, T. E., W. L. Berry, and D. C. Whybark, *Manufacturing Planning and Control Systems,* 3rd ed., Chap. 7. Homewood, IL: Irwin, 1992.

CHAPTER 3

Fogarty, D. W., and T. R. Hoffman, *Production and Inventory Management,* Chap. 3. Cincinnati, OH: South-Western, 1983.

Mather, H., *Competitive Manufacturing.* Englewood Cliffs, NJ: Prentice Hall, 1988.

Plossl, G. W., *Production and Inventory Control, Principles and Techniques,* 2nd ed., Chap. 7. Englewood Cliffs, NJ: Prentice Hall, 1985.

Schonberger, R. J., and E. M. Knod, *Operations Management: Serving the Customer,* 3rd ed., Chap. 6. Plano, TX: Business Publications, 1988.

Vollmann, T. E., W. L. Berry, and D. C. Whybark. *Manufacturing Planning and Control Systems,* 3rd ed., Chaps. 9 and 14. Homewood, IL: Irwin, 1992.

CHAPTER 4

Buffa, E. S., and R. K. Sarin, *Modern Production Operations Management,* 8th ed., Chap. 6. New York: Wiley, 1987.

Fogarty, D. W., and T. R. Hoffmann, *Production and Inventory Management,* Chap. 4. Cincinnati, OH: South-Western, 1983.

Garwood, D., *Bills of Material: Structured for Excellence.* Marietta, GA: Dogwood Publishing, 1988.

Mather, H., *Bills of Materials, Recipes and Formulations.* Atlanta, GA: Wright Publishing, 1982.

Orlicky, J., *Material Requirements Planning.* New York, McGraw-Hill, 1975.

Plossl, G. W., *Production and Inventory Control, Principles and Techniques,* 2nd ed., Chap. 6. Englewood Cliffs, NJ: Prentice Hall, 1985.

Vollmann, T. E., W. L. Berry, and D. C. Whybark, *Manufacturing Planning and Control Systems,* 3rd ed., Chaps. 2 and 11. Homewood, IL: Irwin, 1992.

CHAPTER 5

Blackstone, J. H., Jr., *Capacity Management.* Cincinnati, OH: South-Western, 1989.

Fogarty, D. W., and T. R. Hoffmann, *Production and Inventory Management,* Chaps. 1 and 5. Cincinnati, OH: South-Western, 1983.

Plossl, G. W., *Production and Inventory Control, Principles and Techniques,* 2nd ed., Chap. 9. Englewood Cliffs, NJ: Prentice Hall, 1985.

Vollmann, T. E., W. L. Berry, and D. C. Whybark, *Manufacturing Planning and Control Systems,* 3rd ed., Chap. 4. Homewood, IL: Irwin, 1992.

CHAPTER 6

Fogarty, D. W., and T. R. Hoffman, *Production and Inventory Management,* Chap. 12. Cincinnati, OH: South-Western, 1983.

Goldratt, E. M., and J. Cox, *The Goal,* rev. ed. Croton-on-Hudson, NY: North River Press, 1986.

Goldratt, E. M., and R. E. Fox, *The Race.* Croton-on-Hudson, NY: North River Press, 1986.

Melnyk, S. A., and N. L. Carter, *Production Activity Control, A Practical Guide.* Homewood, IL: Dow Jones-Irwin, 1987.

Plossl, G. W., *Production and Inventory Control, Principles and Techniques,* 2nd ed., Chaps. 10–12. Englewood Cliffs, NJ: Prentice Hall, 1985.

Vollmann, T. E., W. L. Berry, and D. C. Whybark, *Manufacturing Planning and Control Systems,* 3rd ed., Chap. 5. Homewood, IL: Irwin, 1992.

CHAPTER 7

Ammer, D. S., *Materials Management and Purchasing,* 4th ed. Homewood, IL: Irwin, 1980.

Colton, R. R., and W. F. Rohrs, *Industrial Purchasing and Effective Materials Management.* Reston, VA: Reston, 1985.

Dobbler, D. W., L. Lee, and D. Burt, *Purchasing and Materials Management.* New York: McGraw-Hill, 1984.

Zenz, G. J., *Purchasing and the Management of Materials,* 6th ed. New York: Wiley, 1987.

CHAPTER 8

Buffa, E. S., and R. K. Sarin, *Modern Production Operations Management,* 8th ed., Chap. 4. New York: Wiley, 1987.

Fogarty, D. W., and T. R. Hoffmann, *Production and Inventory Management,* Chap. 2. Cincinnati, OH: South-Western, 1983.

Plossl, G. W., *Production and Inventory Control, Principles and Techniques,* 2nd ed., Chaps. 4 and 5. Englewood Cliffs, NJ: Prentice Hall, 1985.

Schonberger, R. J., and E. M. Knod, *Operations Management: Serving the Customer,* 3rd ed., Chap. 4. Plano, TX: Business Publications, 1988.

Vollmann, T. E., W. L. Berry, and D. L. Whybark, *Manufacturing Planning and Control Systems,* 3rd ed., Chap. 16. Homewood, IL: Irwin, 1992.

CHAPTER 9

Buffa, E. S., and R. K. Sarin, *Modern Production Operations Management,* 8th ed., Chap. 5. New York: Wiley, 1987.

Fogarty, D. W., and T. R. Hoffmann, *Production and Inventory Management,* Chaps. 6 and 9. Cincinnati, OH: South-Western, 1983.

Plossl, G. W., *Production and Inventory Control, Principles and Techniques,* 2nd ed., Chap. 2. Englewood Cliffs, NJ: Prentice Hall, 1985.

Schonberger, R. J., and E. M. Knod, *Operations Management: Serving the Customer,* 3rd ed., Chap. 7. Plano, TX: Business Publications, 1988.

Vollmann, T. E., W. L. Berry, and D. C. Whybark, *Manufacturing Planning and Control Systems,* 3rd ed., Chap. 17. Homewood, IL: Irwin, 1992.

CHAPTERS 10 AND 11

Buffa, E. S., and R. K. Sarin, *Modern Production Operations Management,* 8th ed., Chap. 5. New York: Wiley, 1987.

Fogarty, D. W., and T. R. Hoffmann, *Production and Inventory Management,* Chap. 7. Cincinnati, OH: South-Western, 1983.

Martin, A. J., *Distribution Resource Planning.* Essex Junction, VT: Oliver Wight Publications, 1983.

Plossl, G. W., *Production and Inventory Control, Principles and Techniques,* 2nd ed., Chap. 3. Englewood Cliffs, NJ: Prentice Hall, 1985.

CHAPTERS 12 AND 13

Ackerman, K. B., *Warehousing, A Guide for Both Users and Operators.* Washington, DC: Traffic Service Corp., 1977.

Apple, J. M., *Material Handling Systems Design.* New York: Ronald Press, 1972.

Ballou, Ronald H., *Business Logistics Management,* 3rd ed. Englewood Cliffs, NJ: Prentice Hall, 1992.

Blanding, W., *Blanding's Practical Physical Distribution.* Washington, DC: Traffic Service Corp., 1978.

Bowersox, D. J., P. J. Calabro, and G. D. Wagenheim, *Introduction to Transportation.* New York: Macmillan, 1981.

Bowersox, D. J., D. J. Closs, and O. K. Helferich, *Logistical Management,* 3rd ed. New York: Macmillan, 1986.

Coyle, J. J., E. J. Bardi, and C. J. Langley, *The Management of Business Logistics,* 4th ed. St. Paul, MN: West Publishing, 1988.

Johnson, J. L., and D. F. Wood, *Contemporary Physical Distribution and Logistics,* 3rd ed. New York: Macmillan, 1986.

Lieb, R. C., *Transportation, The Domestic System.* Reston, VA: Reston, 1978.

CHAPTER 14

Arnold, J. R., and L. M. Clive, *Introduction to Operations Management.* Qualicum Beach, B C: J. R. Arnold & Associates Ltd., 1996.

Chase, R. B., and N. J. Aquilano, *Production and Operations Management.* Chicago: Richard D. Irwin, 1995.

Schonberger, R. J., and E. M. Knod, *Operations Management.* Chicago: Richard D. Irwin, 1995.

Starr, M. K. *Operations Management.* Danvers, MA: Boyd & Fraser Publishing, 1996.

Turner, W. C., et al. *Introduction to Industrial and Systems Engineering.* Englewood Cliffs, NJ: Prentice Hall, 1993.

CHAPTER 15

Goddard, W. E., *Just-in-Time: Surviving by Breaking Tradition.* Essex Junction, VT: Oliver Wight Publications, 1986.

Hall, R. W., *Zero Inventories.* Homewood, IL: Dow Jones-Irwin, 1983.

Hall, R. W., *Attaining Manufacturing Excellence.* Homewood, IL: Dow Jones-Irwin, 1987.

Schonberger, R. J., *Japanese Manufacturing Techniques: Nine Hidden Lessons in Simplicity.* New York: Free Press, 1982.

Schonberger, R. J., *World Class Manufacturing: The Lessons of Simplicity Applied.* New York: Free Press, 1986.

Suzaki, K., *The New Manufacturing Challenge: Techniques for Continuous Improvement.* New York: Free Press, 1987.

Wemmerlöv, U., *Production Planning and Control Procedures for Cellular Manufacturing.* Falls Church, VA: American Production and Inventory Control Society, 1988.

CHAPTER 16

Arnold, J. R., and L. M. Clive, *Introduction to Operations Management.* Qualicum Beach, B C: J. R. Arnold & Associates Ltd., 1996.

Besterfield, D. H., Besterfield-Michna, C., Besterfield, G. H., Besterfield-Sacre, M. *Total Quality Management.* Englewood Cliffs, NJ: Prentice Hall, 1995.

Chase, R. B., and N. J. Aquilano, *Production and Operations Management.* Chicago: Richard D. Irwin, 1995.

Schonberger, R. J., and E. M. Knod, *Operations Management.* Chicago: Richard D. Irwin, 1995.

Schonberger, R. J., *Japanese Manufacturing Techniques.* New York: Free Press, 1982.

Starr, Martin K. *Operations Management.* Danvers, MA: Boyd & Fraser, 1996.

Index

447